Tribological Characteristics of Bearing System, 2nd Edition

Tribological Characteristics of Bearing System, 2nd Edition

Guest Editor

Yong Chen

Basel • Beijing • Wuhan • Barcelona • Belgrade • Novi Sad • Cluj • Manchester

Guest Editor
Yong Chen
Guangxi University
Nanning
China

Editorial Office
MDPI AG
Grosspeteranlage 5
4052 Basel, Switzerland

This is a reprint of the Special Issue, published open access by the journal *Lubricants* (ISSN 2075-4442), freely accessible at: https://www.mdpi.com/journal/lubricants/special_issues/85923P1892.

For citation purposes, cite each article independently as indicated on the article page online and as indicated below:

Lastname, A.A.; Lastname, B.B. Article Title. *Journal Name* **Year**, *Volume Number*, Page Range.

ISBN 978-3-7258-3085-5 (Hbk)
ISBN 978-3-7258-3086-2 (PDF)
https://doi.org/10.3390/books978-3-7258-3086-2

© 2025 by the authors. Articles in this book are Open Access and distributed under the Creative Commons Attribution (CC BY) license. The book as a whole is distributed by MDPI under the terms and conditions of the Creative Commons Attribution-NonCommercial-NoDerivs (CC BY-NC-ND) license (https://creativecommons.org/licenses/by-nc-nd/4.0/).

Contents

Faisal Rahmani, Emad Makki and Jayant Giri
Influence of Bearing Wear on the Stability and Modal Characteristics of a Flexible Rotor Supported on Powder-Lubricated Journal Bearings
Reprinted from: *Lubricants* **2023**, *11*, 355, https://doi.org/10.3390/lubricants11090355 1

Zhiwei Wang, Shuanglong Mao, Heng Tian, Bing Su and Yongcun Cui
Simulation Analysis and Experimental Research on Electric Thermal Coupling of
Current Bearing
Reprinted from: *Lubricants* **2024**, *12*, 73, https://doi.org/10.3390/lubricants12030073 19

Shuai Li, Yafu Huang, Hechun Yu, Wenbo Wang, Guoqing Zhang, Xinjun Kou, et al.
Calculation and Analysis of Equilibrium Position of Aerostatic Bearings Based on Bivariate Interpolation Method
Reprinted from: *Lubricants* **2024**, *12*, 85, https://doi.org/10.3390/lubricants12030085 36

Bing Su, Han Li, Guangtao Zhang, Fengbo Liu and Yongcun Cui
Study on Cage Stability of Solid-Lubricated Angular Contact Ball Bearings in an Ultra-Low Temperature Environment
Reprinted from: *Lubricants* **2024**, *12*, 124, https://doi.org/10.3390/lubricants12040124 50

Fei Shang, Bo Sun, Shaofeng Wang, Yongquan Han, Wenjing Liu, Ning Kong, et al.
Calibration of Oil Film Thickness Acoustic Reflection Coefficient of Bearing under Multiple Temperature Conditions
Reprinted from: *Lubricants* **2024**, *12*, 125, https://doi.org/10.3390/lubricants12040125 68

Hengdi Wang, Han Li, Zheming Jin, Jiang Lin, Yongcun Cui, Chang Li, et al.
Simulation Analysis and Experimental Study on the Fluid–Solid–Thermal Coupling of Traction Motor Bearings
Reprinted from: *Lubricants* **2024**, *12*, 144, https://doi.org/10.3390/lubricants12050144 89

Omid Safdarzadeh, Alireza Farahi, Andreas Binder, Hikmet Sezen and Jan Philipp Hofmann
WLI, XPS and SEM/FIB/EDS Surface Characterization of an Electrically Fluted
Bearing Raceway
Reprinted from: *Lubricants* **2024**, *12*, 148, https://doi.org/10.3390/lubricants12050148 106

Guozhen Fan, Youhua Li, Yuehua Li, Libin Zang, Ming Zhao, Zhanxin Li, et al.
Research on Design and Optimization of Micro-Hole Aerostatic Bearing in
Vacuum Environment
Reprinted from: *Lubricants* **2024**, *12*, 224, https://doi.org/10.3390/ lubricants12060224 134

Yong Chen, Xiangrun Pu, Lijie Hao, Guangxin Li and Li Luo
Study on the Friction Characteristics and Fatigue Life of Carbonitriding-Treated
Needle Bearings
Reprinted from: *Lubricants* **2024**, *12*, 234, https://doi.org/10.3390/lubricants12070234 153

Yang Yang, Jiayu Wang, Meiling Wang and Baogang Wen
Dynamic Modeling and Behavior of Cylindrical Roller Bearings Considering Roller Skew and the Influence of Eccentric Load
Reprinted from: *Lubricants* **2024**, *12*, 317, https://doi.org/10.3390/lubricants12090317 170

Ioannis Tselios and Pantelis Nikolakopoulos
Combining Artificial Neural Networks and Mathematical Models for Unbalance Estimation in a Rotating System under the Nonlinear Journal Bearing Approach
Reprinted from: *Lubricants* **2024**, *12*, 344, https://doi.org/10.3390/lubricants12100344 . . . . . . **190**

Shengdong Zhang, Dongjiang Yang, Guangming Li, Yongchao Cheng, Jichao Li, Fangqiao Zhao and Wenlong Song
Carrying Capacity of Spherical Hydrostatic Bearings Including Dynamic Pressure
Reprinted from: *Lubricants* **2024**, *12*, 346, https://doi.org/10.3390/lubricants12100346 . . . . . . **213**

Florian de Cadier de Veauce, Yann Marchesse, Thomas Touret, Christophe Changenet, Fabrice Ville, Luc Amar and Charlotte Fossier
Power Losses of Oil-Bath-Lubricated Ball Bearings—A Focus on Churning Losses
Reprinted from: *Lubricants* **2024**, *12*, 362, https://doi.org/10.3390/lubricants12110362 . . . . . . **224**

Yong Chen, Guangxin Li, Anhe Li and Bolin He
End-to-End Intelligent Fault Diagnosis of Transmission Bearings in Electric Vehicles Based on CNN
Reprinted from: *Lubricants* **2024**, *12*, 364, https://doi.org/10.3390/lubricants12110364 . . . . . . **239**

Fangjian Yuan, Hang Xiu, Guohua Cao, Jingran Zhang, Bingshu Chen, Yutang Wang and Xu Zhou
A Computational Fluid Dynamics-Based Study on the Effect of Bionic-Compound Recess Structures in Aerostatic Thrust Bearings
Reprinted from: *Lubricants* **2024**, *12*, 385, https://doi.org/10.3390/lubricants12110385 . . . . . . **261**

Baogang Wen, Yuanyuan Li, Yemin Li, Meiling Wang and Jingyu Zhai
Influence and Optimization of Nozzle Position on Lubricant Distribution in an Angular Contact Ball Bearing Cavity
Reprinted from: *Lubricants* **2024**, *12*, 419, https://doi.org/10.3390/lubricants12120419 . . . . . . **284**

Article

Influence of Bearing Wear on the Stability and Modal Characteristics of a Flexible Rotor Supported on Powder-Lubricated Journal Bearings

Faisal Rahmani [1,*], Emad Makki [2] and Jayant Giri [3,*]

1. Department of Mechanical Engineering, Birla Institute of Technology, Mesra, Ranchi 835215, India
2. Department of Mechanical Engineering, College of Engineering and Islamic Architecture, Umm Al-Qura University, Makkah 24382, Saudi Arabia; eamakki@uqu.edu.sa
3. Department of Mechanical Engineering, Yeshwantrao Chavan College of Engineering, Nagpur 441110, India
* Correspondence: faisalrahmani@bitmesra.ac.in (F.R.); jayantpgiri@gmail.com (J.G.); Tel.: +91-829-456-7477 (F.R.)

Abstract: This study has investigated the influence of journal bearing wear on the dynamic behaviour of a flexible rotor with a central disc. Rotors supported on journal bearings are susceptible to self-excited whirling, leading to unstable conditions. Prior knowledge of the stability limit speed is important to avoid the excessive vibration of rotating machines. For the study in this paper, journal bearings were lubricated with powder owing to high-temperature applications where conventional oil lubricants would fail to perform. The governing equations for lubrication were derived using a simplified grain theory based on the theory of dense gases. The rotor shaft was discretized considering Timoshenko beam elements. Modal analysis was conducted to obtain the system's natural frequencies, mode shapes, damping factors, stability limit speed, and unbalance response. This study has also evaluated the influence of wear depth on the dynamic behaviour of the rotor shaft system and found that bearing wear significantly affects the stiffness and damping characteristics of lubricating film. Consequently, free and forced vibration behaviour is also affected. It has been found that increased wear depth improves stability limit speed but has little influence on the unbalance response.

Keywords: journal bearing; powder lubrication; flexible rotor; modal analysis; stability

Citation: Rahmani, F.; Makki, E.; Giri, J. Influence of Bearing Wear on the Stability and Modal Characteristics of a Flexible Rotor Supported on Powder-Lubricated Journal Bearings. *Lubricants* **2023**, *11*, 355. https://doi.org/10.3390/lubricants11090355

Received: 26 July 2023
Revised: 15 August 2023
Accepted: 18 August 2023
Published: 22 August 2023

Copyright: © 2023 by the authors. Licensee MDPI, Basel, Switzerland. This article is an open access article distributed under the terms and conditions of the Creative Commons Attribution (CC BY) license (https://creativecommons.org/licenses/by/4.0/).

1. Introduction

Journal bearings exhibit excellent lubrication and load-bearing capabilities, due to which it is commonly used to support heavy rotating equipment such as steam turbines and industrial gas turbines. However, when exposed to high temperatures, conventional lubricating oils may deteriorate and become ineffective. Under such circumstances, powder lubricants can be a viable alternative, as they perform well in high-temperature environments [1–3]. Moreover, bearing wear can occur due to extended periods of operation or frequent start/stop cycles, which can influence bearing clearance and ultimately change the dynamic behaviour of rotor-bearing system. The present study explores the influence of bearing wear on the dynamic behaviour of a rotor-bearing system.

Modelling powder flow is complicated for many reasons; thus, several researchers have attempted to develop governing equations using continuum and discrete approaches [4]. In the context of the application of powder flow in bearing lubrication, the simplest yet most powerful theory explaining the general features of grain flow is due to Haff [5], who utilized the theory of dense gases to analyse granular flow, incorporating a microscopic model to relate the parameters present in conservation laws. The performance of a slider bearing was investigated using Haff's model, and analytical solutions were obtained for an infinitely wide slider [6]. For journal bearings lubricated with powders, a closed-form lubrication equation was developed by Tsai and Jeng [7,8], and their numerical results were validated with the experimental findings of Heshmat and Brewe [9]. Higgs and Tichy [10,11] used a slightly different continuum approach to develop a granular kinetic lubrication model with appropriate rheological constitutive equations for stress in a thin shearing flow of granular particles. A robust numerical code using the finite difference method for simple shearing flows has also been developed. Sawyer and Tichy [12] conducted a study involving continuum and particle analyses, comparing their results with those of prior experiment. Although the load-carrying capacity and shear force predicted by both models were higher than those in the experimental data, the trends and orders of magnitude were similar. Recently, the force chain characteristics of shearing granular media have been studied for parallel sliding friction pairs [13], journal bearings [14], and Taylor–Couette geometry [15] using the discrete element method.

Modal analysis is a powerful method to evaluate the natural frequencies and responses of multi-degree freedom systems. Irretier [16] developed a comprehensive mathematical framework for applying modal analysis to rotating systems. Rotating systems have asymmetrical and speed-dependent system matrices, making their modal characteristics speed-dependent as well. Rotors supported on journal bearings require an evaluation of bearing coefficients, which depend on both speed and bearing geometry. These bearing coefficients are necessary for the correct estimation of the dynamics of the machine. The theoretical evaluation of bearing coefficients is restricted to short bearing approximation (length-to-diameter ratios of less than 0.5) and can be found in refs. [17,18]. For a finite bearing, coefficients can be evaluated numerically using the perturbation technique [19,20]. Negatively cross-coupled stiffness tends to destabilize the whirling of rotors due to a tangential force acting in the direction of the whirl orbit. The magnitude of the tangential force will increase with the spin speed; thus, after a certain speed, the rotor will become unstable. El-Shafei et al. [21] conducted an experimental study to explore the factors that contribute to the instability in a journal bearing that supports a flexible rotor.

Bearing wear, caused by prolonged operation or repeated start/stop cycles, affects bearing clearance and dynamic coefficients. A model for wear geometry was developed by Dufrane et al. [22] after they measured the wear in the bearing of a steam turbine. The same model was verified by Hashimoto et al. [23], who studied the effects of wear on the performance parameters of laminar as well as turbulent regimes. A thermo-hydrodynamic analysis of worn journal bearings was carried out by Fillon and Bouyer [24], and it was found that up to 20% of wear had little influence on performance parameters but wear of more than 20% reduced the temperature inside the film. Papadopoulos et al. [25] proposed a method for detecting wear in hydrodynamic bearings based on the analysis of measured rotor responses at specific locations. Gertzos et al. [26] introduced a graphical technique to identify the wear depth associated with the measured performance parameters. Chasalevris et al. [27] found that worn bearings have additional frequency components in the continuous wavelet transform (CWT) of measured responses. Recently, Machado et al. [28,29] examined the influence of wear on the response of a rotor-bearing system using a slightly modified wear model introduced by Dufrane et al. [22]. This wear model considered a slight angular shifting of wear towards the right to match practical conditions. In another study, Machado et al. [30] utilized the response spectrum to diagnose wear. Konig et al. [31] presented a new numerical method to predict wear in journal bearings during steady-state operation. This method considers wear on both the macroscopic and asperity contact scales and shows that reduced asperity interaction due to wearing-in leads to higher accuracy of wear prediction. In recent times, machine learning, along with vibration signals, has emerged as a viable method for condition monitoring and wear fault diagnosis. Gecgel et al. [32] proposed a framework using a deep learning algorithm to classify wear faults in hydrodynamic journal bearings using simulated vibration signals, which can be a promising tool for wear fault diagnostics in journal bearings. Mokhtari et al. [33] presented the use of machine learning algorithms applied to Acoustic Emission (AE) signals for monitoring friction and wear of journal bearings in jet engines containing a gearbox. Konig et al. [34] discussed the use of Acoustic Emission (AE) technique and machine learning methods for wear monitoring in sliding bearing systems. The study achieved high accuracy and sensitivity in detecting and classifying wear failure modes, including running-in, inadequate lubrication, and particle-contaminated oil.

Previous studies have focused on conventional oil lubricants. However, this work aims to investigate the dynamics of rotors supported on journal bearings with powder lubricants. The wear depth influences the bearing coefficients, which further influences the dynamics of the complete system. Thus, the influence of wear on the modal characteristics is also examined. The rotor–shaft system is modelled using Timoshenko beam elements, and each node is assigned four degrees of freedom. The equations of motion, which are obtained by assembling system matrices, are utilized to evaluate eigenvalues, modal damping factors, stability limit speed, and unbalance response.

2. Mathematical Model
2.1. Grain Theory

Haff [5] developed a simple grain theory that treats grain flow as a fluid mechanics problem. The laws of mass, momentum, and energy conservation were employed to examine granular flow. Constitutive equations were obtained from a microscopic model of grain–grain collisions, using the kinetic theory of dense gases. With assumptions such as grain particles being cohesionless identical spheres of diameter d with negligible separation s, and considering a fluctuation velocity $\bar{v}$ of grain particles together with the bulk flow velocity V, it can be shown that the relations for the pressure p, viscosity η, thermal diffusivity K, and energy lost per unit volume per second I can be written as follows (see ref. [5] for details):

$$p = td\rho \frac{\bar{v}^2}{s}; \ \eta = qd^2\rho \frac{\bar{v}}{s}; \ K = rd^2 \frac{\bar{v}}{s}; \ I = \gamma\rho \frac{\bar{v}^3}{s} \tag{1}$$

where, t, q, r, and γ are dimensionless constants. The term $\frac{V}{s}$ is collision rate.

Flow velocities u, v, and w vary gradually from point to point in the flow field. Since viscous forces are due to relative motion between two layers, they will take the same form as in a hydrodynamic system [5]. Continuity, momentum, and pseudo-energy equations for granular flow are written as,

$$\frac{\partial u}{\partial x} + \frac{\partial v}{\partial y} + \frac{\partial w}{\partial z} = 0 \tag{2}$$

$$\rho \frac{Du}{Dt} = -\frac{\partial p}{\partial x} + \frac{\partial}{\partial x}\left[2\eta\left(\frac{\partial u}{\partial x}\right)\right] + \frac{\partial}{\partial y}\left[\eta\left(\frac{\partial u}{\partial y} + \frac{\partial v}{\partial x}\right)\right] + \frac{\partial}{\partial z}\left[\eta\left(\frac{\partial w}{\partial x} + \frac{\partial u}{\partial z}\right)\right] + \rho f_x \tag{3}$$

$$\rho \frac{Dv}{Dt} = -\frac{\partial p}{\partial y} + \frac{\partial}{\partial x}\left[\eta\left(\frac{\partial u}{\partial y} + \frac{\partial v}{\partial x}\right)\right] + \frac{\partial}{\partial y}\left[2\eta\left(\frac{\partial v}{\partial y}\right)\right] + \frac{\partial}{\partial z}\left[\eta\left(\frac{\partial w}{\partial y} + \frac{\partial v}{\partial z}\right)\right] + \rho f_y \tag{4}$$

$$\rho \frac{Dw}{Dt} = -\frac{\partial p}{\partial z} + \frac{\partial}{\partial x}\left[\eta\left(\frac{\partial u}{\partial z} + \frac{\partial w}{\partial x}\right)\right] + \frac{\partial}{\partial y}\left[\eta\left(\frac{\partial v}{\partial z} + \frac{\partial w}{\partial y}\right)\right] + \frac{\partial}{\partial z}\left[2\eta\left(\frac{\partial w}{\partial z}\right)\right] + \rho f_z \tag{5}$$

$$\begin{aligned}\frac{D}{Dt}\left(\frac{\rho\bar{v}^2}{2} + \frac{\rho V^2}{2}\right) &= \frac{\partial}{\partial x}\left[K\frac{\partial}{\partial x}\left(\frac{\rho\bar{v}^2}{2}\right)\right] + \frac{\partial}{\partial y}\left[K\frac{\partial}{\partial y}\left(\frac{\rho\bar{v}^2}{2}\right)\right] + \frac{\partial}{\partial z}\left[K\frac{\partial}{\partial z}\left(\frac{\rho\bar{v}^2}{2}\right)\right] \\ &+ \frac{\partial}{\partial x}\eta\left[2u\frac{\partial u}{\partial x} + v\left(\frac{\partial u}{\partial y} + \frac{\partial v}{\partial x}\right) + w\left(\frac{\partial u}{\partial z} + \frac{\partial w}{\partial x}\right)\right] + \frac{\partial}{\partial y}\eta\left[u\left(\frac{\partial u}{\partial y} + \frac{\partial v}{\partial x}\right) + 2v\frac{\partial v}{\partial y} + w\left(\frac{\partial v}{\partial z} + \frac{\partial w}{\partial y}\right)\right] \\ &+ \frac{\partial}{\partial z}\eta\left[u\left(\frac{\partial u}{\partial z} + \frac{\partial w}{\partial x}\right) + v\left(\frac{\partial v}{\partial z} + \frac{\partial w}{\partial y}\right) + 2w\frac{\partial w}{\partial z}\right] - u\frac{\partial p}{\partial x} - v\frac{\partial p}{\partial y} - w\frac{\partial p}{\partial z} + \rho(uf_x + vf_y + wf_z) - I\end{aligned} \tag{6}$$

In Equations (3)–(6), $\frac{D(\ )}{Dt} = \frac{\partial(\ )}{\partial t} + u\frac{\partial(\ )}{\partial x} + v\frac{\partial(\ )}{\partial y} + w\frac{\partial(\ )}{\partial z}$ represents total derivative.

2.2. Application to Bearing Lubrication

The schematic of a worn journal bearing is shown in Figure 1. Following simplifying assumptions are considered to derive a governing equation similar to the classical Reynolds equation of hydrodynamic lubrication.

1. Inertia forces and body forces can be neglected in comparison with viscous forces.
2. Pressure is considered constant across the film.
3. The terms $\frac{\partial u}{\partial x}, \frac{\partial u}{\partial z}, \frac{\partial v}{\partial x}, \frac{\partial v}{\partial y}, \frac{\partial v}{\partial z}, \frac{\partial w}{\partial x}, \frac{\partial w}{\partial z}$, and their derivatives are negligible in comparison with $\frac{\partial u}{\partial y}, \frac{\partial w}{\partial y}$ and their derivatives.
4. The term $\frac{\partial \bar{v}}{\partial x}, \frac{\partial \bar{v}}{\partial z}$ and their derivatives are negligible in comparison with $\frac{\partial \bar{v}}{\partial y}$.

With these assumptions, Equations (3)–(6) can be reduced to the following:

$$\frac{\partial p}{\partial x} = \frac{\partial}{\partial y}\left(\eta \frac{\partial u}{\partial y}\right) \tag{7}$$

$$\frac{\partial p}{\partial z} = \frac{\partial}{\partial y}\left(\eta \frac{\partial w}{\partial y}\right) \tag{8}$$

$$\frac{\partial^2 \bar{v}}{\partial y^2} - \frac{\gamma}{rd^2}\bar{v} = 0 \tag{9}$$

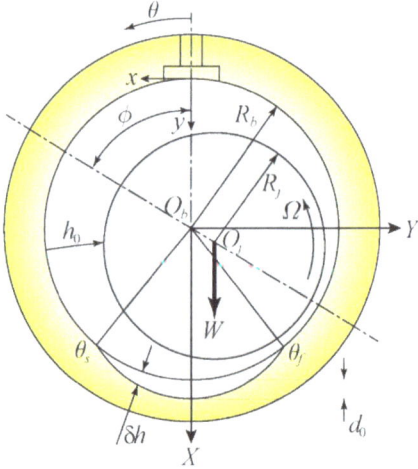

Figure 1. Schematic of a worn journal bearing.

Treating viscosity η in Equations (7) and (8) as the average value across the film [8,35], the modified Reynolds equation is obtained as (see ref. [35] for details) the following:

$$\frac{\partial}{\partial x}\left(\Psi h^2 \frac{1}{p}\frac{\partial p}{\partial x}\right) + \frac{\partial}{\partial z}\left(\Psi h^2 \frac{1}{p}\frac{\partial p}{\partial z}\right) = \frac{U}{2}\frac{\partial h}{\partial x} + \frac{\partial h}{\partial t} \tag{10}$$

where $\Psi = \frac{1}{6}\frac{t}{q}\sqrt{\frac{r}{\gamma}}B\left(\frac{e^{\sigma h}+e^{-\sigma h}-2}{e^{\sigma h}-e^{-\sigma h}}\right), \sigma = \frac{1}{d}\sqrt{\frac{\gamma}{r}}$.

Film forces can be obtained through the integration of pressure over the bearing surface as written below.

$$F_X = \int_0^L \int_0^{2\pi} (p - p_a)\cos\theta\, Rd\theta dz \tag{11}$$

$$F_Y = \int_0^L \int_0^{2\pi} (p - p_a) \sin\theta \, R d\theta dz \qquad (12)$$

The resultant film force on the bearing surface is obtained as the following:

$$F_b = \sqrt{F_X^2 + F_Y^2} \qquad (13)$$

Shearing force on the bearing surface is obtained through the following relation:

$$F_s = \int_0^L \int_0^{2\pi} \left(\frac{h}{2} \frac{\partial p}{\partial x} + \eta \frac{U}{h} \right) R d\theta dz \qquad (14)$$

The coefficient of friction is calculated as the following:

$$f = \frac{F_b}{F_s} \qquad (15)$$

Stiffness and damping coefficients of the powder film are obtained using the finite perturbation technique [19,20]. With very small perturbations of displacements $(\Delta X, \Delta Y)$ and velocities $(\Delta \dot{X}, \Delta \dot{Y})$ about the steady state position (X, Y) of the journal centre, bearing forces (F_X, F_Y) are obtained. Employing central difference, coefficients of stiffness and damping are written as,

$$\begin{aligned}
K_{XX} &= \frac{\partial F_X}{\partial X} = \frac{F_X(X+\Delta X,Y,0,0) - F_X(X-\Delta X,Y,0,0)}{2\Delta X} \\
K_{XY} &= \frac{\partial F_X}{\partial Y} = \frac{F_X(X,Y+\Delta Y,0,0) - F_X(X,Y-\Delta Y,0,0)}{2\Delta Y} \\
K_{YX} &= \frac{\partial F_Y}{\partial X} = \frac{F_Y(X+\Delta X,Y,0,0) - F_Y(X-\Delta X,Y,0,0)}{2\Delta X} \\
K_{YY} &= \frac{\partial F_Y}{\partial Y} = \frac{F_Y(X,Y+\Delta Y,0,0) - F_Y(X,Y-\Delta Y,0,0)}{2\Delta Y} \\
C_{XX} &= \frac{\partial F_X}{\partial \dot{X}} = \frac{F_X(X,Y,\Delta \dot{X},0) - F_X(X,Y,-\Delta \dot{X},0)}{2\Delta \dot{X}} \\
C_{XY} &= \frac{\partial F_X}{\partial \dot{Y}} = \frac{F_X(X,Y,0,\Delta \dot{Y}) - F_X(X,Y,0,-\Delta \dot{Y})}{2\Delta \dot{Y}} \\
C_{YX} &= \frac{\partial F_Y}{\partial \dot{X}} = \frac{F_Y(X,Y,\Delta \dot{X},0) - F_Y(X,Y,-\Delta \dot{X},0)}{2\Delta \dot{X}} \\
C_{YY} &= \frac{\partial F_Y}{\partial \dot{Y}} = \frac{F_Y(X,Y,0,\Delta \dot{Y}) - F_Y(X,Y,0,-\Delta \dot{Y})}{2\Delta \dot{Y}}
\end{aligned} \qquad (16)$$

2.3. Modelling of Bearing Wear

Dufrane et al. [22] proposed a mathematical model for bearing wear after careful experimentation which was validated experimentally by Hashimoto et al. [23]. An additional depth δh in the angular span from θ_s to θ_f (see Figure 1) is incorporated for the film thickness and is written as,

$$\begin{aligned}
\bar{h} &= 1 + \overline{X} \cos\theta + \overline{Y} \sin\theta & \text{for } \theta \leq \theta_s, \theta \geq \theta_f \\
\bar{h} &= (\overline{X} - 1) \cos\theta + \overline{Y} \sin\theta + \bar{d}_0 & \text{for } \theta_s < \theta < \theta_f
\end{aligned} \qquad (17)$$

where, $\overline{X} = X/c$ and $\overline{Y} = Y/c$ are the normalized coordinates in the $X - Y$ coordinate system, and $\bar{d}_0 = d_0/c$ is the normalized maximum wear depth. The range of angular position of the worn region (θ_s and θ_f) is determined by the solution of the equation,

$$\cos\theta = \bar{d}_0 - 1 \qquad (18)$$

2.4. Modal Analysis of Rotors

Modal analysis is a powerful method to decouple equations of motion through a coordinate transformation using a modal matrix. The uncoupled equations are solved independent of each other in the modal coordinates and then solutions are combined to obtain the response of the original system. Unlike non-rotating structures, the rotating systems are Non-Self Adjoint (NSA) systems wherein both right and left eigenvectors are required for modal analysis.

The shaft has distributed mass and stiffness due to flexibility. The shaft is modelled using a two-noded Timoshenko beam element having four degrees of freedom $(x, y, \theta_x, \theta_y)$ at each node. The nodal displacement vector $\mathbf{q}_e$ can be written as,

$$\mathbf{q}_e = \{x_i, \ y_i, \ \theta_{x_i}, \ \theta_{y_i}, \ x_{i+1}, \ y_{i+1}, \ \theta_{x_{i+1}}, \ \theta_{y_{i+1}}\}^T \quad (19)$$

The equations of motion were written for each element and then assembled with boundary conditions to obtain the equations of motion as,

$$\mathbf{M}\ddot{\mathbf{q}}(t) + \mathbf{C}\dot{\mathbf{q}}(t) + \mathbf{K}\mathbf{q}(t) = \mathbf{f}(t) \quad (20)$$

where, $\mathbf{M} = \mathbf{M}_T + \mathbf{M}_R$, $\mathbf{C} = \mathbf{C}_B + \Omega\mathbf{G}$, and $\mathbf{K} = \mathbf{K}_B + \mathbf{K}_S$

The mass matrix $\mathbf{M}$ consists of translatory $\mathbf{M}_T$ and rotary inertia $\mathbf{M}_R$ matrices. The overall damping matrix $\mathbf{C}$ consists of the gyroscopic matrix $\mathbf{G}$, and the bearing damping matrix $\mathbf{C}_B$. The overall stiffness matrix $\mathbf{K}$ is the sum of the bearing stiffness matrix $\mathbf{K}_B$ and the shaft stiffness matrix $\mathbf{K}_S$. Ω denotes the spinning speed. $\mathbf{q}(t)$ and $\mathbf{f}(t)$ are the global displacement and force vectors. The state-space form of Equation (20) is written as [36],

$$\dot{\mathbf{u}}(t) = \mathbf{A}\mathbf{u}(t) + \mathbf{B}\mathbf{f}(t) \quad (21)$$

where $\mathbf{A} = \begin{bmatrix} 0 & \mathbf{I} \\ -\mathbf{M}^{-1}\mathbf{K} & -\mathbf{M}^{-1}\mathbf{C} \end{bmatrix}$; $\mathbf{B} = \begin{bmatrix} 0 \\ \mathbf{M}^{-1} \end{bmatrix}$; $\mathbf{u}(t) = \begin{Bmatrix} \mathbf{q}(t) \\ \dot{\mathbf{q}}(t) \end{Bmatrix}$. By setting $\mathbf{f}(t) = 0$ in Equation (21), the equation of motion for free vibration is obtained as,

$$\dot{\mathbf{u}}(t) = \mathbf{A}\mathbf{u}(t) \quad (22)$$

The solution of Equation (22) results in an eigenvalue problem $\mathbf{A}\mathbf{u}_i = \lambda_i\mathbf{u}_i$, where λ_i is the eigenvalue and $\mathbf{u}_i$ is the corresponding eigenvector. The adjoint eigenvalue problem $\mathbf{A}^T\mathbf{v}_j = \lambda_j\mathbf{v}_j$ has same eigenvalues but different eigenvectors because $\mathbf{A}$ is generally a non-symmetric matrix ($\mathbf{A}^T \neq \mathbf{A}$). Eigenvectors $\mathbf{u}_1, \mathbf{u}_2, \cdots, \mathbf{u}_{2n}$ and $\mathbf{v}_1, \mathbf{v}_2, \cdots, \mathbf{v}_{2n}$ are known as right and left eigenvectors, respectively. Constructing matrices of eigenvalues $\Lambda = diag[\lambda_1 \ \lambda_2 \ \cdots \ \lambda_{2n}]$, right eigenvectors $\mathbf{U} = [\mathbf{u}_1 \ \mathbf{u}_2 \ \cdots \ \mathbf{u}_{2n}]$, and left eigenvectors $\mathbf{V} = [\mathbf{v}_1 \ \mathbf{v}_2 \ \cdots \ \mathbf{v}_{2n}]$, as the biorthonormality condition is expressed as,

$$\begin{aligned} \mathbf{V}^T\mathbf{U} &= \mathbf{I} \\ \mathbf{V}^T\mathbf{A}\mathbf{U} &= \Lambda \end{aligned} \quad (23)$$

The solution of the state vector $\mathbf{u}(t)$ in Equation (21) can be assumed as a linear combination of $\xi_i(t)\mathbf{u}_i$, where $\xi_i(t)$ is the modal coordinate. Thus, $\mathbf{u}(t)$ is written as,

$$\mathbf{u}(t) = \sum_{i=1}^{2n} \xi_i(t)\mathbf{u}_i \quad (24)$$

Substituting Equation (24) into Equation (21), and premultiplying throughout by $\mathbf{V}^T$, the following equation is obtained.

$$\mathbf{V}^T\mathbf{U}\dot{\xi}(t) = \mathbf{V}^T\mathbf{A}\mathbf{U}\xi(t) + \mathbf{V}^T\mathbf{B}\mathbf{f}(t) \quad (25)$$

Using the biorthonormality condition given in Equations (23) and (25) can be transformed to

$$\dot{\xi}(t) = \Lambda \xi(t) + \mathbf{n}(t) \tag{26}$$

where $\mathbf{n}(t) = \mathbf{V}^T \mathbf{B} \mathbf{f}(t)$ is known as the modal excitation vector. Independent modal equations can be extracted from Equation (26) as,

$$\dot{\xi}_i(t) = \lambda_i \xi_i(t) + n_i(t); \; i = 1, 2, \cdots, 2n \tag{27}$$

where $n_i = \mathbf{v}_i^T \mathbf{B} \mathbf{f}(t)$. Considering $\mathbf{f}(t)$ as a harmonic force with frequency ω, the steady state response $\xi_i(t)$ is given below:

$$\xi_i(t) = \frac{\mathbf{v}_i^T \mathbf{B} \mathbf{f}(t)}{\iota \omega - \lambda_i} \tag{28}$$

Substitution of Equation (28) into Equation (24) provides the solution of state vector as,

$$\mathbf{u}(t) = \sum_{i=1}^{2n} \frac{\mathbf{u}_i \mathbf{v}_i^T}{\iota \omega - \lambda_i} \mathbf{B} \mathbf{f}(t) \tag{29}$$

3. Computational Procedure

The modified Reynold's Equation (10) is discretized using finite difference method, and the resulting algebraic equations are solved for pressure iteratively with successive over-relaxation. The criterion for convergence of pressure in the lubricating film is chosen as,

$$\frac{\sum_{i=1}^{m} \sum_{j=1}^{n} \left| (p_{i,j})^{N+1} - (p_{i,j})^N \right|}{\sum_{i=1}^{m} \sum_{j=1}^{n} \left| (p_{i,j})^{N+1} \right|} \leq \epsilon_p \tag{30}$$

where i, j are indices of grid points, and m, n are the total nodes in θ and z directions, respectively. ϵ_p is the error in pressure between successive iterations, and N is the iteration number. The grid size used in the present study is 161×51, which is determined after a grid-independent test.

After computing the pressure field, film forces are obtained through numerical integration of Equations (11) and (12) using Simpson's rule. Once the steady-state data are computed, a finite perturbation of displacement and velocity is provided, and the above steps are followed to obtain film forces with perturbed parameters. Displacement and velocity perturbations of $0.001c$ and $0.001c\Omega$, respectively, are considered in the present work. Bearing coefficients are then evaluated using Equation (16).

After computing the stiffness and damping coefficients of powder film, elemental matrices of shaft and disc elements are computed. After assembling elemental matrices into global matrices and applying the boundary conditions, eigenvalue analysis is carried out to find the natural frequencies, mode shapes, modal damping factors, and unbalance response. A flowchart depicted in Figure 2 outlines the entire computation process.

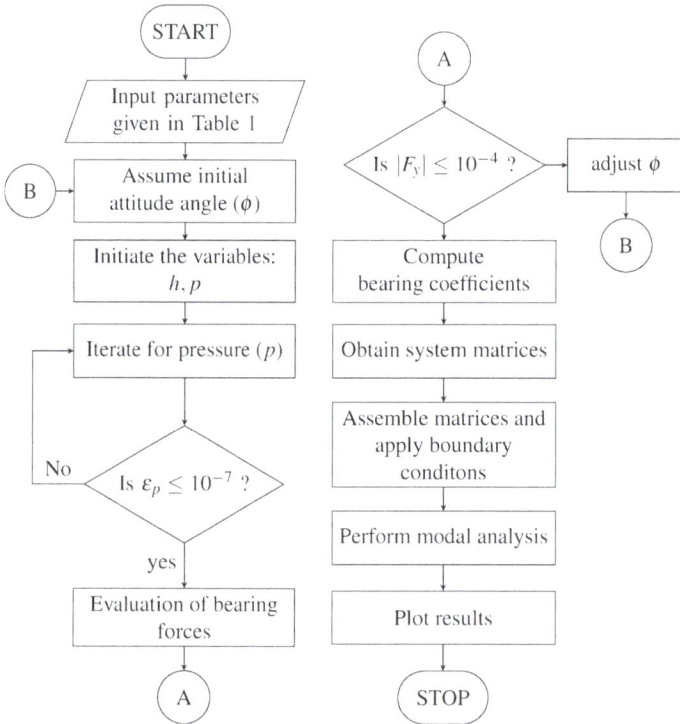

Figure 2. Flowchart of computation.

4. Results and Discussion

Figure 3 shows a simple model of a rotor–shaft system studied in this work. It is a slightly modified rotor obtained from Rao [17]. The model consists of a flexible shaft mounted on identical journal bearings at the ends. A disc of mass m_d is mounted at the centre of the shaft of length L_s. The effect of bearing wear is studied for three different cases of worn bearings with (i) $\bar{d}_0 = 0.1$, (ii) $\bar{d}_0 = 0.2$, and (iii) $\bar{d}_0 = 0.3$, in addition to smooth bearing. The data used in the computation is provided in Table 1. The correctness of the present results is validated with the published work of Tsai and Jeng [8]. A comparison of the coefficient of friction is shown in Figure 4. It can be seen that results are in good agreement, which verifies the correctness of the present results.

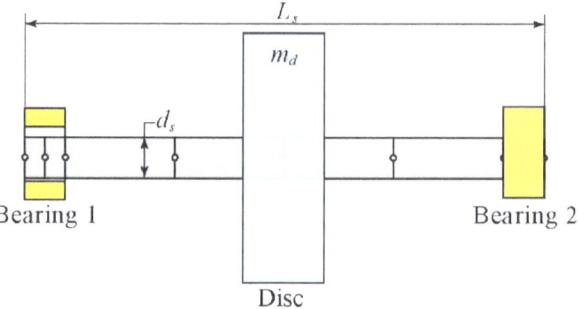

Figure 3. Rotor-bearing system supported on identical journal bearings.

Table 1. Input data.

Parameter	Notation	Unit	Value
Shaft diameter	d_s	mm	34.0
Shaft length	L_s	mm	1000
Disc mass	m_d	kg	20.4
Rotor unbalance	$m_d u_d$	kg·m	0.02
Bearing length	L	mm	34.0
Bearing diameter	D	mm	34.0
Radial clearance	c	mm	0.1
Maximum wear depth	$\bar{d}_0$	-	0.1, 0.2, 0.3
Grain diameter	d	µm	1.0
t/q			1.0
γ/r			0.0004
B/U			4.0

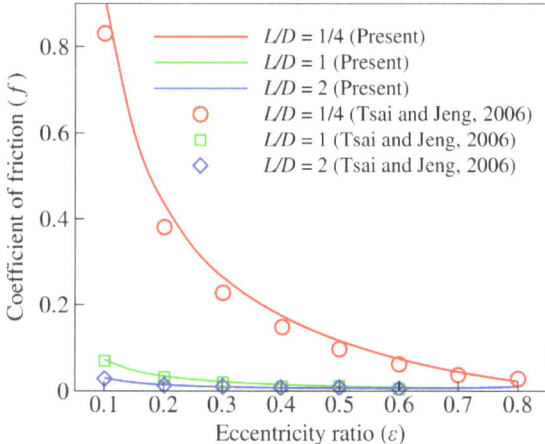

Figure 4. Comparison of present results with published work of Tsai and Jeng [8].

For a particular rotor spin speed under steady-state conditions, the journal centre takes a position inside the bearing, which is quantified by the eccentricity e and the variation in eccentricity ratio ε with rotor spin speed, and is shown in Figure 5. As the speed increases, the journal centre comes closer to the bearing centre, resulting in a smaller eccentricity ratio. This is observed for all the four cases shown in Figure 5. The variation in attitude angle with journal speed is shown in Figure 6. It can be seen that the attitude angle increases with increasing speed implying that journal centre moves away from the vertical load line. Moreover, there is a decrease in attitude angle with bearing wear. Bearing wear has a significant influence on journal eccentricity, and as the wear depth is increased, there is a wider clearance gap and the journal settles farther from the bearing centre. The steady-state positions of the journal centre for different eccentricity ratios and wear depths are depicted in Figure 7. The position of the journal centre shifts towards the vertical line passing through the bearing centre as wear increases. At very high eccentricity ($\varepsilon > 0.8$) (which corresponds to very low speed), the journal tends to settle near the bottom of the bearing. However, with the worn condition, the journal finds some extra space in the worn region, and due to hydrodynamic action, it shifts slightly to the right. This increases with increased wear depth and can be seen in Figure 7. Fillon and Bouyer [24] found similar results for oil-lubricated worn journal bearings.

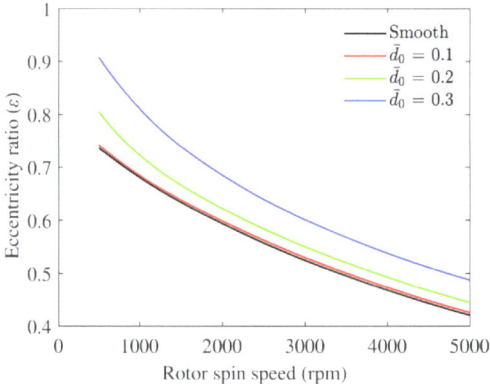

Figure 5. Eccentricity ratio variation with rotor spin speed.

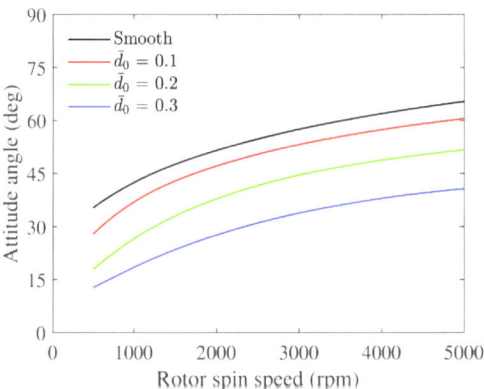

Figure 6. Attitude angle variation with rotor spin speed.

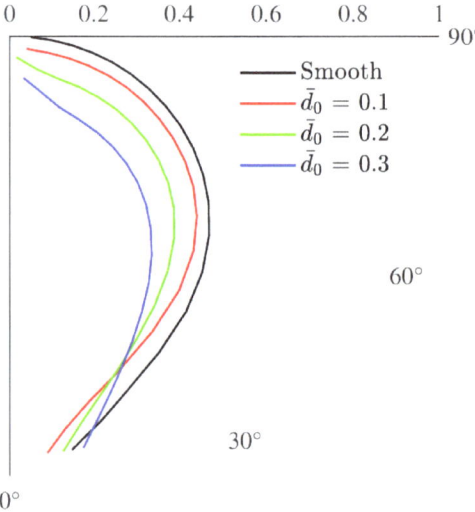

Figure 7. Steady-state position of journal centre.

The film thickness and pressure profile at the bearing mid-plane ($\bar{z} = 0.5$) for different worn conditions and speeds are shown in Figure 8. Film thickness modifies according to wear profile, and pressure profile is subsequently affected resulting in changes in the overall dynamics. At lower speeds, pressure is concentrated near the minimum film thickness. With the increase in speed, pressure is more distributed along the circumference. Figure 9 shows the pressure contours of all the cases to have a better picture of the overall pressure distribution.

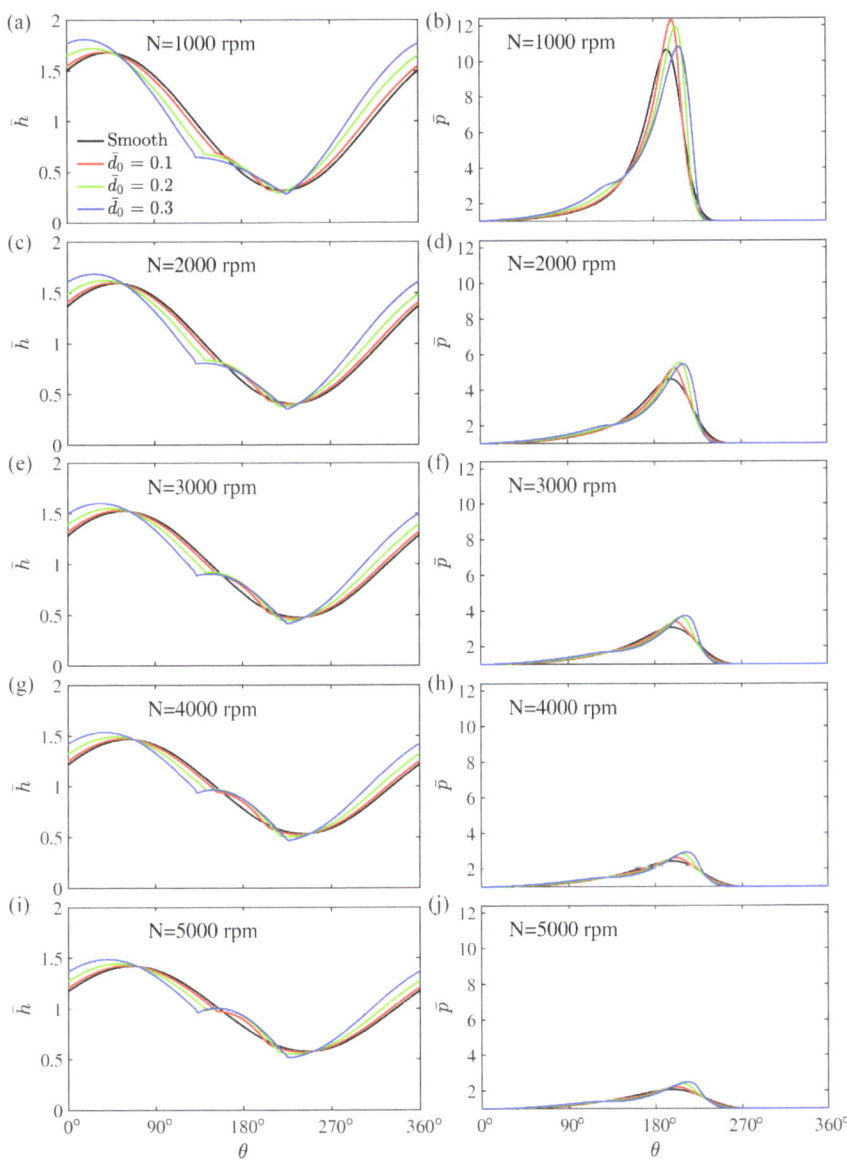

Figure 8. Film thickness and pressure profile at the bearing mid-plane ($\bar{z} = 0.5$). (**a**) $\bar{h}$ vs. θ at $N = 1000$ rpm, (**b**) $\bar{p}$ vs. θ at $N = 1000$ rpm, (**c**) $\bar{h}$ vs. θ at $N = 2000$ rpm, (**d**) $\bar{p}$ vs. θ at $N = 2000$ rpm, (**e**) $\bar{h}$ vs. θ at $N = 3000$ rpm, (**f**) $\bar{p}$ vs. θ at $N = 3000$ rpm, (**g**) $\bar{h}$ vs. θ at $N = 4000$ rpm, (**h**) $\bar{p}$ vs. θ at $N = 4000$ rpm, (**i**) $\bar{h}$ vs. θ at $N = 5000$ rpm, and (**j**) $\bar{p}$ vs. θ at $N = 5000$ rpm.

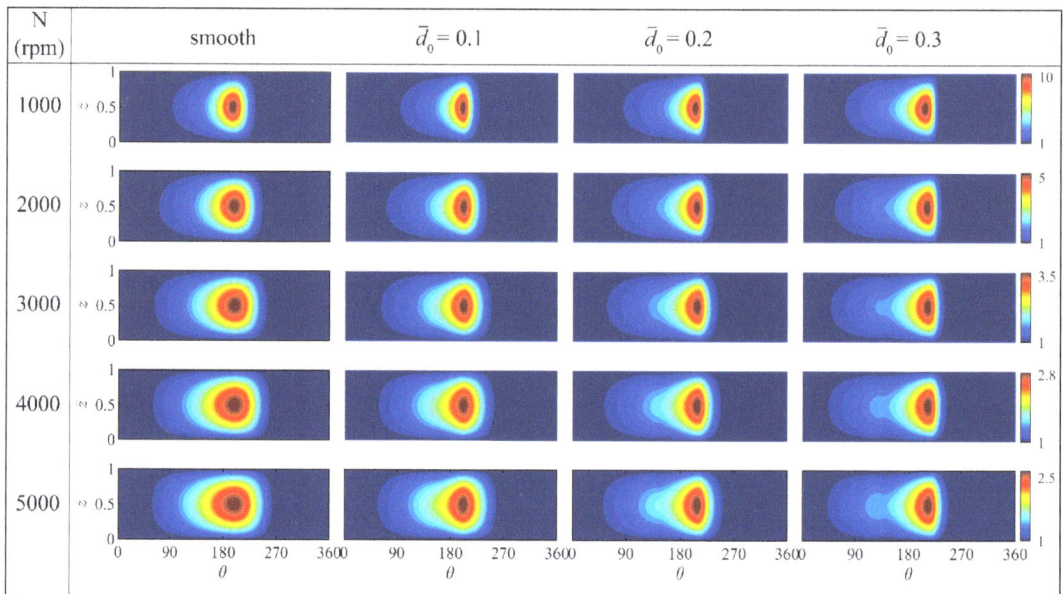

Figure 9. Pressure contours on the unwrapped bearing surface for various speed and worn depth conditions.

To study the dynamic behaviour, it is important to evaluate the bearing coefficients since bearing coefficients play an important role in predicting the stability and vibration response of rotating machinery. The stiffness of a lubricant film primarily relies on its film thickness, where a thinner film results in a higher stiffness. The stiffness coefficient generally decreases as the speed increases due to the increase in film thickness. This is seen in Figure 10. The influence of bearing wear is also substantial because worn bearing influences the film thickness. With the increase in wear depth, the stiffness coefficient is seen to increase. Wear also influences the damping coefficients as shown in Figure 11.

The Campbell diagram of four cases, (i) smooth bearing, (ii) $\bar{d}_0 = 0.1$, (iii) $\bar{d}_0 = 0.2$, and (iv) $\bar{d}_0 = 0.3$, are shown in Figure 12, wherein the whirl frequencies (imaginary part of the eigenvalues) for the first four modes are plotted against rotor spin speed. The synchronous whirl line (SWL) is also shown by a dashed line. The intersection of SWL with the whirl frequencies provides the critical speeds. The values of critical speeds and corresponding whirl frequencies are also indicated. In the subcritical region, which is on the left side of the critical speed, there are two whirl frequencies below the SWL. These are the whirl frequencies due to journal bearing and which increase with rotor spin speed. The whirl frequency corresponding to the first mode is approximately half of the rotor spin speed. The third and fourth whirl frequencies are above the SWL and represent the bending modes of the rotor. Bearing wear influences the whirl frequencies associated with bearing modes (I and II modes before veering, III and IV modes after veering). At a particular speed, there is a slight decrease in the whirl frequency associated with bearing modes. The whirl frequencies associated with shaft-bending modes are unaffected by bearing wear.

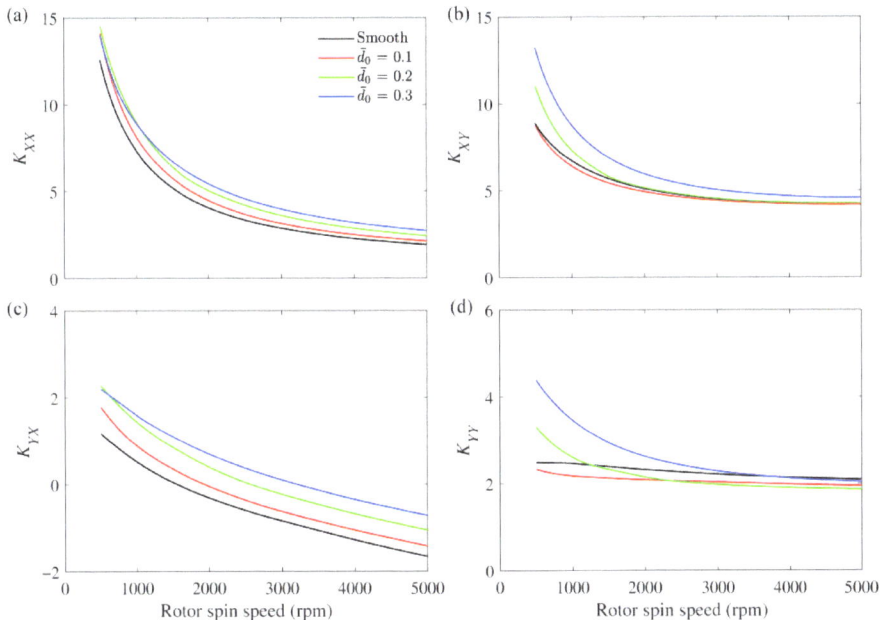

Figure 10. Variation in the dimensionless stiffness coefficients, (**a**) K_{XX}, (**b**) K_{XY}, (**c**) K_{YX}, and (**d**) K_{YY} with rotor spin speed.

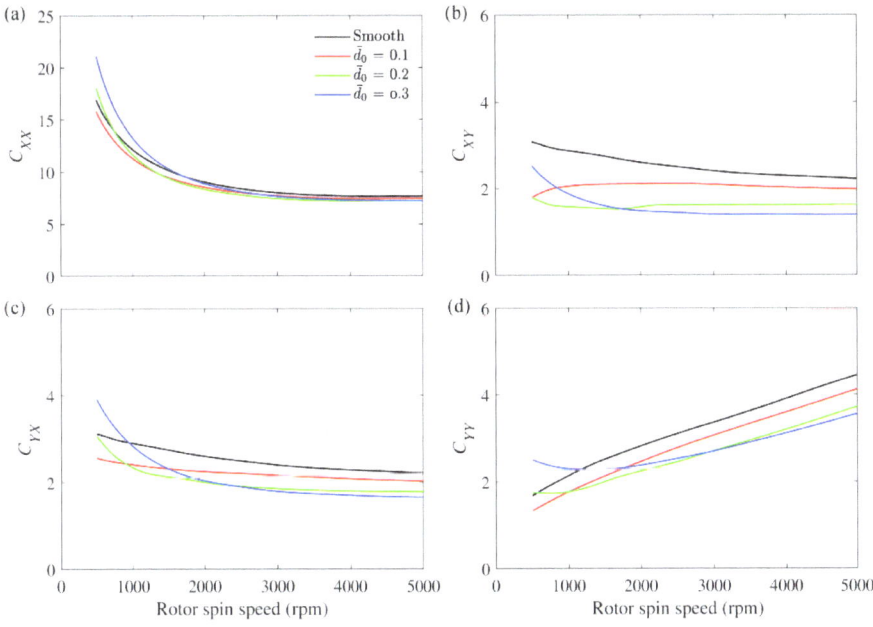

Figure 11. Variation in the dimensionless damping coefficients, (**a**) C_{XX}, (**b**) C_{XY}, (**c**) C_{YX}, and (**d**) C_{YY} with rotor spin speed.

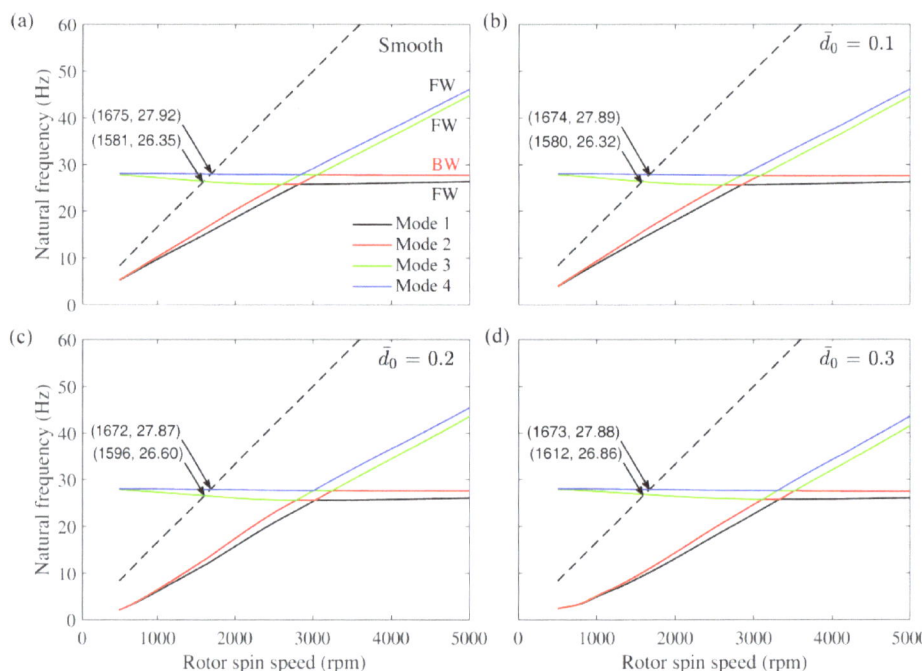

Figure 12. Campbell diagram showing first four whirl frequencies vs. rotor spin speed for (**a**) smooth bearing, (**b**) bearing with $\bar{d}_0 = 0.1$, (**c**) bearing with $\bar{d}_0 = 0.2$, and (**d**) bearing with $\bar{d}_0 = 0.3$. (FW—Forward Whirl, BW—Backward Whirl).

Rotors generally exhibit two different kinds of whirling modes known as Forward Whirl (FW) and Backward Whirl (BW). In the subcritical region, the fourth mode is the BW mode, which is the bending mode of the rotor–shaft structure. The whirl frequency lines seem to cross one another after certain speeds, but in reality, they swap their trends, a phenomenon known as curve veering. The curve veering phenomenon is seen in the post-critical region, near approximately twice the critical speed. The veering near twice the critical speed is also seen in conventional oil-lubricated journal bearings supported rotor [37]. Mode shapes are obtained from eigenvectors obtained through eigenvalue analysis. Nodal values of x and y displacements are extracted from eigenvectors. These are generally complex values implying a rotation. These values are multiplied with coordinates of a unit circle in the transverse plane whose real parts are the coordinates of ellipses shown in Figure 13 [18]. The mode shapes corresponding to the first four modes are shown in Figure 13 for 2500 rpm and 3200 rpm spin speed of rotors, which are before veering and after veering. The swapping of mode shapes can be clearly seen in this figure.

The variation of modal damping factors (MDF) of the first four modes with rotor spin speed is shown in Figure 14. A positive value of MDF implies a stable system as the energy of the system dissipates and a negative value of MDF implies an unstable system as the energy of the system increases through the whirling of the rotor. The onset of instability is defined by the stability limit speed (SLS), which is the rotor spin speed at which the MDF of any of the modes become negative. From Figures 12 and 14, it may be noticed that the SLS is nearly twice the critical speed similar to what is seen in rotor supported by oil-lubricated journal bearings [37]. The modal damping factor in the case of powder-lubricated journal bearings is generally on the higher side as compared with oil-lubricated bearings.

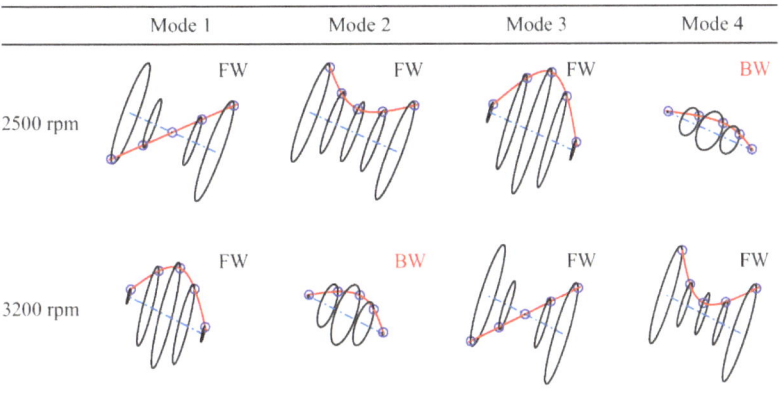

Figure 13. Mode shapes of first four modes at rotor spin speed of 2500 rpm and of 3200 rpm for smooth bearing (FW—Forward Whirl, BW—Backward Whirl).

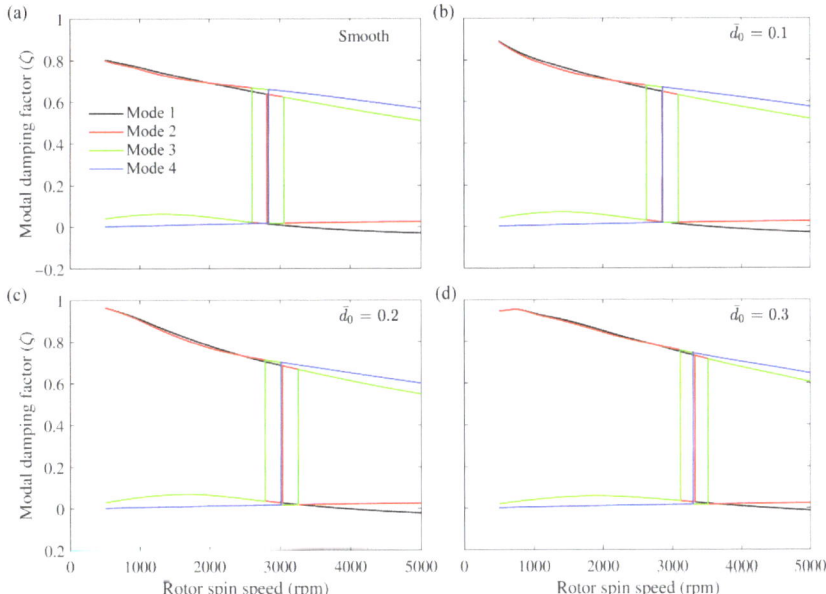

Figure 14. Modal damping factor for first four modes vs. rotor spin speed for (**a**) smooth bearing, (**b**) bearing with $\bar{d}_0 = 0.1$, (**c**) bearing with $\bar{d}_0 = 0.2$, and (**d**) bearing with $\bar{d}_0 = 0.3$.

Figure 15 shows the critical speeds and SLS for the four cases. Critical speeds are almost unaffected by bearing wear because these are the natural frequencies associated with shaft-bending modes. However, SLS is increased with increased wear depth. The negatively cross-coupled stiffness coefficient K_{YX} tends to destabilize the system by providing a force in the direction of whirling. All other coefficients have positive values, which means their forces will act towards equilibrium and bring the journal towards equilibrium. Only K_{YX} becomes negative at higher speeds, which means a force will act on the journal that will move it away from the equilibrium position. Once this force becomes sufficiently high to balance all other forces, the journal will move away from the equilibrium, and thus, an unstable condition will arrive. As previously seen in Figure 10c, the value of K_{YX} becomes negative at a higher speed, for bearing with larger wear depth. Hence, rotors supported on bearings with larger wear depth are more stable.

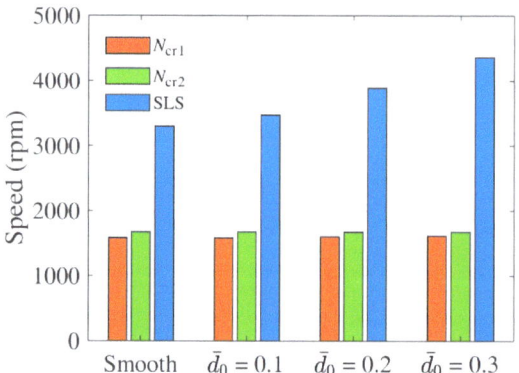

Figure 15. Critical speed and stability limit speed (SLS) for different cases.

The unbalance response (UBR) at the location of disc is shown in Figure 16. There is a clear peak near the critical speed. The bearing wear has very slight influence on the unbalance response, and the amplitude is slightly increased with increased wear depth.

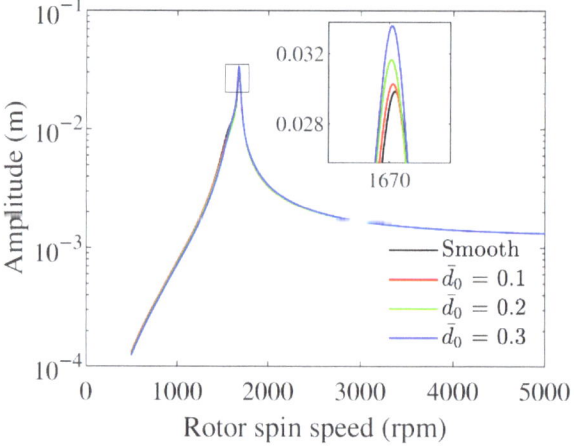

Figure 16. Unbalance response at the location of the disc.

5. Conclusions

In this paper, a rotor supported on worn journal bearings lubricated with powders is analysed for stability and modal characteristics. The rotor–shaft system is discretized using finite element method, and an eigenvalue analysis is performed to determine its modal characteristics. The system matrices are dependent upon rotor spin speed; hence, whirling frequencies and mode shapes vary with speed too. There is a swapping of mode shapes near twice the critical speed. It is also found that with increased wear depth the eccentricity ratio increases, attitude angle decreases, and stiffness coefficients increase. Worn bearing increases the stability limit speed in comparison with smooth bearing. Bearing wear has a very insignificant influence on critical speeds and unbalance response. The amplitude of unbalance response increased only slightly with increased wear depth.

The present study is carried out considering the linearised bearing coefficients, and future studies will consider the non-linearities in the bearing and a transient analysis with varying speeds.

Author Contributions: Conceptualization, F.R.; Data curation, F.R.; Formal analysis, F.R. and E.M.; Investigation, F.R. and J.G.; Methodology, F.R. and E.M.; Resources, E.M. and J.G.; Software, F.R. and J.G.; Validation, F.R.; Writing—original draft, F.R.; Writing—review and editing, F.R., E.M., and J.G. All authors have read and agreed to the published version of the manuscript.

Funding: This research received no external funding.

Data Availability Statement: Not applicable.

Conflicts of Interest: The authors declare no conflict of interest.

References

1. Wornyoh, E.Y.A.; Jasti, V.K.; Higgs III, C.F. A Review of Dry Particulate Lubrication: Powder and Granular. *J. Tribol.* **2007**, *129*, 438–449. [CrossRef]
2. Heshmat, H. *Tribology of Interface Layers*; CRC Press, Taylor and Francis Group: Boca Raton, FL, USA, 2010.
3. Rahmani, F.; Dutt, J.K.; Pandey, R.K. Performance Behaviour of Elliptical-Bore Journal Bearings Lubricated with Solid Granular Particulates. *Particuology* **2016**, *27*, 51–60. [CrossRef]
4. Iordanoff, I.; Seve, B.; Berthier, Y. Solid Third Body Analysis Using a Discrete Approach: Influence of Adhesion and Particle Size on Macroscopic Properties. *J. Tribol.* **2002**, *124*, 530–538. [CrossRef]
5. Haff, P.K. Grain Flow as a Fluid-Mechanical Phenomenon. *J. Fluid Mech.* **1983**, *134*, 401. [CrossRef]
6. Dai, F.; Khonsari, M.M.; Lu, Z.Y. On the Lubrication Mechanism of Grain Flows. *Tribol. Trans.* **1994**, *37*, 516–524. [CrossRef]
7. Tsai, H.-J.; Jeng, Y.-R. An Average Lubrication Equation for Thin Film Grain Flow With Surface Roughness Effects. *J. Tribol.* **2002**, *124*, 736–742. [CrossRef]
8. Tsai, H.-J.; Jeng, Y.-R. Characteristics of Powder Lubricated Finite-Width Journal Bearings: A Hydrodynamic Analysis. *J. Tribol.* **2006**, *128*, 351–357. [CrossRef]
9. Heshmat, H.; Brewe, D.E. Performance of a Powder Lubricated Journal Bearing With WS2 Powder: Experimental Study. *J. Tribol.* **1996**, *118*, 484–491. [CrossRef]
10. Higgs III, C.F.; Tichy, J. Effect of Particle and Surface Properties on Granular Lubrication Flow. *Proc. Inst. Mech. Eng. Part J J. Eng. Tribol.* **2008**, *222*, 703–713. [CrossRef]
11. Higgs III, C.F.; Tichy, J. Granular Flow Lubrication: Continuum Modeling of Shear Behavior. *J. Tribol.* **2004**, *126*, 499–510. [CrossRef]
12. Sawyer, W.G.; Tichy, J. Lubrication With Granular Flow: Continuum Theory, Particle Simulations, Comparison With Experiment. *J. Tribol.* **2001**, *123*, 777–784. [CrossRef]
13. Meng, F.; Liu, K.; Wang, W. The Force Chains and Dynamic States of Granular Flow Lubrication. *Tribol. Trans.* **2015**, *58*, 70–78. [CrossRef]
14. Wang, W.; Gu, W.; Liu, K.; Wang, F.; Tang, Z. DEM Simulation on the Startup Dynamic Process of a Plain Journal Bearing Lubricated by Granular Media. *Tribol. Trans.* **2014**, *57*, 198–205. [CrossRef]
15. Wang, W.; Gu, W.; Liu, K. Force Chain Evolution and Force Characteristics of Shearing Granular Media in Taylor-Couette Geometry by DEM. *Tribol. Trans.* **2015**, *58*, 197–206. [CrossRef]
16. Irretier, H. Mathematical Foundations of Experimental Modal Analysis in Rotor Dynamics. *Mech. Syst. Signal Process.* **1999**, *13*, 183–191. [CrossRef]
17. Rao, J.S. *Rotor Dynamics*; New Age International (P) Ltd.: Delhi, India, 1996.
18. Friswell, M.I.; Penny, J.E.; Garvey, S.D.; Lees, A.W. *Dynamics of Rotating Machines*; Cambridge University Press: Cambridge, UK, 2010.
19. Qiu, Z.L.; Tieu, A.K. The Effect of Perturbation Amplitudes on Eight Force Coefficients of Journal Bearings. *Tribol. Trans.* **1996**, *39*, 469–475. [CrossRef]
20. Rahmani, F.; Dutt, J.K.; Pandey, R.K. Stability of Rotor Supported on Powder Lubricated Journal Bearings with Surface Pockets. *Proc. IMechE Part C J. Mech. Eng. Sci.* **2021**, *235*, 2317–2329. [CrossRef]
21. El-Shafei, A.; Tawfick, S.H.; Raafat, M.S.; Aziz, G.M. Some Experiments on Oil Whirl and Oil Whip. In *ASME Turbo Expo 2004: Power for Land, Sea, and Air, Proceedings of the Turbo Expo 2004, Vienna, Austria, 14–17 June 2004*; ASME: New York, NY, USA, 2008; Volume 6, pp. 701–710. [CrossRef]
22. Dufrane, K.F.; Kannel, J.W.; McCloskey, T.H. Wear of Steam Turbine Journal Bearings at Low Operating Speeds. *J. Lubr. Technol.* **1983**, *105*, 313–317. [CrossRef]
23. Hashimoto, H.; Wada, S.; Nojima, K. Performance Characteristics of Worn Journal Bearings in Both Laminar and Turbulent Regimes. Part I: Steady-State Characteristics. *ASLE Trans.* **1986**, *29*, 565–571. [CrossRef]
24. Fillon, M.; Bouyer, J. Thermohydrodynamic Analysis of a Worn Plain Journal Bearing. *Tribol. Int.* **2004**, *37*, 129–136. [CrossRef]

25. Papadopoulos, C.A.; Nikolakopoulos, P.G.; Gounaris, G.D. Identification of Clearances and Stability Analysis for a Rotor-Journal Bearing System. *Mech. Mach. Theory* **2008**, *43*, 411–426. [CrossRef]
26. Gertzos, K.P.; Nikolakopoulos, P.G.; Chasalevris, A.C.; Papadopoulos, C.A. Wear Identification in Rotor-Bearing Systems by Measurements of Dynamic Bearing Characteristics. *Comput. Struct.* **2011**, *89*, 55–66. [CrossRef]
27. Chasalevris, A.C.; Nikolakopoulos, P.G.; Papadopoulos, C.A. Dynamic Effect of Bearing Wear on Rotor-Bearing System Response. *J. Tribol.* **2013**, *135*, 011008. [CrossRef]
28. Machado, T.H.; Cavalca, K.L. Modeling of Hydrodynamic Bearing Wear in Rotor-Bearing Systems. *Mech. Res. Commun.* **2015**, *69*, 15–23. [CrossRef]
29. Machado, T.H.; Mendes, R.U.; Cavalca, K.L. Directional Frequency Response Applied to Wear Identification in Hydrodynamic Bearings. *Mech. Res. Commun.* **2016**, *74*, 60–71. [CrossRef]
30. Machado, T.H.; Alves, D.S.; Cavalca, K.L. Investigation about Journal Bearing Wear Effect on Rotating System Dynamic Response in Time Domain. *Tribol. Int.* **2019**, *129*, 124–136. [CrossRef]
31. König, F.; Ould Chaib, A.; Jacobs, G.; Sous, C. A Multiscale-Approach for Wear Prediction in Journal Bearing Systems—From Wearing-in towards Steady-State Wear. *Wear* **2019**, *426–427*, 1203–1211. [CrossRef]
32. Gecgel, O.; Dias, J.P.; Ekwaro-Osire, S.; Alves, D.S.; Machado, T.H.; Daniel, G.B.; de Castro, H.F.; Cavalca, K.L. Simulation-Driven Deep Learning Approach for Wear Diagnostics in Hydrodynamic Journal Bearings. *J. Tribol.* **2020**, *143*, 084501. [CrossRef]
33. Mokhtari, N.; Pelham, J.G.; Nowoisky, S.; Bote-Garcia, J.L.; Gühmann, C. Friction and Wear Monitoring Methods for Journal Bearings of Geared Turbofans Based on Acoustic Emission Signals and Machine Learning. *Lubricants* **2020**, *8*, 29. [CrossRef]
34. König, F.; Sous, C.; Ould Chaib, A.; Jacobs, G. Machine Learning Based Anomaly Detection and Classification of Acoustic Emission Events for Wear Monitoring in Sliding Bearing Systems. *Tribol. Int.* **2021**, *155*, 106811. [CrossRef]
35. Rahmani, F.; Pandey, R.K.; Dutt, J.K. Performance Studies of Powder-Lubricated Journal Bearing Having Different Pocket Shapes at Cylindrical Bore Surface. *J. Tribol.* **2018**, *140*, 031704. [CrossRef]
36. Meirovitch, L. *Fundamentals of Vibrations*; Waveland Press: Long Grove, IL, USA, 2010.
37. Chouksey, M.; Dutt, J.K.; Modak, S.V. Modal Analysis of Rotor-Shaft System under the Influence of Rotor-Shaft Material Damping and Fluid Film Forces. *Mech. Mach. Theory* **2012**, *48*, 81–93. [CrossRef]

Disclaimer/Publisher's Note: The statements, opinions and data contained in all publications are solely those of the individual author(s) and contributor(s) and not of MDPI and/or the editor(s). MDPI and/or the editor(s) disclaim responsibility for any injury to people or property resulting from any ideas, methods, instructions or products referred to in the content.

Article

Simulation Analysis and Experimental Research on Electric Thermal Coupling of Current Bearing

Zhiwei Wang, Shuanglong Mao, Heng Tian, Bing Su and Yongcun Cui *

School of Mechatronics Engineering, Henan University of Science and Technology, Luoyang 471003, China; wangzw87@yeah.net (Z.W.); 15716670075@163.com (S.M.); tianheng_1988@163.com (H.T.); subing@haust.edu.cn (B.S.)
* Correspondence: 9906172@haust.edu.cn; Tel.: +86-0379-64231479

Abstract: With the advancement of industries such as high-speed railways, new energy vehicles, and wind power, bearings are frequently exposed to various electric field environments, leading to the need for lubricating oil films of bearings to withstand voltage. One of the major issues caused by the breakdown discharge process of the lubricating oil film in bearings is the generation of local instantaneous high temperatures. This temperature rise is a key factor contributing to problems such as high operating temperature of bearings, surface damage in the contact area, and degradation of lubrication performance. This research article focuses on the comprehensive influence of bearing friction and electrical factors. It establishes a heat source calculation model and a temperature field simulation model specifically for current-carrying bearings. This study analyzes both the overall temperature rise of bearings and the local temperature rise resulting from breakdown discharge. Furthermore, the accuracy of the simulation analysis is verified through experiments. The temperature field simulation and experimental results consistently indicate that electrical environmental factors can cause an increase in the overall temperature rise of a bearing. Additionally, the breakdown and discharge of the lubricating oil film generate local instantaneous high temperatures in the contact area of the bearing.

Keywords: bearings; electrical erosion; equivalent circuit; electric thermal coupling

1. Introduction

With the development of industries such as high-speed railways, new energy vehicles, and wind power, bearings are increasingly exposed to various electric field environments. This electrical intervention can result in a new type of damage known as electrical corrosion in bearings [1]. In sectors like high-speed railways and new energy vehicles, AC motors are widely utilized due to their excellent operational and control characteristics when driven by a variable-frequency power supply, which has led to significant economic benefits. However, despite the substantial economic advantages brought about by variable-frequency drive technology, it also has negative consequences, with electrical corrosion of bearings being a particularly severe issue [2–4]. When the shaft voltage remains below the threshold voltage of the bearing oil film, the bearing functions as a capacitor, and the dv/dt current represents the charging and discharging current of the bearing capacitor. Typically, this value is minimal and does not cause any damage to the bearing. However, when the shaft voltage exceeds the threshold voltage of the lubricating grease film, the electrical energy stored in the parasitic capacitor will generate a discharge pulse current, which is referred to as the discharge bearing current [5,6]. The amplitude of the change in electric field strength in the lubricating oil film is a crucial factor in determining the level of bearing discharge. The breakdown field strength is influenced by several factors, including the viscosity of the base oil, the use of additives, and humidity. The breakdown strength of the lubricating oil film typically falls within the range of 15–50 kV/mm [7]. In the Hertz contact area of bearings, the thickness of the oil film center is generally around 0.2–0.8 μm [8]. If the shaft

Citation: Wang, Z.; Mao, S.; Tian, H.; Su, B.; Cui, Y. Simulation Analysis and Experimental Research on Electric Thermal Coupling of Current Bearing. *Lubricants* **2024**, *12*, 73. https://doi.org/10.3390/lubricants12030073

Received: 30 January 2024
Revised: 20 February 2024
Accepted: 22 February 2024
Published: 26 February 2024

Copyright: © 2024 by the authors. Licensee MDPI, Basel, Switzerland. This article is an open access article distributed under the terms and conditions of the Creative Commons Attribution (CC BY) license (https://creativecommons.org/licenses/by/4.0/).

voltage exceeds 3 V, it can result in breakdown discharge in the Hertz contact area, leading to local high temperatures and potential electrical corrosion damage to the bearing.

The issue of motor shaft current caused by frequency converters has become increasingly prominent with their widespread use in driving AC motors. Since the 1990s, this topic has received significant attention in academic reports [9,10]. Scholars have conducted research on the phenomenon of bearing electrical corrosion during this period. Tischmacher et al. [11] demonstrated through experiments that the thickness of the lubricating oil film and its breakdown are influenced by the magnitude of the applied voltage, conductivity of lubricating grease, and bearing speed. El Hadj Miliani [12] highlighted the detrimental effects of axle current on the service life of bearings in railway locomotives and electric vehicles. Romanenko et al. [13] analyzed the changes in the dielectric strength and chemical properties of various lubricating greases under the influence of current, providing an explanation for bearing damage. Niu K. et al. [14] described the mechanism of shaft voltage and shaft current generation, and analyzed the associated hazards such as material weakening, surface ablation, and lubrication deterioration, along with their relative mechanisms. The calculation of the capacitance of the bearing lubricating oil film and the shaft current is a crucial step in analyzing the phenomenon of bearing electrical corrosion. Binder et al. [15] considered the bearing as a flat plate capacitor, with the Hertz contact surface as the polar plate and the lubricating oil film center thickness as the polar plate spacing. They analyzed the bearing based on this model. Lu F.M. [16] developed circuit models for ordinary bearings and insulated bearings to analyze performance indicators such as the insulation resistance, AC voltage value, and coating capacitance of insulated bearing coatings. Liu R.F. et al. [17] proposed a bearing capacitance calculation method based on electromagnetic field and determined the center thickness of the bearing oil film using the theory of elastohydrodynamic lubrication. They established a finite element analysis model and calculated the equivalent capacitance of the bearing. Xiong F. et al. [18] took into account the variation in the radial clearance of bearings and the eccentricity of motor air gaps. They divided the equivalent capacitance of bearings into Hertz contact capacitance and non-Hertz contact capacitance based on Hertz contact theory, and established a calculation model for the equivalent capacitance of bearings. From a microscopic perspective, the bearing raceway and rolling element act as the electrodes, with the medium being an oil film. Bearing breakdown is considered a form of fine electric spark breakdown. The breakdown mechanism of liquid media in general electric spark breakdown is usually explained using the electron avalanche theory [19]. Subsequently, J M. Meek and L.B. Loeb proposed the streamer theory to address the limitations of the electron avalanche theory in terms of breakdown time [20]. When the lubricating oil film of a bearing is disrupted, it can lead to local instantaneous high temperature, resulting in partial melting of the contact area. Analyzing the temperature field of the bearing is an important step in evaluating its performance indicators. Lei J. et al. [21] developed a steady-state calculation model for bearings using the thermal network method. They considered various factors that affect the friction power consumption of bearing components. Liu Y.Y. et al. [22] established the dynamic differential equations of the double-row angular contact ball bearing based on the rolling bearing dynamics theory. Li J.S. et al. [23] used a local method to construct a calculation model for friction power consumption in bearings. They established a heat transfer model and analyzed the impact of rotational speed, axial force, and lubricating oil temperature on the temperature rise in the bearing system. Xie Y.B. et al. [24] developed a mathematical model for heat generation in bearings at different speeds based on friction and heat transfer theories. They used finite element software to simulate and calculate the temperature field distribution of bearings and bearing seats without cooling structures. Wang Q.Q. et al. [25] established a model for bearing discharge breakdown and conducted electric–thermal coupling simulations to determine the temperature rise at the breakdown channel and point over time. They also proposed a method for determining the critical current of bearing electrical corrosion. While there are numerous comprehensive theories and methods available for analyzing the temperature fields of conventional bearings, there is a noticeable lack of research on the analysis of temperature fields in bearings operating in electrical environments.

This article presents a calculation model for two types of heat sources: comprehensive friction power consumption and electric heating power consumption. It provides simulation analysis models and methods for the temperature rise in the outer ring of the bearing and the local temperature rise in the contact area, which are then validated through experiments.

2. Simulation Model of Heat Generation in Current-Carrying Bearings

In previous studies on temperature field analysis of bearings, researchers have often focused solely on the power loss caused by friction. However, in the case of current-carrying bearings, it is necessary to consider both the power consumption due to friction and the power consumption due to electric heating. Therefore, this section presents a heat generation model that accounts for both factors in current-carrying bearings.

2.1. Equivalent Circuit of Current Bearing

During the operation of a current bearing, a shaft current is generated. As the current passes through the bearing, it encounters resistance, leading to energy loss. This lost energy is converted into heat, causing the bearing temperature to increase. Hence, it is crucial to construct an equivalent circuit for current-carrying bearings, establish a model for heat generation due to current, and analyze it.

The insulation coating (ceramic layer attached to the outer diameter surface of the bearing and the two outer ring end faces) capacitance, inner and outer ring resistance, steel ball resistance, and oil film capacitance of the current-carrying bearing collectively form a hybrid circuit during operation. In the non-load-bearing area, the surface of the steel ball is covered by a thicker liquid film compared to the lubricating oil film in the load-bearing area, and the capacitance can be disregarded. Hence, when constructing an equivalent circuit, only the load-bearing area needs to be taken into account.

In the case of bearings with multiple steel balls, the capacitance and resistance between the steel balls and the inner and outer raceways are connected in series. However, the steel balls themselves are connected in parallel. When a bearing is coated with a ceramic coating, the capacitance of the ceramic coating is connected in series with the equivalent capacitance of the bearing. The equivalent circuit of a single steel ball in a bearing consists of a combination of a capacitor and an oil film resistance. This oil film is formed by the lubricating oil between the steel ball and the inner and outer rings. The schematic diagram of the bearing is depicted in Figure 1.

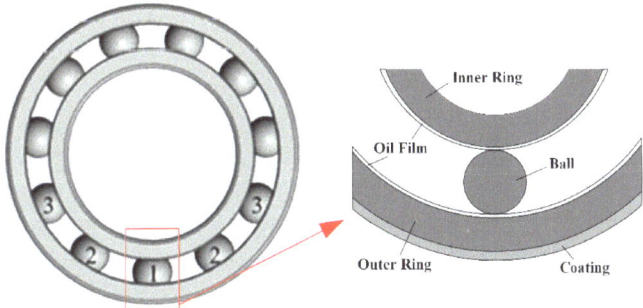

Figure 1. Bearing schematic diagram (without cage).

According to Figure 2, the circuit model of the bearing can be represented as a parallel connection of n steel ball units, combined with the inner and outer ring resistors, coating capacitors, and resistors. From a circuit equivalence perspective, when the bearing is intact, the steel ball, inner and outer raceways, lubricating oil film, and coating together form the equivalent capacitance of the bearing. The lubricating oil film primarily carries the voltage in this state. However, when the lubricating oil film breaks down, the steel ball and inner and outer raceways enter a resistive state with a very small resistance value. In this case, the majority of the voltage is borne by the coating.

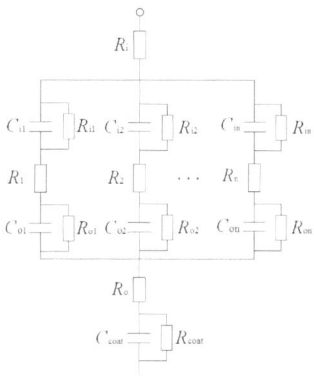

Figure 2. Equivalent model of bearing circuit.

2.2. Calculation of Current-Carrying Bearing Capacitance

Constructing an equivalent circuit and calculating the equivalent capacitance of current-carrying bearings are important steps in the analysis. Figure 3 illustrates the structure of the bearing and the Hertz contact deformation diagram. It can be observed that the bearing capacitance in the bearing area can be considered as a parallel connection of two types of capacitors. The first type is the Hertz contact capacitance, denoted as C_{Hz}, which is formed in the Hertz contact region. The second type is the non-Hertz contact capacitance, denoted as C_{noHz}, which is formed in the non-Hertz contact region. These two capacitances are connected in series to obtain the total capacitance, C_{in} and C_{on}, of the inner and outer rings of a single steel ball. The capacitance formed by the steel ball and the inner and outer raceway oil film is also connected in series to obtain the capacitance of a single steel ball. The capacitance formed by the steel balls in the bearing area is connected in parallel, ultimately resulting in the total bearing capacitance, C_B.

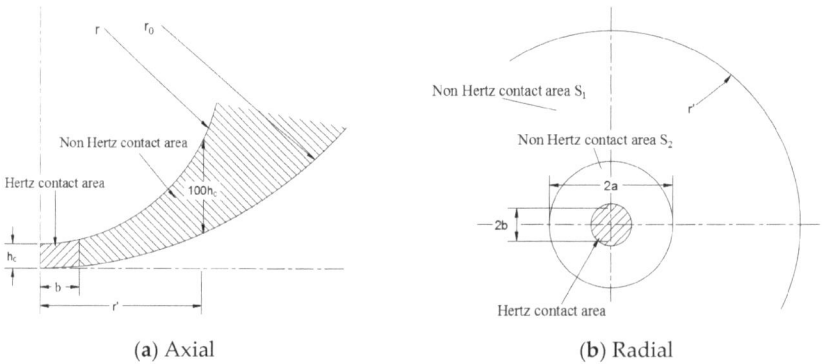

(a) Axial   (b) Radial

Figure 3. Schematic diagram of contact area between steel ball and raceway.

The steel ball and raceway experience elastic deformation upon contact, transitioning from point contact to surface contact and eventually creating an elliptical Hertz contact area. The Hertz contact capacitance between the steel ball and the inner or outer raceway is as follows:

$$C_{Hz} = \frac{\varepsilon_0 \times \varepsilon_r \times A_h}{h_c} \qquad (1)$$

In the formula, ε_0 is the vacuum dielectric constant, ε_r is the relative dielectric constant, A_h is the Hertz contact area, and h_c is the central oil film thickness.

When a bearing is only subjected to radial force, the central oil film thickness between a single steel ball and raceway in the load zone can be calculated using the theory of elastohydrodynamic lubrication and by considering the structure of the bearing steel ball and raceway. The formula for this calculation is provided in reference [26].

$$h_c = 2.69 \frac{\alpha^{0.53}(\rho v u_0)^{0.68} R_x^{0.464}}{E_0^{0.073} Q_{max}^{0.067}} (1 - 0.61 e^{-0.72K}) \qquad (2)$$

In the formula, α is the viscosity pressure index of the lubricating grease; ρ is the density of the lubricating grease; v is the viscosity of the lubricating grease base oil, which is related to the temperature of the working environment of the lubricating grease; u_0 is the average velocity of the contact surface between the steel ball and the raceway; R_x is the equivalent curvature radius along the rolling direction of the steel ball; R_y is the equivalent curvature radius in the axial plane; K is the ellipticity of the stress area; E_0 is the equivalent elastic modulus of the steel ball and ring; and Q_{max} is the maximum load borne by the steel ball.

According to Hertz contact theory, the contact area between the steel ball and raceway of deep groove ball bearings is elliptical [27]. Therefore, the formula for calculating the area of the ellipse formed by contact deformation at the contact point is as follows:

$$A_h = \pi ab \qquad (3)$$

In the formula, a and b are the major and minor axes of the elliptical surface, respectively.

Figure 3 illustrates the plane projection of the contact deformation between the steel ball and the raceway in the non-Hertz contact area of the bearing. When calculating the capacitance in this area for a ball bearing, only the portion that has a notable influence on the capacitance is taken into account. The capacitance between the steel ball and the inner or outer raceway, within the axial and radial contact points, is determined as follows [8]:

$$C_{noHz} = 4\varepsilon_0 \varepsilon_r \int_b^{r\prime} \frac{\sqrt{\frac{r_0}{r_0 - r}}}{h_c + \frac{1}{2r}x^2} dx \qquad (4)$$

In the formula, $r\prime$ is the horizontal distance projected from this point to the surface of the raceway when the gap thickness between the steel ball and the raceway is $100 h_c$, r is the radius of the steel ball, and r_0 is the raceway radius.

For both the inner and outer raceways of the bearing, there are C_{Hz} and C_{noHz}. The total capacitance between each steel ball in the bearing area and the inner raceway is C_{Bi}. This capacitance is the parallel connection of the Hertz contact area capacitance of the inner raceway and the non-Hertz contact area capacitance of the inner raceway. It can be represented as follows:

$$\begin{cases} C_{in} = C_{Hzi} + C_{noHzi} \\ C_{on} = C_{Hze} + C_{noHze} \end{cases} \qquad (5)$$

The formula for calculating the equivalent capacitance of bearings is as follows:

$$C_B = C_{B1} + 2C_{B2} + 2C_{B3} = \frac{C_{i1}C_{o1}}{C_{i1} + C_{o1}} + 2\frac{C_{i2}C_{o2}}{C_{i2} + C_{o2}} + 2\frac{C_{i3}C_{o3}}{C_{i3} + C_{o3}} \qquad (6)$$

After applying an insulation coating to the outer ring of the bearing, the capacitance of the insulation coating C_T is connected in series with the capacitance of the lubricating oil film of the bearing. The calculation formula for the total capacitance of the bearing C_{Total} is as follows:

$$C_{Total} = \frac{C_B C_T}{C_B + C_T} \qquad (7)$$

The electric thermal energy loss caused by the presence of a shaft current in the current-carrying bearing can be determined by calculating the shaft current I and the equivalent impedance of the bearing X_C, based on the established equivalent circuit. As stated in [28], the formula for calculating the current through the capacitor and the equivalent impedance of the capacitor is as follows:

$$I = C_{Total} \frac{dU}{dt} \qquad (8)$$

$$X_C = \frac{1}{2\pi f C_{Total}} \qquad (9)$$

In the formula, f is the frequency value of the applied excitation signal.

2.3. Calculation Model of Bearing Heat Source

In the case of grease-lubricated current-carrying bearings guided by steel balls, the power consumption generated during operation is a combination of various factors. These include friction power consumption caused by elastic hysteresis between steel balls and raceways, friction power consumption caused by differential sliding between steel balls and raceways, friction power consumption caused by spin sliding of steel balls, friction power consumption between steel balls and cages, and friction power consumption caused by oil film viscosity loss [21]. Additionally, current-carrying bearings may experience electrical energy losses due to the shaft current.

Combining Equations (8) and (9), the formula for calculating the electric thermal energy loss of current bearings due to the presence of a shaft current is as follows:

$$P = I^2 X_C \qquad (10)$$

The formula for calculating the total friction power consumption of ball bearings is as follows:

$$H_{total} = H_E + H_D + H_S + H_{cb} + H_L + P \qquad (11)$$

In the formula, H_E is the friction power consumption caused by the elastic hysteresis of the steel ball, H_D is the friction power consumption caused by the differential sliding between the steel ball and the raceway, H_S is the friction power consumption caused by the spin sliding of the steel ball, H_{cb} is the friction power consumption between the steel ball and the cage, H_L is the friction power consumption caused by the loss of oil film viscosity, and H_{total} is the total friction power consumption of the bearing.

3. Simulation and Analysis of Temperature Field in Current-Carrying Bearings

In this article, we first simulate the temperature field to study the frictional heat generation in current-carrying bearings. The simulation helps us to determine the temperature in the contact area between the bearing steel ball and the inner and outer rings. Subsequently, we use this temperature as the initial temperature to perform electric thermal coupling simulation on the bearings.

3.1. Temperature Field Simulation Model

This article focuses on the modeling and analysis of 6215 bearings with insulation coatings for motors. The structural parameters of the bearing are shown in Table 1, and the relevant parameters of the lubricating grease are shown in Table 2.

A simplified model of the bearing is illustrated in Figure 4. It comprises an air fluid domain, a simplified model of the bearing seat, a simplified model of the bearing's inner and outer rings, and a lubricating grease fluid domain (consisting of the bearing cavity and grease storage chamber). The outermost layer is the air layer, with the outer air wall set to room temperature. Heat exchange occurs between the air and the bearing seat through convection. The bearing seat wraps around the bearing and the lubricating grease storage chamber. The inner ring comes into contact with the main shaft, and the bearing seat exhibits better heat dissipation performance compared to the shaft. The issue of shaft heat dissipation is not taken into consideration. The lubricating grease outlet is located in the red section on the side of the bearing, and the outlet is assumed to dissipate heat through air convection.

Table 1. Structural parameters of bearings.

Parameter	Meaning	Value
d	Inside diameter	75 mm
B	Width	25 mm
f_i	Coefficient of curvature radius of inner raceway groove	0.5097
D_W	Steel ball diameter	17.462 mm
λ	Radial clearance	46~71 μm
E_0	The elastic modulus of steel	2.08×10^{11} Pa
ε_0	Vacuum dielectric constant	8.85×10^{-12} F/m
D	Outside diameter	130 mm
d_m	Bearing pitch diameter	102.5 mm
f_e	Coefficient of curvature radius of outer raceway groove	0.5268
N	Number of steel balls	11
γ	Dimensionless geometric parameters	0.17
M	Viscosity–pressure coefficient	2.08×10^{-8} Pa^{-1}
ε_r	Dielectric constant of lubricating grease	2.5

Table 2. Lubricating grease-related parameters.

Parameter	Value
40 °C base oil viscosity	220 mm^2/s
Density	880 kg/m^3
100 °C base oil viscosity	19 mm^2/s

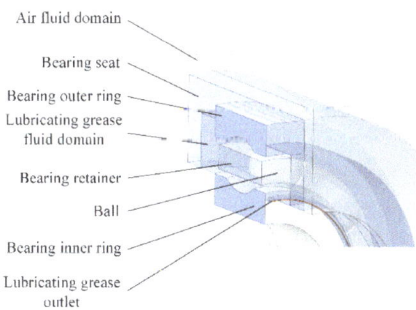

Figure 4. Simplified analysis diagram of temperature field simulation bearings.

The erosion pit area between the current bearing steel ball and the raceway is very small. In order to improve calculation efficiency without compromising simulation accuracy, it is necessary to reasonably simplify the bearing electric thermal coupling simulation model. When the bearing is subjected to radial load, the steel ball directly beneath the bearing area bears the maximum load. The lubricating oil film thickness in this contact area is the smallest, making it more susceptible to breakdown. Therefore, the steel ball and the inner and outer local raceways in contact with it were selected as the objects of the electric thermal coupling simulation (refer to Figure 5a,b). Since the breakdown of the bearing oil film only occurs locally, only a small portion of the oil film is considered when simulating the temperature field of the bearing's electric thermal coupling. Consequently, the simulation model will be further simplified by focusing on a cube centered on the Hertz contact part between the steel ball and the raceway (refer to Figure 5c). Finally, the remaining parts were removed, and the intercepted steel balls and raceways were locally enlarged to obtain the corresponding bearing electric thermal coupling simulation model, as shown in Figure 5d.

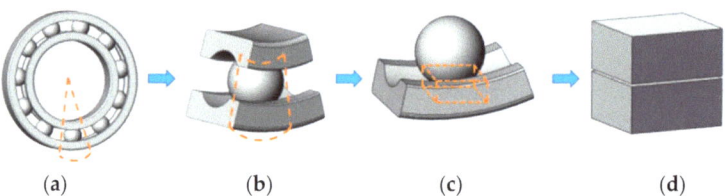

(a)   (b)   (c)   (d)

Figure 5. Simplification of bearing electric thermal coupling model.

In order to analyze the breakdown of the bearing's lubricating oil film involving electromagnetics and thermodynamics, a simulation analysis was conducted using Comsol. The analysis involved adding the current physical field and transient temperature field. The temperature and its variation patterns at different positions in the current channel of the bearing's lubricating oil film were analyzed. Figure 6 shows the points selected for analysis in the current channel. Point a, located at the center of the plane where the excitation current is applied, is the breakdown point within the current channel and serves as the connection between the outer raceway of the bearing and the current channel. Point B, located at the junction of the current channel and the lubricating oil film, belongs to the outer layer of the current channel. Point c is located at the center of the current channel. Point d, located on the other side of the penetration point that is symmetrical with point a in the horizontal plane of the model center, serves as the connection between the current channel and the steel ball.

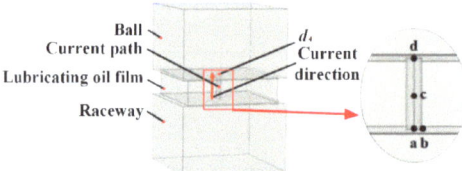

Figure 6. Electric thermal coupling simulation model.

During the operation of the bearing, the outer ring remains stationary while the steel ball and inner raceway rotate. The contact between the steel ball and any point on the inner and outer raceway is dynamic. This contact area experiences electrical erosion and breakdown, which involves the establishment of a breakdown channel, breakdown, and elimination processes. The duration of this process is determined by the relative motion speed. Additionally, it is important to calculate the time it takes for the steel ball to roll over the breakdown area and simulate the resulting temperature rise in the current channel during this time.

The calculation formula for the time required for the steel ball to roll over the current path area is as follows:

$$t = \frac{b}{\Delta V} \tag{12}$$

In the formula, b is the cross-sectional diameter of the current channel, which is the short half axis of the contact elliptical surface; ΔV is the relative velocity of the steel ball relative to the inner and outer raceways, which can be obtained from [26].

3.2. Simulation Results of Frictional Heat Generation Temperature Field

According to Section 2.1, the power consumption resulting from friction heat in the bearing under the specified working conditions was calculated to be 220 W. This power was then allocated and used to simulate the temperature distribution using the Ansys Fluent module, as shown in Figure 7. From the figure, it can be observed that the outer ring of the bearing reached a maximum temperature of 58.6 °C, whereas the contact area between the steel ball and the inner raceway reached a maximum temperature of 106 °C, and the contact area between the steel ball and the outer raceway reached a maximum temperature of 104 °C.

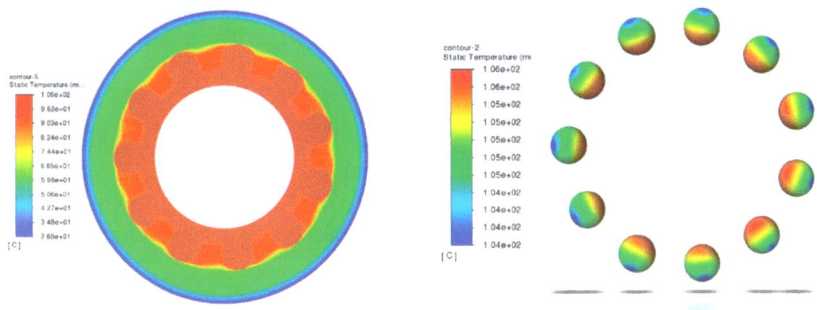

(**a**) Overall temperature distribution of bearings.　　(**b**) Temperature distribution of steel balls.

Figure 7. Temperature distribution diagram of bearings before turning power on.

After applying a peak voltage of 90 V and a frequency of 100 kHz to the bearing, the electric heating generated a power consumption of 18.76 W. The temperature distribution was obtained through temperature field simulation, as shown in Figure 8. From the figure, it can be observed that the outer ring of the bearing reached a maximum temperature of 62 °C after applying the voltage. The contact area between the steel ball and the inner raceway reached a maximum temperature of 110 °C, while the contact area between the steel ball and the outer raceway reached a maximum temperature of 105 °C.

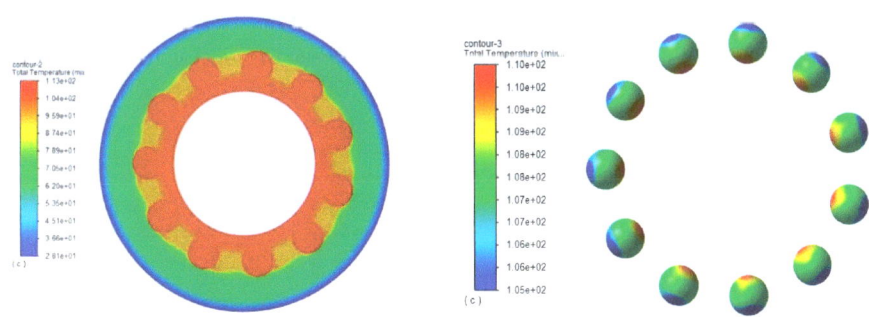

(**a**) Overall temperature distribution of bearings.　　(**b**) Temperature distribution of steel balls.

Figure 8. Temperature distribution diagram of bearings after turning power on.

Under specific operating conditions, the temperature in the contact area between the steel ball and the inner ring of the bearing surpassed that of the outer ring. This was attributed to the higher relative speed of the steel ball to the inner ring compared to the outer ring. Following power activation, the temperature of all bearing components increased, aligning with the electric heating phenomenon.

3.3. Simulation Analysis of Local Temperature Rise in the Contact Area of Current Bearing

After conducting measurements, the capacitance range of the insulation coating of the insulated ball bearing used in the experiment was found to be between 1.3 nF and 1.6 nF. For the purpose of this article, a value of 1.45 nF was considered. When the operating conditions included an inner ring speed of 4500 r/min and a radial load of 4.6 kN, the calculated bearing capacitance value was as shown in Table 3.

Table 3. Calculation results of bearing capacitance.

Steel Ball Number	1	2	3
C_{in}/pF	93.36	79.95	45.35
C_{en}/pF	79.12	67.96	38.83
$C_{B1\sim3}/\mathrm{pF}$	42.83	36.73	20.92
C_B/pF		158.13	
C_{Total}/pF		142.58	

A sinusoidal AC signal with a peak-to-peak voltage of 60 V and 90 V, as well as a voltage frequency of either 100 kHz or 130 kHz, were applied to the bearing. The shaft current values before and after the breakdown of the lubricating oil film were calculated, as indicated in Table 4.

Table 4. Axis currents at different voltages and frequencies.

Voltage Peak to Peak Value (V)	Frequency (kHz)	Unbreakable Shaft Current (mA)	Breakdown Rear Axle Current (mA)
60	100	2.69	27.33
90	100	4.03	41.00
60	130	3.49	35.53
90	130	5.24	53.31

The simulation model of the electric thermal coupling temperature field, as described in Section 3.1, was established. The Comsol 6.0 software was used to simulate the electric thermal coupling temperature field, and the relevant parameters were set according to Figure 9.

Name	Expression	Value	Describe
L_con	0.43e-6[m]	4.3E-7 m	Path length
R_con	0.2e-6[m]	2E-7 m	Path radius
Effective	0.0029[A]	0.0029 A	
Maximum	0.0041[A]	0.0041 A	Maximum current value
Fre	100000[Hz]	1E5 Hz	Current frequency

Figure 9. Parameter settings for electric thermal coupling simulation.

The lubrication micro zone of the bearing experiences Joule heat loss due to current breakdown. This heat can lead to changes in the film-forming characteristics between the bearing steel ball and the raceway, resulting in degradation of lubrication conditions. In severe cases, the local temperature can reach the melting temperature of the bearing steel, causing mechanical damage and damaging the raceway and steel ball surface. To study this phenomenon, a simulation was conducted by coupling multiple physical fields to calculate the transient heat generated by current loss. The temperature changes at different times in the current path and breakdown point of the bearing model were obtained from the temperature field. Figures 10–13 show the simulation results of electric thermal coupling at different positions under different frequencies and voltages before and after the breakdown of the lubricating oil film. As shown in the figures, the heat in the current path continuously increased over time, with the temperature at each point gradually rising. The highest temperature was observed at point c, which was located at the center of the path. Point a, which was the current excitation input point, had a temperature equal to the symmetric point d at the other end of the path. Both temperatures were lower than point c, but higher than point b. Point b, located outside the current path, had the smallest current density, and the heat generated there dissipated into the lubricating oil film. As a result, the temperature at point b was the lowest. Eventually, the temperature at each point stabilized at around 1000 ns. The calculation shows that it took 4415 ns for the steel ball to roll over the outer ring channel area and 2956 ns to roll over the inner ring channel area. The black vertical dashed line represents the inner ring, while the red vertical dashed line represents the outer ring.

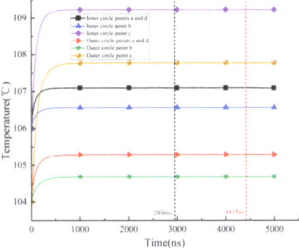 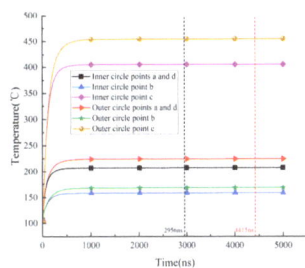

(**a**) Before lubricating oil film breakdown.   (**b**) After lubricating oil film breakdown.

Figure 10. Time variation curve of channel temperature at voltage of 60 V and frequency of 100 kHz.

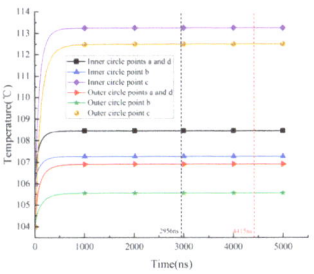 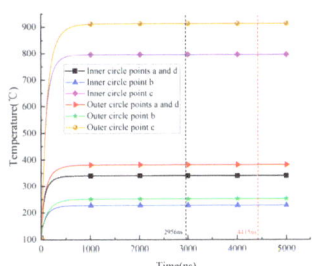

(**a**) Before lubricating oil film breakdown.   (**b**) After lubricating oil film breakdown.

Figure 11. Time variation curve of channel temperature at voltage of 90 V and frequency of 100 kHz.

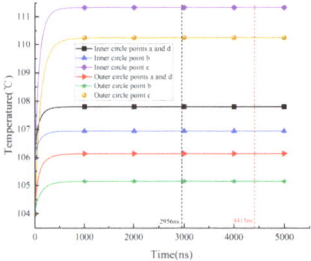
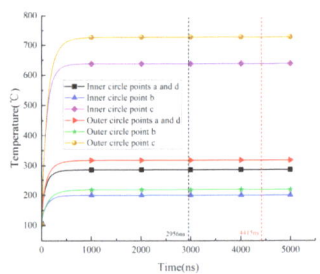

(a) Before lubricating oil film breakdown.　　(b) After lubricating oil film breakdown.

Figure 12. Time variation curve of channel temperature at voltage of 60 V and frequency of 130 kHz.

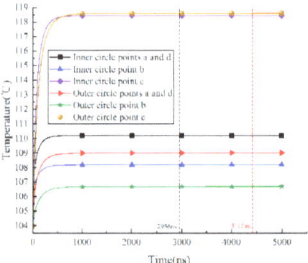
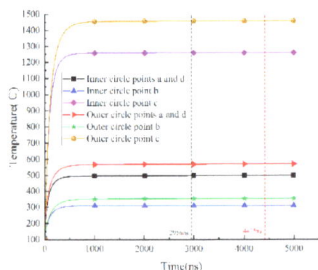

(a) Before lubricating oil film breakdown.　　(b) After lubricating oil film breakdown.

Figure 13. Time variation curve of channel temperature at voltage of 90 V and frequency of 130 kHz.

The variation curve of the channel temperature with time is shown in Figure 10, before and after the breakdown of the lubricating oil film under the conditions of a voltage of 60 V and a frequency of 100 kHz. For the inner ring, prior to breakdown, the temperature at point c reached 109 °C, while the temperatures at points a and d reached 107.2 °C, and at point b it reached 106.6 °C. After breakdown, the temperature at point c increased up to 405 °C, while the temperatures at points a and d rose up to 207 °C, and at point b, it reached up to 160 °C. As for the outer ring, before breakdown, the temperature at point c reached 107.5 °C, the temperatures at points a and d reached 105.2 °C, and at point b, it reached 104.6 °C. After breakdown, the temperature at point c rose up to 455 °C, while the temperatures at points a and d reached up to 225 °C, and at point b, it reached up to 170 °C.

The variation curve of the channel temperature with time is shown in Figure 11, before and after the breakdown of the lubricating oil film under the conditions of a voltage of 90 V and a frequency of 100 kHz. For the inner ring, prior to breakdown, the temperature at point c reached 113 °C, while the temperatures at points a and d reached 108.2 °C, and at point b, it reached 107.2 °C. After breakdown, the temperature at point c increased up to 797 °C, while the temperatures at points a and d rose up to 340 °C, and at point b, it reached up to 229 °C. As for the outer ring, before breakdown, the temperature at point c reached 112 °C, the temperatures at points a and d reached 106.6 °C, and at point b, it reached 105.4 °C. After breakdown, the temperature at point c rose up to 915 °C, while the temperatures at points a and d reached up to 383 °C, and at point b, it reached up to 253 °C.

The variation curve of the channel temperature with time is shown in Figure 12, before and after the breakdown of the lubricating oil film under the condition of a voltage of 60 V and a frequency of 130 kHz. For the inner ring, prior to breakdown, the temperature at point c reached 111 °C, while the temperatures at points a and d reached 107.7 °C, and at point b, it reached 106.8 °C. After breakdown, the temperature at point c increased up to 638 °C, while the temperatures at points a and d rose up to 287 °C, and at point b, it reached up to 200 °C. As for the outer ring, before breakdown, the temperature at point c reached 110 °C, the temperatures at points a and d reached 106 °C, and at point b, it reached 105 °C. After breakdown, the temperature at point c rose up to 728 °C, while the temperatures at points a and d reached up to 319 °C, and at point b, it reached up to 219 °C.

The variation curve of the channel temperature with time is shown in Figure 13, before and after the breakdown of the lubricating oil film under the condition of a voltage of 90 V and a frequency of 130 kHz. For the inner ring, prior to breakdown, the temperature at point c reached 118.2 °C, while the temperatures at points a and d reached 109.8 °C, and at point b, it reached 108 °C. After breakdown, the temperature at point c increased up to 1260 °C, while the temperatures at points a and d rose up to 498 °C, and at point b, it reached up to 311 °C. As for the outer ring, before breakdown, the temperature at point c reached 118 °C, the temperatures at points a and d reached 108.5 °C, and at point b, it reached 106.5 °C. After breakdown, the temperature at point c rose up to 1458 °C, while the temperatures at points a and d reached up to 570 °C, and at point b, it reached up to 354 °C.

Based on the simulated results, it is evident that the application of current to the bearing led to an increase in temperature across different parts of the bearing. Specifically, as the voltage and frequency of the current increased, the temperature in the contact area between the bearing and the inner and outer rings also increased. Moreover, when the lubricating oil film broke down, localized high temperatures occurred in the contact area.

4. Experimental Verification of Electric Thermal Coupling Temperature Field for Current Bearing

This article uses a self-developed current bearing testing machine to conduct electric thermal coupling temperature field tests on the current bearing, verifying the accuracy of the above simulation model.

4.1. Test Conditions

Figure 14 presents a schematic diagram illustrating the overall structural layout of the testing machine. The testing machine utilized in this article comprised a testing device, a control system, a power supply system, and a measurement device. These system modules collaborated with each other to conduct comprehensive tests on the performance of bearings under current-carrying conditions.

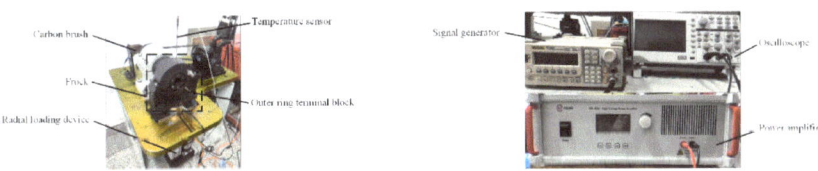

(a) Test equipment.   (b) Power supply and detection device.

Figure 14. Testing and measurement equipment.

This article describes the use of a signal generator and a power amplifier to apply excitation signals to the test bearing. The excitation current was input from the main shaft of the testing machine through a carbon brush and output from the outer ring voltage terminal via the main shaft inner ring steel ball outer ring coating bearing seat. This created a closed current circuit, as shown in Figure 15, to power the test bearing. The voltage signal changes between different parts of the bearing were measured using an oscilloscope. Channel 1 displays the total voltage change applied to the bearing, channel 2 shows the voltage change at both ends of the bearing coating, and channel 3 shows the voltage change at both ends of the bearing oil film.

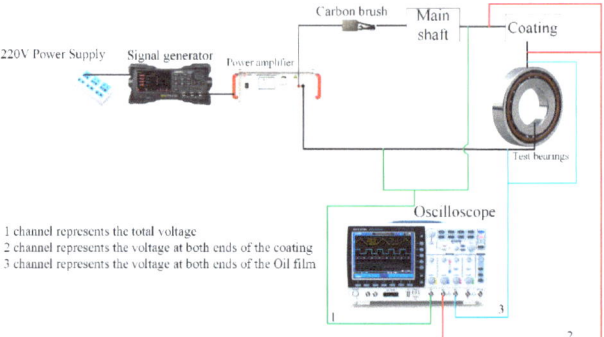

Figure 15. Experimental circuit diagram.

4.2. Results and Discussion of Outer Ring Temperature Rise Test

The current-carrying bearing performance testing machine described in Section 4.1 was utilized to conduct a current-carrying test on the 6215-type test bearing. The test was conducted under the following conditions: the inner ring speed of the bearing was set at 3500 rpm, the radial load applied was 3.5 kN, and an excitation signal with a voltage Vpp = 90 V and frequency f = 100 kHz was applied to the bearing.

As depicted in Figure 16a, the temperature rise in the outer ring of the bearing was shown before power was applied. The figure illustrates that, after the test bearing began running, the temperature of the outer ring gradually stabilized from the initial room temperature of 25 °C to 55 °C. Similarly, Figure 16b presents the temperature rise in the outer ring of the bearing after power was applied. It can be observed from Figure 16 that an excitation signal was applied to the bearing once it began running. However, due to the interference of the excitation signal on the temperature sensor, the measured outer ring temperature fluctuated within a certain range. Between 0 and 25,000 s, the outer ring temperature exhibited an upward trend and eventually fluctuated within the range of 54 °C–62 °C. To determine the average value, the excitation signal was applied, causing the temperature of the outer ring to gradually stabilize from the initial room temperature of 25 °C to 58 °C. Before the excitation signal was applied, the temperature of the outer ring remained stable at 55 °C at the same room temperature, with an error of 6.14% compared to the simulation results. After the excitation signal was applied, the temperature rise of the outer ring stabilized at 58 °C, with an error of 6.45% compared to the simulation results. Within the acceptable error range, the accuracy of the simulation model was verified.

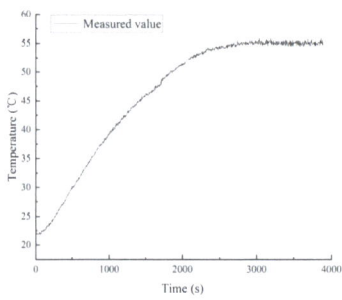 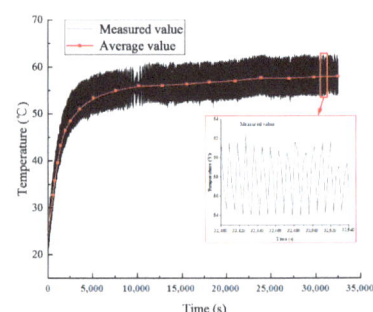

(**a**) Before power was applied.   (**b**) After power was applied.

Figure 16. Temperature rise and partial enlargement of the outer ring before and after turning power on.

4.3. Verification of Temperature Rise Test in Contact Area

The slice diagram in Figure 17 illustrates the condition of the bearing after 25 h of operation under the specified working conditions mentioned in Section 4.2. When the lubricating oil film of the bearing was penetrated, it resulted in an immediate increase in temperature, causing the lubricating grease to darken and some of it to adhere to the raceway. This, in turn, led to pitting corrosion in the contact area of the bearing. The figure indicates that the corrosion marks on the outer ring of the bearing were more severe compared to those on the inner ring. Therefore, it can be inferred that the temperature rise in the contact area of the outer ring of the bearing was higher than that of the inner ring after the lubricating oil film was breached, which aligns with the simulation results.

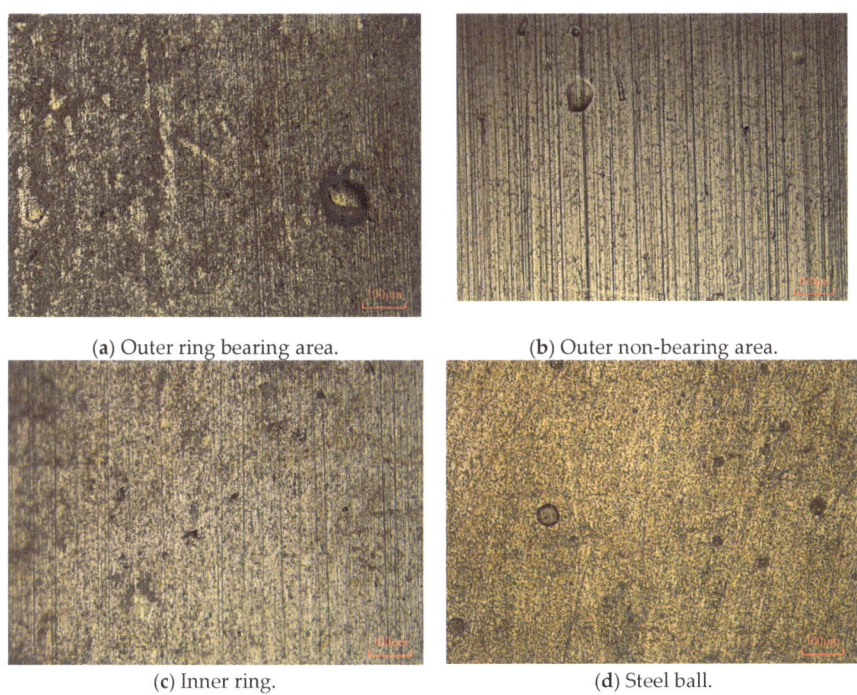

(**a**) Outer ring bearing area.　　(**b**) Outer non-bearing area.

(**c**) Inner ring.　　(**d**) Steel ball.

Figure 17. Slice diagram of test bearings.

5. Conclusions

This article presents a simulation analysis of the temperature field of deep groove ball bearings under an electric environment, considering electric thermal coupling. The study also includes experimental verification. The following conclusions can be drawn:

(1) Electrical environmental factors can cause an increase in the overall temperature rise in the bearing. Specifically, the temperature of the outer ring of the bearing can increase by approximately 3 °C compared to the temperature before powering on.
(2) In an electrical environment, the contact areas of bearings experience localized high temperatures due to the breakdown and discharge of the lubricating oil film. The temperature rise in the contact area between the rolling element and the outer raceway is particularly significant, increasing the likelihood of electrical corrosion damage.
(3) The consistency between the experimental results and simulation results is good, which verifies the accuracy of the simulation model.

Author Contributions: Conceptualization, Z.W., S.M. and H.T.; data curation, S.M. and H.T.; formal analysis, S.M. and Z.W.; funding acquisition, Y.C.; investigation, S.M., B.S. and Z.W.; methodology,

Z.W. and Y.C.; project administration, Z.W. and Y.C.; resources, Y.C.; software, S.M.; supervision, Y.C. and Z.W.; validation, H.T., B.S. and S.M.; writing—original draft, H.T. and Z.W.; writing—review and editing, S.M. and Z.W. All authors have read and agreed to the published version of the manuscript.

Funding: This research was funded by the National Natural Science Foundation of China (52105182) and Zhejiang Province Key R&D Plan (2021C01095).

Data Availability Statement: The data used to support the findings of this study are available from the corresponding author upon request.

Acknowledgments: The authors would like to thank the National Natural Science Foundation of China (52005158) and Zhejiang Province Key R&D Plan (2021C01095) for the financial support.

Conflicts of Interest: The authors declare no conflicts of interest.

References

1. Lou, Z.F. The Analysis and Suppression of Bearing Current in Varied Frequency Supply AC Motor. Master's Thesis, Beijing Jiaotong University, Beijing, China, 2015.
2. Ma, J.J.; Xue, Y.J.; Han, Q.K.; Li, X.J.; Yu, C.X. Motor Bearing Damage Induced by Bearing Current: A Review. *Machines* **2022**, *10*, 1167. [CrossRef]
3. Liu, R.F.; Sang, B.Q.; Cao, J.C. Investigation on the Influence of Motors Grounding States on Bearing Voltage in Inverter Drive System. *Proc. CSEE* **2015**, *35*, 177–183. (In Chinese)
4. He, F.; Xie, G.X.; Luo, J.B. Electrical bearing failures in electric vehicles. *Friction* **2020**, *8*, 4–28. [CrossRef]
5. Wang, Y.; Bai, B.D.; Liu, W.F. Determination of Key Parameters in Distribution Parameters Common Mode Equivalent Circuit and Calculation of the Bearing Currents. *Trans. China Electrotech. Soc.* **2014**, *29*, 124–131. (In Chinese)
6. Ma, S.Y.; Wang, Z.W.; Li, C. Computational Analysis and Experimental Verification of Equivalent Capacitance of Cylindrical Roller Bearings Under Current-Carrying Condition. *Bearing* **2023**, *9*, 11–18. (In Chinese)
7. Busse, D.; Erdman, J.; Kerkman, R.; Schlegel, D.; Skibinski, G. The effects of PWM voltage source inverters on the mechanical performance of rolling bearings. *IEEE Trans.* **1997**, *33*, 567–576. [CrossRef]
8. Wang, Q.Q. Research on Capacitances Calculation and the Bearing Damage Degree in Bearing Currents of AC Motors. Master's Thesis, Beijing Jiaotong University, Beijing, China, 2021.
9. Puchtler, S.; Maier, R.; Kuhn, M.; Burkhardt, Y. The Influence of Load and Speed on the Initial Breakdown of Rolling Bearings Exposed to Electrical Currents. *Lubricants* **2024**, *12*, 1. [CrossRef]
10. Li, Y.; Qiu, L.; Zhi, Y.J.; Gao, Z.F.; Ma, J.E.; Zhang, J.; Fang, Y.T. An overview of bearing voltages and currents in rail transportation traction motors. *Appl. Phys. Eng.* **2023**, *24*, 226–242. [CrossRef]
11. Tischmacher, H.; Gattermann, S. Investigations on Bearing Currents in Converter-Fed Electrical Motors. In Proceedings of the 20th International Conference on Electrical Machines, Marseille, France, 2–5 September 2012; pp. 1764–1770.
12. Miliani, E.H. Leakage current and commutation losses reduction in electric drives for Hybrid Electric Vehicle. *J. Power Sources* **2014**, *255*, 266–273. [CrossRef]
13. Romanenko, A.; Ahola, J.; Muetze, A. Influence of electric discharge activity on bearing lubricating grease degradation. In Proceedings of the IEEE Energy Conversion Congress and Exposition, Montreal, QC, Canada, 20–24 September 2015; SEP: Montreal, QC, Canada, 2015; Volume 18–24, pp. 4851–4856.
14. Niu, K.; Zeng, Z.X.; Chen, T.H. Research Progress of the Mechanism of Shaft Current Corrosion in Bearing and the Protection Technology. *Lubr. Eng.* **2023**, *48*, 179–188. (In Chinese)
15. Binder, A.; Muetze, A. Scaling Effects of Inverter-Induced Bearing Currents in AC Machines. *IEEE Ind. Appl.* **2008**, *44*, 769–776. [CrossRef]
16. Lu, F.M. Principle of Electrical Erosion and Insulation Protection Indexes of PWM Frequency Modulation Motor Bearing. *Bearing* **2019**, *5*, 6–9. (In Chinese)
17. Liu, R.F.; Sang, B.Q.; Li, W.L. Calculations and Measurements of Bearing Capacitance in AC Motor Bearings. *Proc. CSEE* **2017**, *37*, 2986–2993. (In Chinese)
18. Xiong, F.; Shi, W.; Liao, A.H. Calculation of Equivalent Capacitance of Traction Motor Bearings for Rail Vehicle Considering Radial Clearance Variation of Bearings and Air Gap Eccentricity of Motor. *Bearing* **2023**, *10*, 49–56. (In Chinese)
19. Suo, L.C.; Zhao, W.S.; Liang, L.P. Research on the Mechanism of Dielectric Breakdown in Powder Mixing Electric Discharge Machining. *Electromachining Mould.* **2001**, *5*, 10–13. (In Chinese)
20. Nakao, Y.; Hamano, N.; Naito, T.; Nakagami, Y.; Shimizu, R.; Sakai, Y.; Tagashira, H. Influence of Molecular Structure on Propagation of Positive Streamer Discharge in Dielectric Liquids. *Electr. Eng. Jpn.* **2004**, *149*, 15–20. [CrossRef]
21. Lei, J.T.; Su, B.; Zhang, S.L. Dynamics-Based Thermal Analysis of High-Speed Angular Contact Ball Bearings with Under-Race Lubrication. *Machines* **2023**, *11*, 691. [CrossRef]
22. Liu, Y.Y.; Yang, H.S.; Su, B. Simulation Analysis of the Friction Power Consumption of Double Row Angular Contact Ball Bearings for Aero-engine. *J. Mech. Transm.* **2021**, *45*, 129–135. (In Chinese)

23. Li, J.S.; Yuan, Y.; Hao, D.Q. Analysis and Experimental Study on Bearing Temperature Rise of High Speed Machine Tools. *J. Mech. Transm.* **2023**, *47*, 129–136. (In Chinese)
24. Xie, Y.B.; Zhang, M.Z.; Wang, D.F. Simulation Analysis on Temperature Field of Grease-Lubricated Bearings for Motorized Spindles with Cooling Systems of Different Structures. *Bearing* **2023**, *11*, 22–26. (In Chinese)
25. Wang, Q.Q.; Liu, R.F.; Ren, X.J. The Motor Bearing Discharge Breakdown Based on the Multi-Physics Field Analysis. *Trans. China Electrotech. Soc.* **2020**, *35*, 4251–4257. (In Chinese)
26. Deng, S.E.; Jia, Q.Y.; Xue, J.X. *Principles of Rolling Bearing Design*, 2nd ed.; China Standard Press: Beijing, China, 2014; pp. 138–139.
27. Muetze, A.; Binder, A. Practical Rules for Assessment of Inverter-Induced Bearing Currents in Inverter-Fed AC Motors up to 500 kW. *IEEE T. Ind. Electron.* **2007**, *54*, 1614–1622. [CrossRef]
28. Tang, J. *Electrotechnics*, 2nd ed.; Higher Education Press: Beijing, China, 2005; pp. 64–65.

Disclaimer/Publisher's Note: The statements, opinions and data contained in all publications are solely those of the individual author(s) and contributor(s) and not of MDPI and/or the editor(s). MDPI and/or the editor(s) disclaim responsibility for any injury to people or property resulting from any ideas, methods, instructions or products referred to in the content.

Article

Calculation and Analysis of Equilibrium Position of Aerostatic Bearings Based on Bivariate Interpolation Method

Shuai Li [1], Yafu Huang [1], Hechun Yu [1,*], Wenbo Wang [1,2], Guoqing Zhang [1], Xinjun Kou [1], Suxiang Zhang [1] and Youhua Li [1]

[1] School of Mechatronics Engineering, Zhongyuan University of Technology, Zhengzhou 451191, China; lishuai23@zut.edu.cn (S.L.); 2022104113@zut.edu.cn (Y.H.)
[2] Faculty of Mechanical and Electrical Engineering, Kunming University of Science and Technology, Kunming 650500, China
* Correspondence: 6222@zut.edu.cn

Abstract: The solution of equilibrium positions is a critical component in the calculation of the dynamic characteristic coefficients of aerostatic bearings. The movement of the rotor in one direction leads to bidirectional variations in the air film force, resulting in low efficiency when using conventional calculation methods. It can even lead to iterative divergence if the initial value is improperly selected. This study concentrates on the orifice throttling aerostatic bearings and proposes a novel method called the bivariate interpolation method (BIM) to calculate the equilibrium position. The equilibrium equation for the rotor under the combined influence of air film forces, gravity, and external loads is established. A calculation program based on the finite difference method is developed to determine the equilibrium position. The process of solving the equilibrium position and the convergence is compared with the secant method and the search method. Furthermore, the variation trend of the equilibrium position and stiffness when the external loads changes are studied based on the BIM. Finally, the correctness of the BIM to solve the equilibrium position is proved by comparing it with the experiment results. The calculation results indicate that the BIM successfully resolves the problem of initial value selection and exhibits superior computational efficiency and accuracy. The equilibrium position initially moves away from the direction of the external load as the load increases, and then this gradually approaches the load direction. The main stiffness increases with increases in the external load, while the variation in cross stiffness depends on the direction of the external load.

Keywords: equilibrium position; bivariate interpolation method; aerostatic bearing; external load; stiffness

Citation: Li, S.; Huang, Y.; Yu, H.; Wang, W.; Zhang, G.; Kou, X.; Zhang, S.; Li, Y. Calculation and Analysis of Equilibrium Position of Aerostatic Bearings Based on Bivariate Interpolation Method. *Lubricants* 2024, 12, 85.
https://doi.org/10.3390/lubricants12030085

Received: 22 January 2024
Revised: 17 February 2024
Accepted: 5 March 2024
Published: 7 March 2024

Copyright: © 2024 by the authors. Licensee MDPI, Basel, Switzerland. This article is an open access article distributed under the terms and conditions of the Creative Commons Attribution (CC BY) license (https://creativecommons.org/licenses/by/4.0/).

1. Introduction

Aerostatic bearings utilize compressed air from external air sources to form an air film that supports the movement of the parts. Compared to precision rolling bearings, aerostatic bearings offer significant advantages in terms of motion accuracy, friction, rotation speed, and environmental impact [1]. Consequently, aerostatic bearings have found widespread applications in aerospace, precision machine tools, electronics and semiconductors, and medical equipment [2–4]. Commonly used orifice throttling aerostatic bearings materials are steel, brass, and aluminum [5]. In addition, porous materials also have more applications because of their superior characteristics [6], and the commonly used material is graphite. Research on aerostatic bearings primarily focuses on topics such as bearing capacity, stiffness, stability, and flow rate [7,8]. The equilibrium position is an important basis for judging the stability of aerostatic bearings [9], and it is also the premise of analyzing the dynamic characteristic coefficients of aerostatic bearings under normal operation. While the static load of practical bearings is typically known, the equilibrium position remains unknown [10]. The accurate and efficient solution of the equilibrium position of a bearing under a given load is a necessary prerequisite for studying the relevant characteristics of aerostatic bearings.

Solving the Reynolds equation is the first step to study the relevant characteristics of aerostatic bearings, which can be divided into the finite difference method (FDM), the finite element method (FEM), and the finite volume method (FVM) according to different solving methods [11]. Due to the nonlinearity of the equilibrium equation, multiple iterations are necessary to determine the equilibrium position of the bearing. Scholars have proposed numerical calculation methods to solve the equilibrium position of the rotor. Shuai [12] used the coordinate rotation method to solve the equilibrium position of the three-lobe journal bearing and proved the feasibility of the numerical calculation method to solve the equilibrium position of the three-lobe journal bearing. However, this method is not suitable for the cylindrical sliding bearing. Qian [13] determined the equilibrium position of the three-lobe journal bearing using the secant method. Wang [14] analyzed the influence of unbalance force on the vibration characteristics of the narrow slit throttling aerostatic bearing and observed that as the speed increases, the rotor's equilibrium position gradually moves towards the center of the bearing. However, no detailed method was provided for solving the equilibrium position. Yu [15] used the search method to solve the equilibrium position of the tilting pad journal bearing. By calculating the direction of the fastest convergence of the air film force based on a given trial calculation point, the equilibrium position of the bearing can be obtained. Results show that the method effectively solves the divergence problem, although its calculation efficiency is low due to the frequent calculation of the air film force. Wan [16] used the Newton iteration method to solve the equilibrium position of the bearing and proposed a load approaching method to solve the problem of the Newton iteration method, getting stuck in a dead loop due to inappropriate initial value selection. Zheng [17] used the fast convergence speed of the Newton–Raphson method to solve the equilibrium position of the bearing and subsequently determined the dynamic characteristic coefficients. The accuracy of the method was proven by comparing it with relevant references. Pokorny [18] optimized the Newton iteration method and compared it with the Newton–Raphson method. The equilibrium position of the tilting pad bearing was then determined, demonstrating the superiority of the optimized Newton method over the Newton–Raphson method. Yang [19] has demonstrated through numerical analysis that the twofold secant method converges faster than both Newton's method and Newton's method with the P.C. format. Zhou [20] used the twofold secant method to solve the static equilibrium position. The results show that the twofold secant method significantly narrows down the search range in the initial iteration and achieves faster convergence compared to the dichotomy method and the secant method. All of the aforementioned methods for solving the equilibrium position used the static Reynolds equation. Furthermore, the dynamic trajectory method can be employed to determine the equilibrium position, although it often necessitates two to three days to ensure accurate calculations, making it relatively time-consuming [21].

When the rotor is in a stable state, the dynamic characteristic coefficients of the bearing during normal operation can be solved and analyzed. The methods for solving the dynamic characteristic coefficients of the bearing include the difference method, the partial derivative method, the small parameter method, and the finite element method [22–25]. Meng [26] used the difference method to solve the stiffness of the oil-lubricated journal bearing and analyzed the influence of bearing parameters. The accuracy of the calculated values was proved by comparing the theoretical calculation values with the experimental data. Qi [27] used the partial derivative method to solve the Reynolds equation defined in the complex range and focused on investigating the effect of rotor disturbance frequency on stiffness and damping coefficients. Li [28] solved the Reynolds equation using the partial derivative method. The calculation results indicate that as the rotor position gradually approaches the equilibrium position, the dynamic characteristic coefficients gradually converge to a stable value.

In summary, many existing references calculate bearing-related characteristics using a fixed eccentricity, but there is a lack of research on solving the equilibrium position and conducting characteristic analysis under given loads. The current methods for solving the equilibrium position are primarily used for the liquid sliding bearing, and varying levels of efficiency and convergence problems exist. This paper focuses on the orifice throttling radial aerostatic bearing as the subject of research and presents a new method, called the BIM, for determining the equilibrium position. The influence of external load on the equilibrium position and dynamic characteristic coefficients is studied, and the calculation results are discussed and compared with references and experimental data.

2. Building and Calculation of the BIM

2.1. Establishment of Mathematical Model of Bearing

Figure 1 depicts the structural model of the orifice throttling aerostatic bearing studied in this paper. Table 1 provides the fundamental parameters of the bearing. O_1 is the center of the bearing, and O_2 is the center of the journal. When there are no external loads present, the rotor experiences air film forces Fx and Fy, in addition to its own weight Mg, at the equilibrium position.

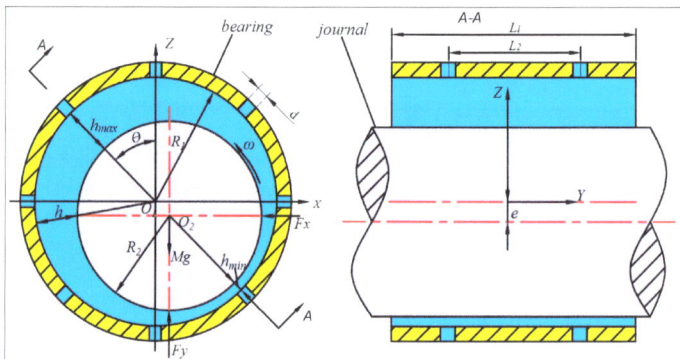

Figure 1. Structure model of orifice throttling aerostatic bearing.

Table 1. Bearing structure parameters and environmental parameters.

Parameter	Value
Journal length L1 (mm)	160
Journal radius R2 (mm)	25
Mean air film thickness c (μm)	10
Diameter of orifice d (mm)	0.2
Row number of Orifice on journal bearing	2
Orifice number of each row	8
Rotor quality M (kg)	6.5
Density of air ρ (kg·m^{-3})	1.204
Supply pressure Ps (MPa)	0.5
Environment pressure P (MPa)	0.1
Viscosity of air μ (N·s·m^{-2})	1.8×10^{-5}

To determine the capacity of the aerostatic bearing, it is necessary to first obtain the air film pressure distribution of the bearing. The bearing is at the equilibrium position, with a stable air film pressure distribution. Because the influence of temperature on aerostatic bearing is revlatively small, the flow process duration is short, so it can be isothermal flow [29]. The commonly used Reynolds equation for compressible gas lubrication can be expressed as

$$\frac{\partial}{\partial x}\left(Ph^3 \frac{\partial P}{\partial x}\right) + \frac{\partial}{\partial y}\left(Ph^3 \frac{\partial P}{\partial y}\right) + Q\delta = 6\mu U \left(\frac{\partial (Ph)}{\partial x}\right) \quad (1)$$

The air film thickness can be expressed as $h = c(1 + \varepsilon \cos \theta)$, where ε is the eccentricity, and θ is the attitude angle.

It is noted that $\delta_k = 1$ at the orifice entrance, and that $\delta_k = 0$ at the orifice exit. Q is the air mass flow factor:

$$Q = 12\mu \frac{P_a}{\rho_a} \cdot \rho v \tag{2}$$

Orifice flow equation:

$$m = \varphi P_s A \sqrt{\frac{2\rho_a}{P_a}} \psi = \rho v dx dy \tag{3}$$

and

$$\psi = \begin{cases} \left[\frac{K}{2}\left(\frac{2}{K+1}\right)^{\frac{K+1}{K-1}}\right]^{0.5}, \frac{P}{P_s} \leq \beta_k \\ \left[\frac{K}{K-1}\left(\left(\frac{P}{P_s}\right)^{\frac{2}{K}} - \left(\frac{P}{P_s}\right)^{\frac{K+1}{K}}\right)\right]^{0.5}, \frac{P}{P_s} > \beta_k \end{cases} \tag{4}$$

where φ is the flow coefficient, A is the restriction area, K is the air constant, and β_k is the critical pressure ratio.

By utilizing the FDM to solve Equation (1), the air film pressure can be determined for a given rotor position. Then, integrating the air film pressure using Equation (5) yields the horizontal and vertical air film forces of the bearing.

$$\begin{aligned} Fx = \int_0^L \int_0^{2\pi} PR \sin\theta d\theta dy \\ Fy = \int_0^L \int_0^{2\pi} PR \cos\theta d\theta dy \end{aligned} \tag{5}$$

Assuming that the initial position of the rotor is vertically oriented, as shown in Figure 2a, the rotor is then horizontally displaced by a distance of $+\Delta x$, resulting in the position shown in Figure 2b. The horizontal movement causes a change in the attitude angle θ, subsequently impacting the distribution of the air film pressure P. According to Equation (5), both the horizontal and vertical bearing capacity are consequently altered.

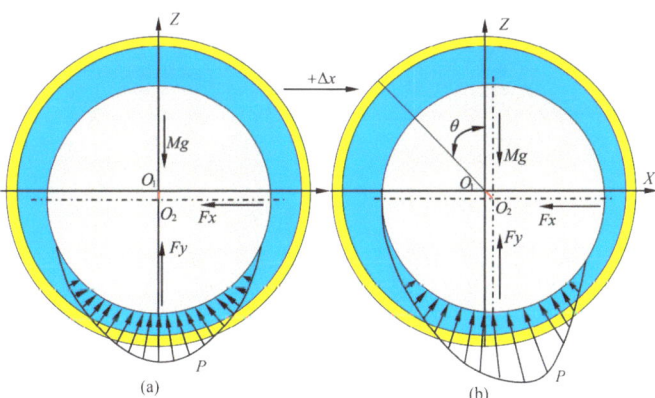

Figure 2. The influence of single direction displacement changes on bearing capacity. (**a**) Initial position. (**b**) Position after moving horizontally by $+\Delta x$.

2.2. Mathematical Model of the BIM

During the motion of the rotor, irrespective of the effects of other factors, the rotor is affected by the combined action of the air film force of the bearing and the gravity of the rotor. When the rotor reaches an equilibrium position, Equation (6) need to be satisfied.

$$\begin{cases} FX = Fx = 0 \\ FY = Fy - Mg = 0 \end{cases} \tag{6}$$

When solving the equilibrium position using the secant method and the Newton method, the horizontal and vertical coordinates are successively iterated in one direction, as shown in Figure 3. When solving the process, the iterative process of abscissa is embedded in the iterative process for the ordinate. The correction of the ordinate can only begin once the horizontal bearing capacity satisfies the requirements. It can be seen from Figure 2 that the movement of the rotor in one direction causes the bidirectional variations in the air film force. Therefore, there are the following problems:

(1) After completing a calculation in the vertical direction, the horizontal bearing capacity fails to satisfy the convergence requirements, necessitating a recalculation of the abscissa. Therefore, until the ordinate satisfies the convergence requirements, each correction of the ordinate requires iterative adjustments of the abscissa. This process results in a reduction in computational efficiency.

(2) In the iterative calculation process, it is essential to determine the influence factors. Influence factors are used to correct the coordinates. The influence factors vary depending on the rotor's position. If the influence factors are excessively large, it can result in iterative divergence, whereas if they are excessively small, it can impact computational efficiency.

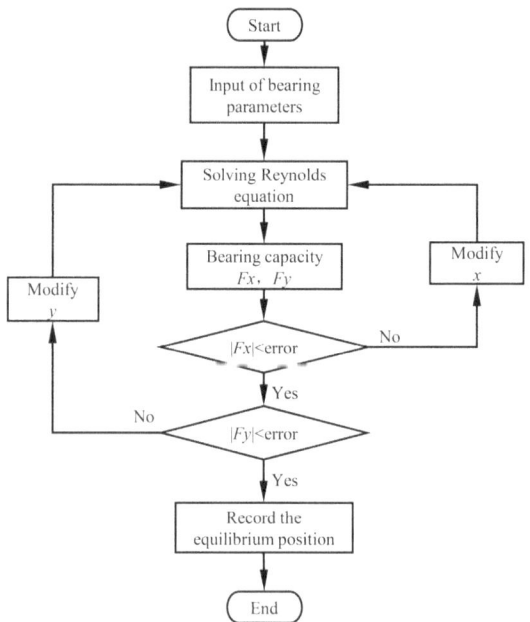

Figure 3. Common methods to solve the process.

The proposed BIM in this paper takes into account the simultaneous effect of horizontal and vertical coordinates on the air film force. There are relationships between the position to be solved and the four known positions surrounding it. Subsequently, the unknown position is determined through two consecutive interpolation steps. Figure 4 illustrates the principle diagram of the BIM, while Equation (7) represents the interpolation equation.

$$F = ax + by + cxy + d \tag{7}$$

The coefficient a controls the coordinate x. The abscissa x of the equilibrium position can be corrected by adjusting a. The coefficient b controls the coordinate y. The ordinate y of the equilibrium position can be corrected by adjusting b. The coefficient c controls the cross-term between the coordinates x and y and describes the interaction relationship between the horizontal and vertical coordinates. It can adjust the iteration direction. The coefficient d is a constant term used to adjust the overall offset of the function.

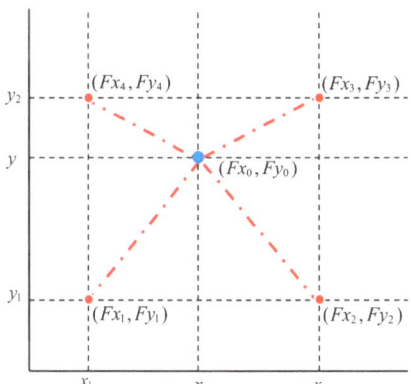

Figure 4. Principle diagram of the BIM.

The interpolation conditions are as follows:

$$Fx(x_1, y_1) = Fx_1, Fy(x_1, y_1) = Fy_1$$
$$Fx(x_2, y_1) = Fx_2, Fy(x_2, y_1) = Fy_2$$
$$Fx(x_2, y_2) = Fx_3, Fy(x_2, y_2) = Fy_3$$
$$Fx(x_1, y_2) = Fx_4, Fy(x_1, y_2) = Fy_4$$

To determine the equilibrium position and account for the impact of rotor position changes on the horizontal and vertical air film forces, the interpolation equation can be simplified to Equation (8) by using the interpolation conditions, where Fx_{load} and Fy_{load} represent the magnitude of the external load applied in the respective horizontal and vertical directions.

$$Fx_{load} = Fx_1(1-u)(1-v) + Fx_2 u(1-v) + Fx_3 uv + Fx_4(1-u)v \qquad (8)$$
$$Fy_{load} = Fy_1(1-u)(1-v) + Fy_2 u(1-v) + Fy_3 uv + Fy_4(1-u)v$$

where

$$u = \frac{x - x_1}{x_2 - x_1}, \quad v = \frac{y - y_1}{y_2 - y_1}$$

During the iterative calculation process, it is necessary to determine the modified coordinate value based on the magnitudes of the influence factors a and b. If $a > b$, the abscissa x is corrected; otherwise, the ordinate y is corrected. Adjusting the coordinates based on the influence factors helps prevent incorrect iteration direction and improves computational efficiency. By combining Equations (7) and (8), the influence factors a and b are obtained.

$$a = \frac{y_2(Fy_2 - Fy_1) + y_1(Fy_4 - Fy_3)}{(x_2 - x_1)(y_2 - y_1)}, b = \frac{x_1(Fy_2 - Fy_1) + x_2(Fy_4 - Fy_3)}{(x_2 - x_1)(y_2 - y_1)}$$

2.3. Calculation Process of the BIM

Figure 5 illustrates the flowchart of the BIM used to calculate the equilibrium position of the bearing. The bearing capacity is determined through the Reynolds equation. The database is used to temporarily store and transfer data between the journal position obtained through the BIM and the bearing capacity calculated through the Reynolds equation.

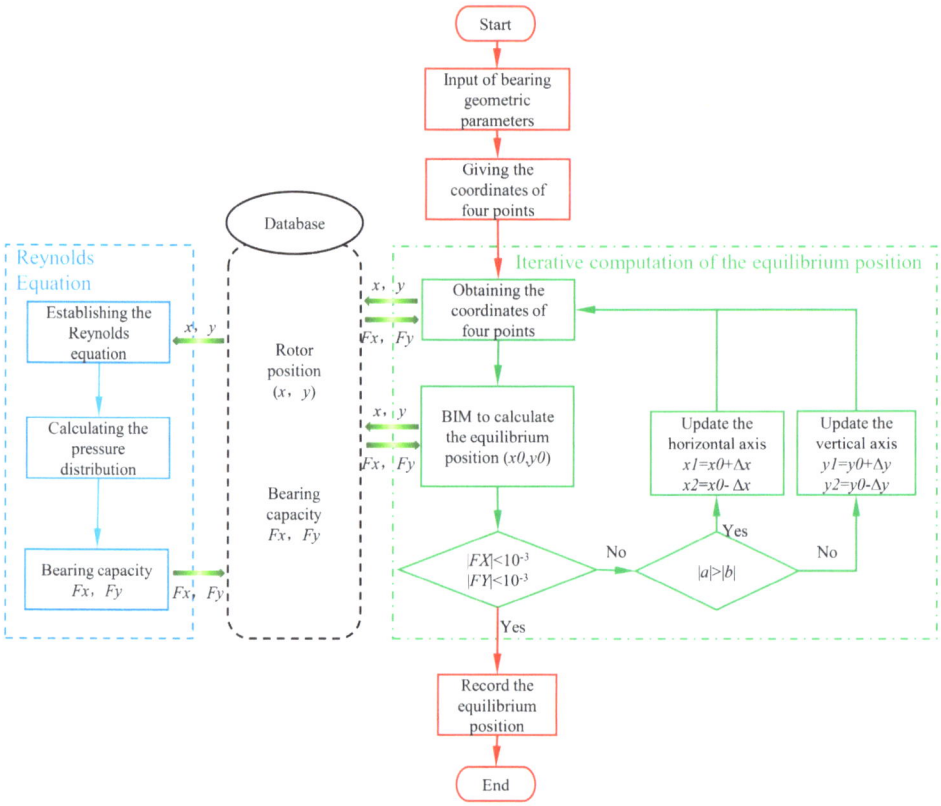

Figure 5. Flowchart for solving equilibrium positions using the BIM.

The iterative process is as follows:

(1) Given convergence criteria (resultant force $|FX| < 10^{-3}$ N, $|FY| < 10^{-3}$ N, 0.001 N force can be ignored).
(2) According to Figure 4, four points (x_1, y_1), (x_2, y_1), (x_2, y_2), (x_1, y_2) in the bearing range are selected, and u, v are calculated.
(3) The air film forces (Fx_1, Fy_1), (Fx_2, Fy_2), (Fx_3, Fy_3), (Fx_4, Fy_4) at the corresponding positions were calculated by the Reynolds equation.
(4) Through Equation (8), coordinate points (x, y) are obtained after two consecutive interpolation calculations.
(5) The air film forces (Fx, Fy) at the position of the fourth step are obtained by the Reynolds equation and compared with the convergence condition. If it is satisfied, the iteration is terminated. If it is not satisfied, proceed to step 6.
(6) Determine the size of the impact factors a and b and determine the coordinate value that has the greatest influence on the air film force. If $a > b$, modify x; otherwise, modify y. Repeat steps 2 to 5 after updating the data.

3. Results and Discussion

3.1. Comparison of Different Methods

Figure 6 illustrates the iterative convergence process of the BIM, the secant method, and the search method for solving the bearing equilibrium position under the same parameters. It can be observed from Figure 6 that the equilibrium positions obtained by the three methods are almost coincident. The main reason is that different calculation methods have different convergence. The convergence condition of the secant method is $F < 1 \times 10^{-2}$ N, the search method is $F < 5 \times 10^{-2}$ N, and the BIM is $F < 1 \times 10^{-3}$ N. The BIM exhibits the highest calculation accuracy. When compared with the secant method, the BIM has errors of 0.03% and 0.02% in the horizontal and vertical coordinates of the equilibrium position, respectively. When compared with the search method, the BIM has errors of 0.15% and 0.06% in the horizontal and vertical coordinates of the equilibrium position, respectively. The maximum error is 0.15%, while the minimum error is 0.02%. Table 2 provides a comparison of the results obtained using the three methods. The BIM requires 4 iterations, the secant method requires 22 iterations, and the search method requires 15 iterations. The BIM exhibits superior calculation efficiency compared with the other methods. In the initial iteration calculation, the convergence curve experiences substantial variations due to the significant initial distance from the equilibrium point. The convergence speed gradually slows down after approaching the equilibrium point. The secant method exhibits poor convergence performance near the equilibrium position due to the high density of points, resulting in reduced calculation efficiency.

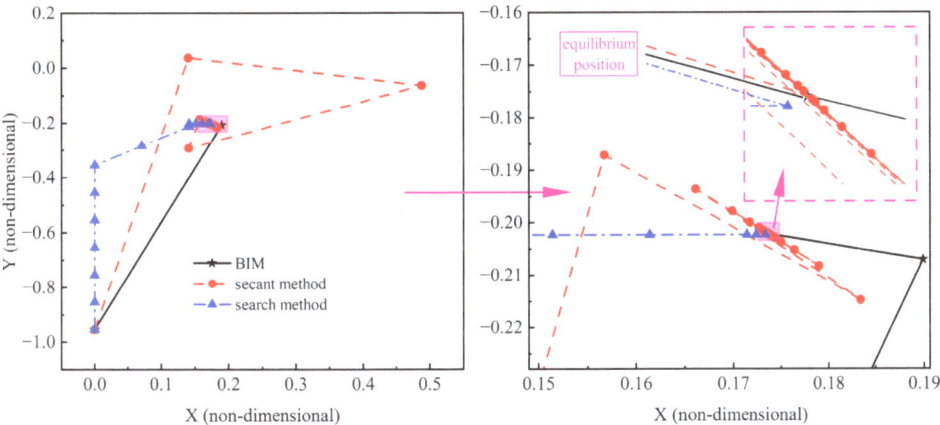

Figure 6. Comparison of convergence process of different methods.

Table 2. Comparison of calculation results. ('▲' represents the calculation baseline for relative error).

Solution Method	Condition of Convergence	Iteration Step	Equilibrium Position	Relative Error	
Secant method	1×10^{-2}	22	X = 0.17369 Y = −0.20219	▲ ▲	\ \
Search method	5×10^{-2}	15	X = 0.17342 Y = −0.20228	\ \	▲ ▲
BIM	1×10^{-3}	4	X = 0.17364 Y = −0.20215	0.03% 0.02%	0.15% 0.06%

3.2. Calculation of Equilibrium Position Based on the BIM

The equilibrium position in this study is primarily influenced by the air film pressure, which is closely associated with the supply pressure and rotor speed. Figure 7a illustrates the variations in the equilibrium position as the supply pressure and rotational speed. As shown in Figure 7a, the equilibrium position gradually moves towards the center of the bearing with increasing rotational speed and supply pressure. This rule is consistent with the results of reference [8].

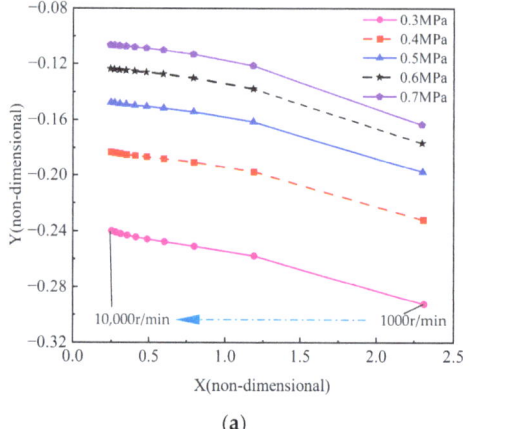
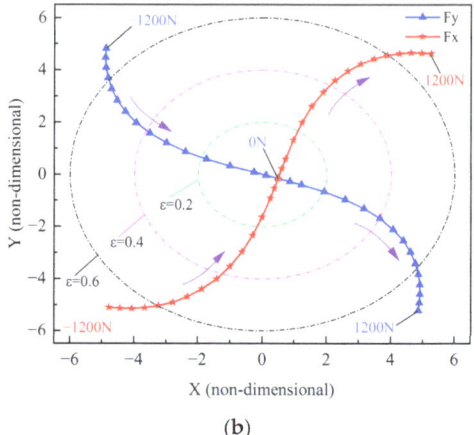

(a) (b)

Figure 7. (**a**) Equilibrium position under different pressures and rotating speeds. (**b**) Equilibrium position under different external loads.

Taking a speed of 5000 r/min and a supply pressure of 0.5 MPa as an example, the variation in the equilibrium position is shown in Figure 7b when the horizontal force Fx and the vertical force Fy ranging from -1200 N to 1200 N with an increment of 100 N are applied to the rotor. When subjected to external loads, the eccentricity ε gradually increases, and the equilibrium position of the axis will first move away from the load direction. As the absolute value of the external load increases, the equilibrium position progressively aligns with the direction of the external load, while the extent of the change decreases.

3.3. Calculation of Stiffness of Equilibrium Position Based on the BIM

The stiffness of aerostatic bearing is solved by the difference method. After obtaining the equilibrium position, a horizontal disturbance $+\Delta x$ is applied at the equilibrium position, and the air film forces Fx_1 and Fy_1 are solved under the new geometric relationship. Similarly, the disturbance $-\Delta x$ is taken, and Fx_2, Fy_2 can be obtained. Then, the stiffness values are then obtained as Kxx and Kyx.

$$Kxx = \frac{Fx_1 - Fx_2}{2\Delta x} \tag{9}$$

$$Kyx = \frac{Fy_1 - Fy_2}{2\Delta x} \tag{10}$$

Similarly, the stiffness values Kxy, Kyy can be obtained by adding disturbance $+\Delta y$ and $-\Delta y$ in the vertical direction.

$$Kxy = \frac{Fx_3 - Fx_4}{2\Delta y} \tag{11}$$

$$Kyy = \frac{Fy_3 - Fy_4}{2\Delta y} \tag{12}$$

Figure 8 illustrates that the direct stiffness and the absolute values of cross stiffness at the equilibrium position gradually rise with increasing rotational speed and supply pressure. Moreover, stiffness values are more responsive to changes in rotational speed. This phenomenon can be attributed to the growing dynamic pressure effect of the bearing as the rotational speed increases, leading to an overall increase in stiffness. The main stiffness Kxx is approximately equal to Kyy, and the cross stiffness Kxy is approximately inversely proportional to Kyx. This rule is consistent with the conclusion of reference [16].

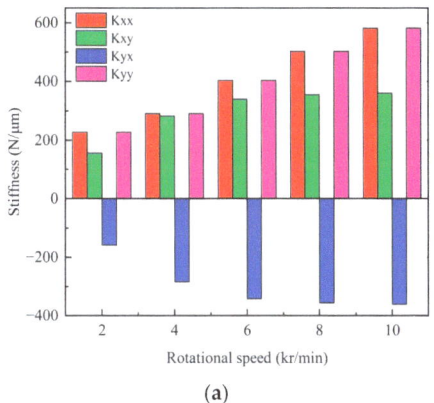
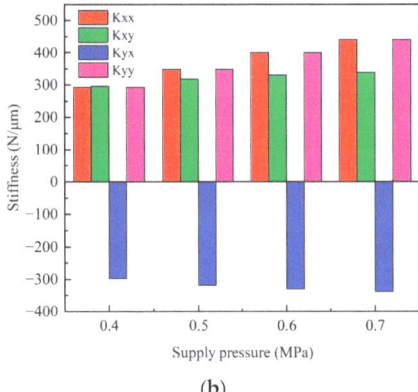

Figure 8. (**a**) The influence of rotational speed on the stiffness at the equilibrium position. (**b**) The influence of supply pressure on the stiffness at the equilibrium position.

The increase in both rotational speed and supply pressure leads to an increase in air film stiffness. The change in rotational speed has a greater influence on stiffness compared to supply pressure. This also explains that the influence of rotational speed on the change in the equilibrium position in Figure 7a is greater than that of supply pressure. With the increase in both speed and supply pressure, the equilibrium position shifts towards the center of the bearing.

Figure 9a,b illustrate the variations in stiffness at the equilibrium position caused by the external loads Fx and Fy. The direct stiffness increases as the absolute value of the external load increases, while the variation in cross stiffness depends on the direction of the applied load. If the absolute value of the external load Fx increases gradually, the corresponding Kxy increases, while the absolute value of Kyx decreases. If the absolute value of the external load Fy increases gradually, the corresponding Kxy decreases, while the absolute value of Kyx increases. This can also explain the trend observed in Figure 7b regarding the influence of external loads on the change in equilibrium position.

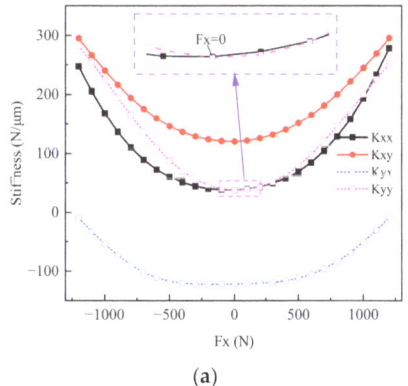
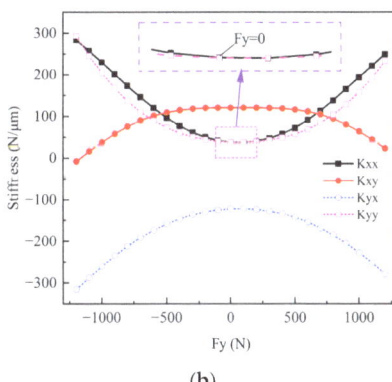

Figure 9. Stiffness changes at the equilibrium position under external load. (**a**) Horizontal load Fx. (**b**) Vertical load Fy.

4. Experiments and Comparison

4.1. Introduction of Experimental Platform

Figure 10 shows the experiment setup used to measure the equilibrium position, comprising the air supply system, the tested motorized spindle, and the TESA TT80 inductance micrometer. The air supply system is constructed by an air compressor (1), filter (3), globe value (4), solenoid throttle value (5), flowmeter (6), and pressure gauge (7). The inductance micrometer (accuracy 10 nm, range ±5 μm) is constructed with an inductive probe (9) and a digital display unit (11). The rotor diameter is measured with the Mitutoyo MDE-75MX (resolution 0.001 mm, allowable error 2 μm), while the bearing diameter is measured with the Mitutoyo CG-D100 (resolution 0.001 mm, allowable error 3 μm). Table 3 gives three sets of measured data, with a mean air film thickness of 12.2 μm. The experimental platform is designed with the mean air film thickness of 10 μm. Although the measured value is 12 μm, accounting for the influence of manufacturing and measurement errors, 10 μm is still used for calculation and comparison.

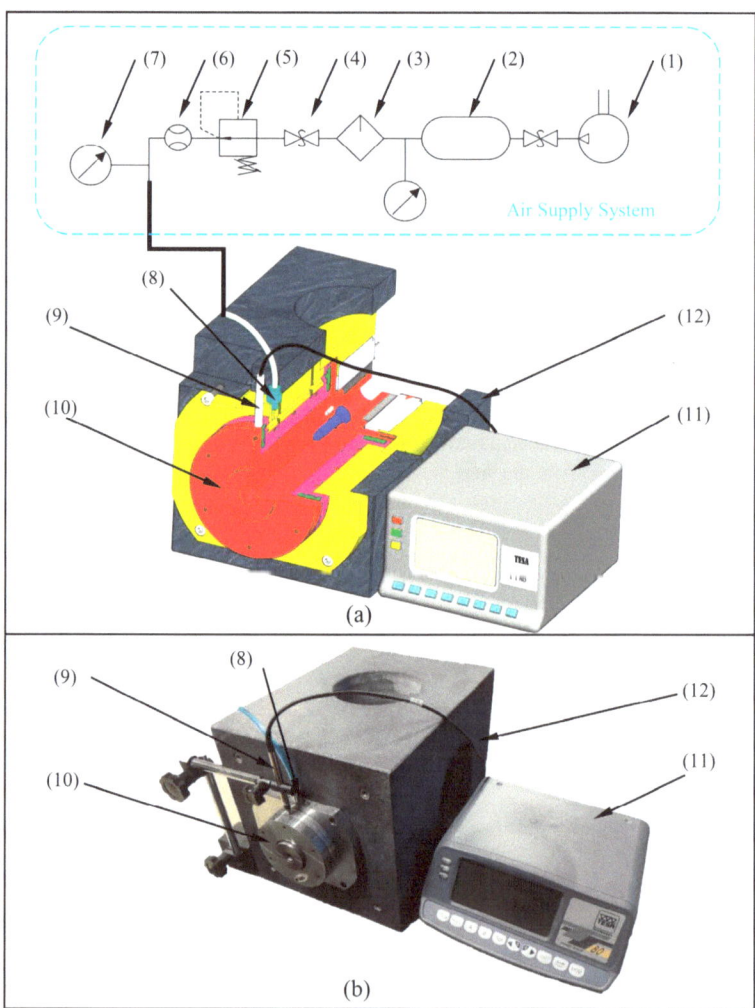

Figure 10. Equilibrium position measurement experimental platform. (**a**) Schematic diagram. (**b**) Physical diagram. (1) air compressor, (2) air tank, (3) filter, (4) globe value, (5) solenoid throttle value, (6) flowmeter, (7) pressure gauge, (8) air inlet, (9) inductive probe, (10) motorized spindle, (11) digital display unit, (12) marble platform.

Table 3. Mean air film thickness.

	Journal Diameter (mm)	Bearing Diameter (mm)	Mean Air Film Thickness (μm)
1	49.997	50.021	12
2	49.998	50.023	12.5
3	49.996	50.020	12
Average	\	\	12.2

4.2. Experiment Process

The measured motorized spindle (10) is fixed in the marble platform (12) and connected to the air supply system via the air inlet (8). During the measurement of the equilibrium position, the spindle speed is set to 0 r/min, and the initial supply pressure is set to 0 MPa. The inductance probe is positioned at the vertical baseline of the motorized spindle and calibrated to zero. The supply pressure is gradually adjusted from 0.2 to 0.6 MPa, and data are collected at intervals of 0.5 MPa. The collected data are processed and displayed using a digital display unit. A set of data is measured every 60° of rotor rotation, resulting in a total of six sets of data, from which the average value is calculated. The experimental data obtained are presented in Figure 11.

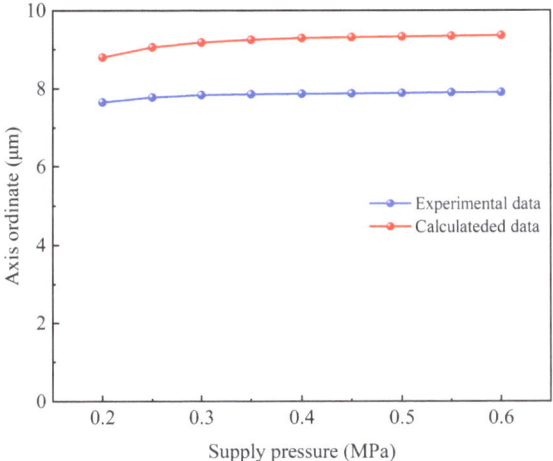

Figure 11. Experimental data and calculated data under different supply pressures.

4.3. Discussion of Experimental Results

It can be seen from Figure 11 that the calculated data and the experimental data exhibit similar trends, although the calculated results are generally higher than the experimental data. The minimum error is 12.9%, and the maximum error is 15.4%. The reasons for the difference between the experimental data and the simulated data are as follows: there are errors in the machining and assembly of the spindle in the experiment, the calculation model is simplified, and ideal conditions are used in the simulation. Therefore, the simulation calculation yields an excessive bearing capacity, resulting in a rotor position that exceeds the experimental value.

5. Conclusions

(1) The BIM effectively solves the problem of iteration divergence caused by inappropriate initial value selection. Compared with the calculation results of the secant method and search method, the maximum error of the equilibrium position is 0.15%, the required

iterative steps are only 1/4 of those of other methods, and the BIM convergence is better.

(2) When the direction of external load on the rotor is unchanged and the amplitude continues to increase, the eccentricity increases nonlinearly. The equilibrium position initially moves away from the direction of the load and later moves closer to it. The main stiffness of the bearing increases with the increase in the external load, independent of the direction of the external load. When the horizontal external load increases gradually, the absolute value of the cross stiffness Kyx decreases, while the Kxy increases. Conversely, when the vertical external load gradually increases, the absolute value of the cross stiffness Kyx increases, while the Kxy decreases.

(3) Without external load, the change in the equilibrium position and its stiffness with rotational speed and supply pressure is consistent with the conclusions of the references. And the reliability of the BIM is proved by the comparative analysis of the experiment and calculation, and the maximum error between them is 15.4%.

Author Contributions: Conceptualization, Y.H. and W.W.; methodology, H.Y. and S.Z.; software, G.Z. and Y.H.; validation, H.Y. and W.W.; formal analysis, S.L. and Y.L.; investigation, S.Z. and Y.L.; data curation, Y.H. and X.K.; writing—original draft preparation, Y.H. and X.K.; writing—review and editing, H.Y. and W.W.; visualization, S.L. and G.Z.; funding acquisition, H.Y. and S.L. All authors have read and agreed to the published version of the manuscript.

Funding: This research was funded by the National Nature Science Foundation of China, grant number 51875586; the Key Research Project of Higher Education Institutions in Henan Province, grant number 24A460030.

Data Availability Statement: Data are contained within the article.

Conflicts of Interest: The authors declare no conflicts of interest.

References

1. Gao, Q.; Chen, W.; Lu, L.; Huo, D.; Cheng, K. Aerostatic bearings design and analysis with the application to precision engineering: State-of-the-art and future perspectives. *Tribol. Int.* **2019**, *135*, 1–17. [CrossRef]
2. Hwang, J.; Park, C.H.; Kim, S.W. Estimation method for errors of an aerostatic planar XY stage based on measured profiles errors. *Int. J. Adv. Manuf. Technol.* **2010**, *46*, 877–883. [CrossRef]
3. Abele, E.; Altintas, Y.; Brecher, C. Machine tool spindle units. *CIRP Ann.* **2010**, *59*, 781–802. [CrossRef]
4. Tsai, M.H.; Hsu, T.Y.; Pai, K.R.; Shih, M.C. Precision position control of pneumatic servo table embedded with aerostatic bearing. *J. Syst. Des. Dyn.* **2008**, *2*, 940–949. [CrossRef]
5. Pandey, N.P.; Tiwari, A. CFD Simulation of Air Bearing Material. *Int. J. Sci. Res. Dev.* **2010**, *3*, 126–129.
6. Vainio, V.; Miettinen, M.; Majuri, J.; Theska, R.; Viitala, R. Manufacturing and static performance of porous aerostatic bearings. *Precis. Eng.* **2023**, *84*, 177–190. [CrossRef]
7. Cui, W.; Li, S.; Zhu, B.; Yang, F.; Chen, B. Research on the influence of a micro-groove-orifice structure and its layout form on the static characteristics of aerostatic journal bearings under a high gas supply pressure. *Adv. Mech. Eng.* **2023**, *15*, 16878132231153263. [CrossRef]
8. Zhang, J.; Deng, Z.; Zhang, K.; Jin, H.; Yuan, T.; Chen, C.; Su, Z.; Cao, Y.; Xie, Z.; Wu, D.; et al. The Influences of Different Parameters on the Static and Dynamic Performances of the Aerostatic Bearing. *Lubricants* **2023**, *11*, 130. [CrossRef]
9. Ene, N.M.; Dimofte, F.; Keith, T.G. A stability analysis for a hydrodynamic three-wave journal bearing. *Tribol. Int.* **2008**, *41*, 434–442. [CrossRef]
10. Li, Y.; Yu, M.; Ao, L.; Yue, Z. Thermohydrodynamic lubrication analysis on equilibrium position and dynamic coefficient of journal bearing. *Trans. Nanjing Univ. Aeronaut. Astronaut.* **2012**, *29*, 227–236. [CrossRef]
11. Chakraborty, M.B.; Charraborti, D.P. Design and Development of Different Applications of PATB (Porous Aerostatic Thrust Bearing): A Review. *Tribol.—Finn. J. Tribol.* **2023**, *40*, 18–28. [CrossRef]
12. Shuai, Q.; Wu, L.; Chen, Z. Solution of the three lobe's journal center orbit under steady-state force. *Mech. Eng. Autom.* **2006**, *2*, 119–121.
13. Qian, K.; Li, X. Lubrication simulation of three lobe journal bearing and solution for the equilibrium position problem. *Lubr. Eng.* **2008**, *10*, 52–54. [CrossRef]
14. Wang, W.; Song, P.; Yu, H.; Zhang, G. Research on vibration amplitude of ultra-precision aerostatic motorized spindle under the combined action of rotor unbalance and hydrodynamic effect. *Sensors* **2023**, *23*, 496. [CrossRef] [PubMed]
15. Yu, S.; Chen, S. Calculation of steady-state equilibrium position of tilting pad journal bearing in motorized spindle. *China Mech. Eng.* **2012**, *23*, 2492. [CrossRef]

16. Wan, Z.; Wang, W.; Long, X.; Meng, G. On the calculation of journal static equilibrium position of oil film bearings. *Lubr. Eng.* **2019**, *44*, 11–16. [CrossRef]
17. Zheng, T.; Hasebe, N. Calculation of equilibrium position and dynamic coefficients of a journal bearing using free boundary theory. *J. Tribol.* **2000**, *122*, 616–621. [CrossRef]
18. Pokorný, J. An efficient method for establishing the static equilibrium position of the hydrodynamic tilting-pad journal bearings. *Tribol. Int.* **2021**, *153*, 106641. [CrossRef]
19. Yang, M.; Xu, C. Twofold secant method of solving nonlinear equation. *J. Henan Norm. Univ.* **2010**, *38*, 14–16. [CrossRef]
20. Zhou, W.; Wei, X.; Wang, L.; Wu, G. A super-linear iteration method for calculation of finite length journal bearing's static equilibrium position. *R. Soc. Open Sci.* **2017**, *4*, 161059. [CrossRef]
21. Deng, Z.; Cheng, W.; Cao, G. Solution of quiescent point in gas foil bearings. *Bearing* **2021**, *9*, 20–28. [CrossRef]
22. Wen, B.; Gu, J.; Xia, S. *Advanced Rotor Dynamics: Theory, Technology and Application*; China Machine Press: Beijing, China, 1999; pp. 86–87.
23. Jia, C.; Pang, H.; Ma, W.; Qiu, M. Analysis of dynamic characteristics and stability prediction of gas bearings. *Ind. Lubr. Tribol.* **2017**, *69*, 123–130. [CrossRef]
24. Yang, L.; Li, H.; Yu, L. Dynamic stiffness and damping coefficients of aerodynamic tilting-pad journal bearings. *Tribol. Int.* **2007**, *40*, 1399–1410. [CrossRef]
25. Wang, D.; Zhu, J. A finite element method for computing dynamic coefficient of hydrodynamic journal bearing. *J. Aerosp. Power* **1995**, *69–71*, 110. [CrossRef]
26. Meng, S. Research on the rotary accuracy of the hydrostatic-dynamic spindle affected by journal geometric error and electromagnetic eccentricity. Ph.D. Thesis, Hunan University, Changsha, China, 2016.
27. Qi, S.; Geng, H.; Lu, L. Dynamic Stiffness and Dynamic Damping Coefficients of Aerodynamic Bearings. *J. Mech. Eng.* **2007**, *5*, 91–98. [CrossRef]
28. Li, Y.; Ao, L.; Li, L.; Yue, Z. Dynamic analysis method of dynamic character coefficient of hydrodynamic journal bearing. *J. Mech. Eng.* **2010**, *46*, 48–53. [CrossRef]
29. Lo, C.; Wang, C.; Lee, Y. Performance analysis of high-speed spindle aerostatic bearings. *Tribol. Int.* **2005**, *38*, 5–14. [CrossRef]

Disclaimer/Publisher's Note: The statements, opinions and data contained in all publications are solely those of the individual author(s) and contributor(s) and not of MDPI and/or the editor(s). MDPI and/or the editor(s) disclaim responsibility for any injury to people or property resulting from any ideas, methods, instructions or products referred to in the content.

Article

Study on Cage Stability of Solid-Lubricated Angular Contact Ball Bearings in an Ultra-Low Temperature Environment

Bing Su [1,*], Han Li [1], Guangtao Zhang [2], Fengbo Liu [2] and Yongcun Cui [1]

[1] School of Mechatronics Engineering, Henan University of Science and Technology, Luoyang 471003, China; li_han0013@163.com (H.L.); 9906172@haust.edu.cn (Y.C.)

[2] China Aviation Optical-Electrical Technology Co., Ltd., Luoyang 471003, China; hkd_zhanggt@163.com (G.Z.); hnly_lfb@163.com (F.L.)

* Correspondence: subing@haust.edu.cn; Tel.: +86-0379-6423-1479

Abstract: In the ultra-low temperature environment, the material properties of the bearing change, which puts forward higher requirements for the dynamic performance of the bearing cage. The bearings operating in ultra-low temperature environments commonly use solid lubricants. This study first focused on measuring the traction coefficients of molybdenum disulfide (MoS_2) solid lubricant in a nitrogen atmosphere, and the Gupta fitting model is constructed to derive the traction equation. Subsequently, the dynamic differential equation of angular contact ball bearings was established, and the stability of the bearing cage in a nitrogen environment was simulated and analyzed based on the dynamic model. The accuracy of the simulation model was verified through comparison. The results show that less than 10% of errors exist between the experimental data and the traction curve fitted by the Gupta model, and the stability of the cage is closely related to operating parameters and bearing structure parameters. Cage stability increases with axial load but decreases with radial load. The cage stability is optimal when the radial internal clearance of the bearing is approximately 0.06 mm. When other conditions remain unchanged and the ratio of the cage pocket hole gap to the cage guide surface gap is 0.2, the cage stability is the best. The research results will provide a foundation for the design and application of solid-lubricated angular contact ball bearings in ultra-low temperature environments.

Keywords: cage stability; ultra-low temperature; solid lubrication; angular contact ball bearings

Citation: Su, B.; Li, H.; Zhang, G.; Liu, F.; Cui, Y. Study on Cage Stability of Solid-Lubricated Angular Contact Ball Bearings in an Ultra-Low Temperature Environment. *Lubricants* **2024**, *12*, 124. https://doi.org/10.3390/lubricants12040124

Received: 5 March 2024
Revised: 29 March 2024
Accepted: 5 April 2024
Published: 7 April 2024

Copyright: © 2024 by the authors. Licensee MDPI, Basel, Switzerland. This article is an open access article distributed under the terms and conditions of the Creative Commons Attribution (CC BY) license (https://creativecommons.org/licenses/by/4.0/).

1. Introduction

Due to the limitations imposed by ultra-low temperature environments, conventional lubricants such as oil or grease cannot be utilized [1,2]. Instead, solid lubrication substances are employed through solid coating technology to effectively lubricate the friction pairs [3]. Solid lubricating materials possess characteristics such as a wide temperature range, low evaporation rates, and corrosion resistance [4–6]. Commonly used solid lubricants in ultra-low temperature environments include silver, PTFE, and MoS_2. PTFE is particularly well suited for extreme environments, enhancing the wear resistance of contact pairs and reducing the traction coefficient by modifying the friction surface [7]. On the other hand, MoS_2 solid lubricants possess a layered structure, excellent wear resistance, and perform effectively at low temperatures, while also offering a wide temperature range [8].

Solid lubricants are widely used in the bearing field. Kwak, et al. [9] conducted a ball and disk experiment using silver and PTFE to study the traction curve of the lubricant. They also verified the hydrodynamic traction model, taking into account the low-temperature hydrodynamic effect. The wear resistance of most solid lubricating materials in ultra-low temperature environments deteriorates [10,11]. Zhang, et al. [12] tested the frictional moments of PTFE-coated and MoS_2-coated solid-lubricated bearings in liquid nitrogen. Even the same material can produce conflicting results due to differences in environment and preparation methods. Different operating conditions also have different effects on the friction behavior of materials [13–15]. Gradt, et al. [16] found that the hardness of polymer materials increased at ultra-low temperatures. Zhang, et al. [17] found that the friction coefficient of composites in liquid nitrogen and liquid hydrogen environment was lower than that at room temperature. The reason for this phenomenon may be related to the relaxation of internal stress caused by the lateral base flows [18].

During bearing operation, a collision occurs between the cage and the bearing elements. This can lead to cage instability and affect friction–wear characteristics. The main cause of bearing failure is that fatigue failure is no longer the cause [19]. Li, et al. [20] believe that cage instability increases cage wear. The wear loss of the cage increased with the increase of the mass imbalance [21]. Gao, et al. [22] believed that the frequent impulse collisions and wear between the ball and cage pocket not only affect the bearing stability but also significantly impact the deterioration of the bearing's service life.

Ghaisas, et al. [23] established a model of the six-degree-of-freedom motion of the bearing and analyzed the influence of the rotation speed and clearance ratio on the trajectory of the center of mass of the cage. Pederson, et al. [24] studied the dynamic performance of flexible cages and rigid cages for deep-groove ball bearings. Chen, et al. [25] studied the effects of cage guidance and oil film thickness on cage stability. Nogi, et al. [26] studied the motion of a ball-bearing cage by using a dynamic analysis program and concluded that an increase in traction coefficient could result in the unstable motion of the cage. Ryu, et al. [27] analyzed the stability of the cage by the Fourier transform of the coefficient of friction and sound vibration. Wen, et al. [28] developed a calculation model to analyze the dynamics of bearings, taking into account non-Newtonian fluids. Another study by Ma [29] revealed that the collision probability between the balls and the cage increased in the bearing area, resulting in greater instability of the cage. Zhang, et al. [30] analyzed three states of cage vortex.

There are a few scholars who study the traction characteristics of solid-lubricated bearings in ultra-low temperature environments. However, we have not seen a similar paper on the stability of bearing cages in ultra-low temperature environments in combination with solid lubrication traction tests. This study aims to analyze the force of the cage under ultra-low temperature and high-speed conditions, using the actual working state of solid-lubricated ball bearings and experimental results of the traction coefficient as boundary conditions. This study also investigates the influence of working conditions and structural parameters on the trajectory of the cage centroid, the deviation of centroid eddy velocity, and the collision force between the cage and bearing elements. It provides a theoretical basis for the stability research of solid-lubricated angular contact ball-bearing cages in ultra-low temperature environments.

2. Calculation of the Traction Coefficient of Solid Lubricant

To further investigate the ultra-low temperature traction characteristics of bearing materials, the ball–disc friction and wear testing machine developed by Henan University of Science and Technology, as shown in Figure 1. The structure diagram of the testing machine is shown in Figure 2. The Gupta solid slip model [31] was used to fit the traction coefficient. Equation (1) represents the traction coefficient as a function of the sliding speed:

$$\mu = (A + B\Delta U)e^{-C\Delta U} + D \tag{1}$$

where μ is the traction coefficient; A, B, C, and D are the constant coefficients that have no physical significance; and ΔU is sliding speed.

Prior to this, other scholars conducted relevant research [28], and, on this basis, I selected different working conditions to study the traction coefficient of solid-lubricated bearings in a nitrogen environment. The ball–disc testing machine was employed to determine the traction coefficient (μ) of MoS_2 solid lubricant under various working conditions. The disc samples are covered with solid lubricating coatings, as shown in Figure 3. The accuracy level of the ball was G10. A complete bearing coated with MoS_2 is shown in

Figure 4, where the MoS$_2$ film has a jet-black color, no matte and a soft texture [12]. When the spraying process is completed, the outer ring is heated to assemble the bearing parts.

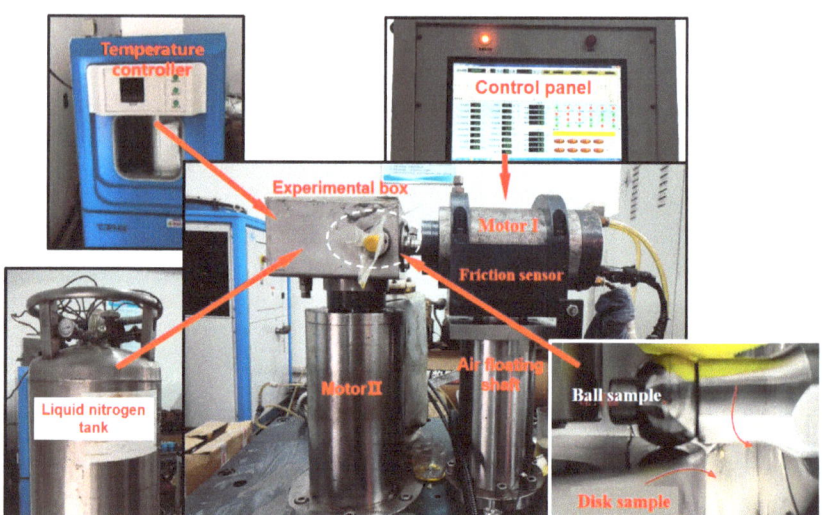

Figure 1. The ultra-low temperature solid lubrication traction force testing machine.

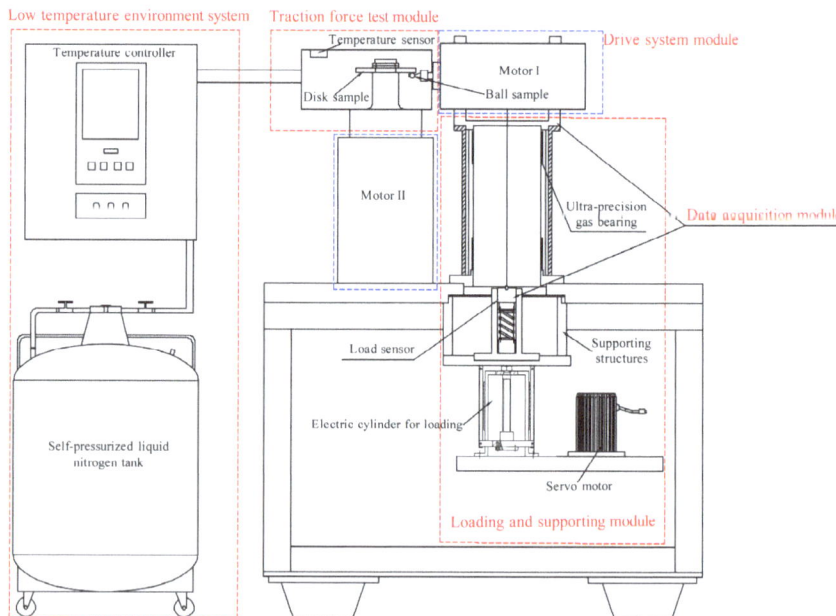

Figure 2. Structure diagram of testing machine.

The ambient temperature of the ball and disc working area is −175 °C to −170 °C. The nominal load between the ball and disk sample is 70 N, 140 N, 230 N, and 390 N, and the sliding speed is $\Delta U = 0 \sim 4$ m/s. Select the ball and disk sample sliding velocity for 0.0 m/s, 0.16 m/s, 0.32 m/s, 0.48 m/s, 0.64 m/s, 0.8 m/s, 1.6 m/s, and 4 m/s.

Figure 3. Specimens with MoS$_2$ coating.

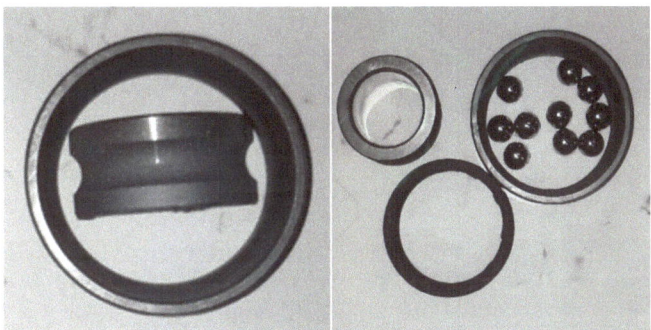

Figure 4. Bearing with MoS$_2$ coating.

As shown in Figure 5, the traction coefficient of the MoS$_2$ coating changes with the sliding speed. The overall trend of the traction coefficient is that it first increases and then slightly decreases with the increase in sliding speed. The traction coefficient decreases with the increase in load.

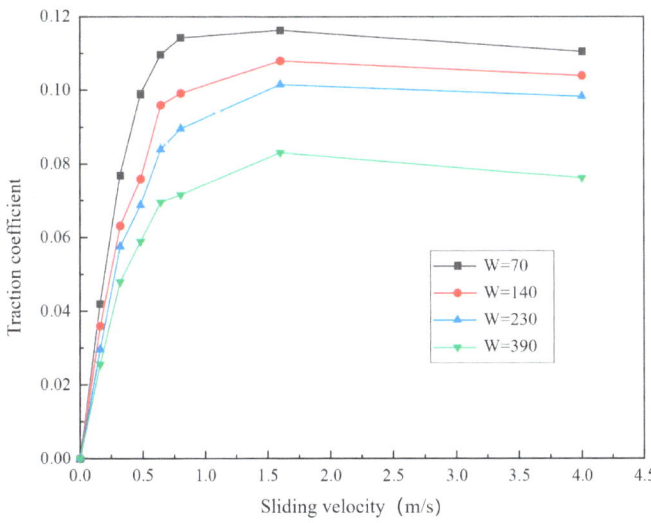

Figure 5. Traction coefficient of MoS$_2$ coating changes with sliding velocity.

To facilitate the calculation of traction parameters, the traction equation is obtained by introducing the dimensionless parameters, such as Equation (2).

$$\overline{W} = \overline{W}/(E^*R^2) \begin{cases} A = A_0 \overline{W}^{A_1} \\ B = B_0 \overline{W}^{B_1} \\ C = C_0 \overline{W}^{C_1} \\ D = D_0 \overline{W}^{D_1} \end{cases} \quad (2)$$

where $\overline{W}$ is a dimensionless parameter; E^* is the equivalent elastic modulus; R is the equivalent radius of curvature. The coefficients of A, B, C, and D and the dimensionless parameter $\overline{W}$ with the approximate exponential function relationship between them.

The Gupta solid slip model was fitted using the least squares method. By compiling the Matlab program of least squares, the optimal values of the mathematical model coefficients A, B, C, and D are listed in Table 1.

Table 1. The fitting values of coefficients of Equation (1).

Coating Materials	Load N	A	B	C	D	Residual Error	Correlation Coefficient
MoS$_2$	70	−0.0299	0.0982	5.1642	0.0299	0.0005	0.9992
	140	−0.0238	0.0787	4.2853	0.0239	0.0008	0.9965
	230	−0.0215	0.0315	6.0399	0.0215	0.0004	0.9985
	390	−0.0176	0.0520	3.3555	0.0178	0.0009	0.9922

The Gupta solid slip model was used to fit the experimental data, and the fitting values of corresponding coefficients with a strong correlation were selected and brought into Equation (2) for regression analysis. The correlation coefficients are greater than 0.95, indicating good fitting accuracy. The comparison between the traction coefficient fitting value and the test value is in Figure 6.

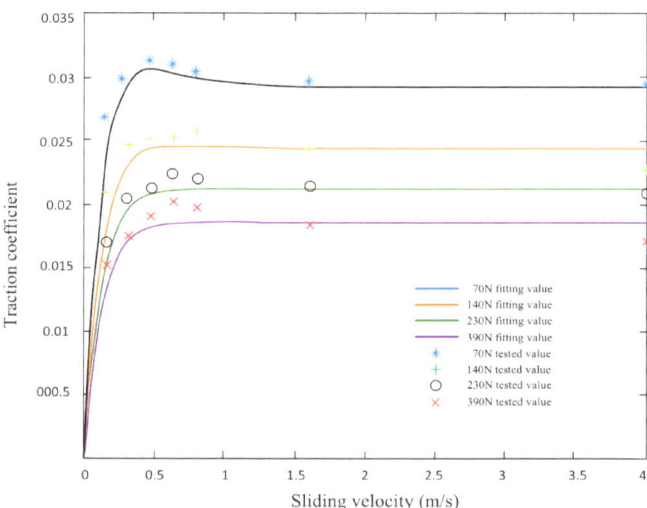

Figure 6. Comparison of traction coefficient fitting value and tested value.

The results indicate that the calculated values of the traction coefficient equation derived from the Gupta model closely match the actual test values, with an error range of less than 10%. This suggests that the Gupta solid slip model accurately predicts the ultra-low temperature traction coefficient.

3. Bearing Dynamics Modeling

Based on the dynamic theory of rolling bearings, this section analyzes the forces and motion states of balls and cages of solid-lubricated ball bearings. Combined with the traction coefficient model, the nonlinear dynamic differential equations of solid-lubricated angular contact ball bearings are derived and established to analyze the dynamic performance of the cage.

3.1. Establish the Bearing Coordinate System

In order to accurately and clearly describe the motion and force state of each component, the azimuth, force, torque, velocity, acceleration, and other parameters of each element are transformed into the inertial coordinate system by coordinates to establish a bearing dynamics model, as shown in Figure 7.

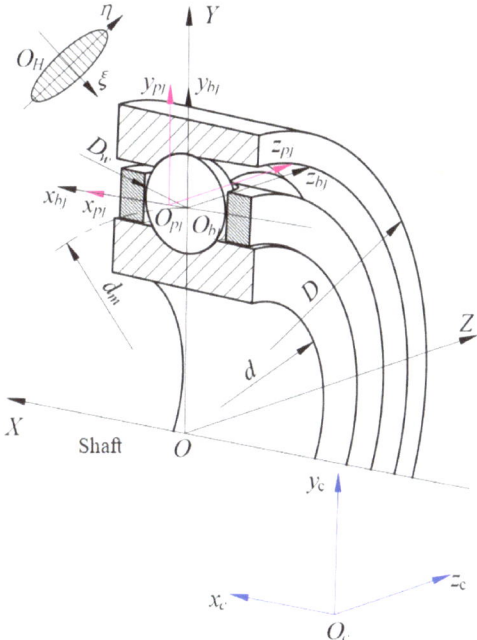

Figure 7. Coordinate systems of the bearing.

$S = \{O; X, Y, Z\}$ is a coordinate system fixed in space, and point O coincides with the center of mass of the cage. The X-axis is parallel to the axial direction of the bearing.

$S_{bj} = \{O_{bj}; x_{bj}, y_{bj}, z_{bj}\}$ is the centroid coordinate system of the ball, and point O_{bj} coincides with the center of mass of the ball. The x_{bj}-axis is parallel to the X-axis.

$S_{Hi(e)j} = \{O_H; \xi, \eta\}_{i(e)j}$ is the local coordinate system of the contact surface, and point O_H is located in the center of the contact area. The ξ-axis is the contact elliptical minor axis, which follows the rolling direction of the steel ball. The η-axis is the contact elliptical major axis, which points to the inside of the compressed part. The i represents contact with the inner ring, and the e represents contact with the outer ring.

$S_c = \{O_c; x_c, y_c, z_c\}$ is the centroid coordinate system of the cage, and point O_c coincides with the center of mass of the cage. The x_c-axis is parallel to the axial direction of the bearing.

$S_{pj} = \{O_{pj}; x_{pj}, y_{pj}, z_{pj}\}$ is a coordinate system of cage pocket holes, and point O_{pj} coincides with the center of the hole where the j-th ball is located. The x_{pj}-axis is parallel to the x_{bj}-axis.

3.2. Cage Force Analysis

3.2.1. Normal Force between Ball and Cage Pocket

Figure 8 illustrates the three motion states of the ball and cage. The calculation process for determining the normal force associated with the center displacement is described in reference [32].

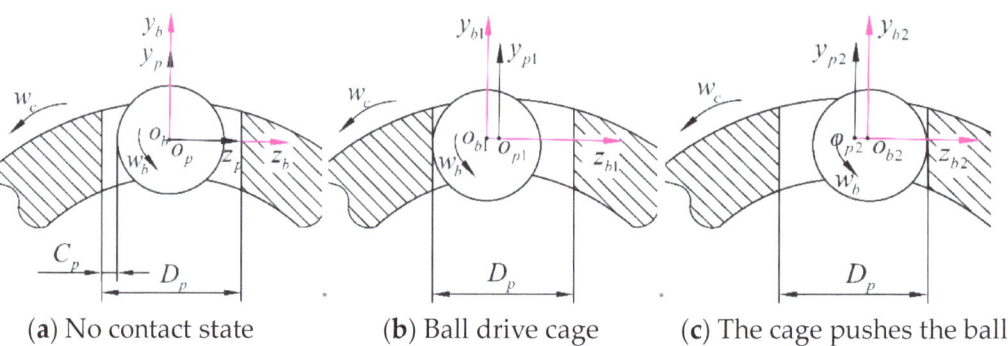

(a) No contact state (b) Ball drive cage (c) The cage pushes the ball

Figure 8. Relation between ball and cage pocket center.

Where w_b is the rotation of the ball, w_c is the rotation of the cage, C_p is the cage pocket hole clearance, and D_p is the diameter of the cage pocket hole. When O_p coincides with O_b, the steel ball is in no contact with the cage. When O_{b1} is ahead of O_{p1}, the ball drives the cage to rotate. When O_{b2} lags behind O_{p2}, the cage pushes the ball to rotate.

3.2.2. Friction at the Contact Surface between the Ball and the Pocket

When the bearing is operating, it will generate rolling friction ($P_{R\xi(\eta)j}$) and sliding friction ($P_{S\xi(\eta)j}$) in the contact zone. Part of the calculation equation is as follows, and the detailed calculation method is shown in the reference [33].

$$\begin{cases} P_{R\xi j} = 0.5 C_{opj} \overline{P_{Rj}} \cos \theta_{pj} \\ P_{R\eta j} = 0.5 C_{opj} \overline{P_{Rj}} \sqrt{\frac{R_{p\xi}}{R_{p\eta}}} \sin \theta_{pj} \\ P_{S\xi j} = \overline{P_{Sj}} \eta_0 u_{Sp\xi j} \sqrt{R_{p\xi} R_{p\eta}} \\ P_{S\eta j} = \overline{P_{Sj}} \eta_0 u_{Sp\eta j} \sqrt{R_{p\xi} R_{p\eta}} \\ \overline{P_{Rj}} = 34.74 \ln \rho_{1Rj} - 27.6 \\ \overline{P_{Sj}} = 0.26 \overline{P_{Rj}} + 10.9 \\ C_{opj} = \eta_0 u_{p\xi j} \sqrt{R_{p\xi} R_{p\eta} \left[(3 + 2k_p)^{-2} + u_{p\eta}^2 \left(3 + 2k_p^{-1}\right)^{-2} k_p^{-1} / u_{p\xi}^2 \right]} \\ k_p = R_{p\xi} / R_{p\eta} \end{cases} \quad (3)$$

where C_{op} is the auxiliary parameter, $u_{p\xi}$ is the average velocity of the ball and the surface of the cage pocket hole in the ξ direction, $u_{p\eta}$ is the average velocity of the ball and the surface of the cage pocket hole in the η direction, $R_{p\xi}$ is the radius of curvature of the ball and the surface of the cage pocket hole in the ξ direction, $R_{p\eta}$ is the radius of curvature of the ball and the surface of the cage pocket hole in the η direction, and $\overline{P_{Sj}}$ and $\overline{P_{Rj}}$ are auxiliary parameters, which can be solved by $\ln \rho_{1Rj}$.

3.2.3. Force between Cage and Ring Guide Surface

During the operation of the solid-lubricated ball bearing, the gap between the cage's inner cylinder surface and the ring guide surface will create pressure. Figure 9 shows the interaction of the guide surface and the cage, and the movement guidance of the cage is the inner ring guidance.

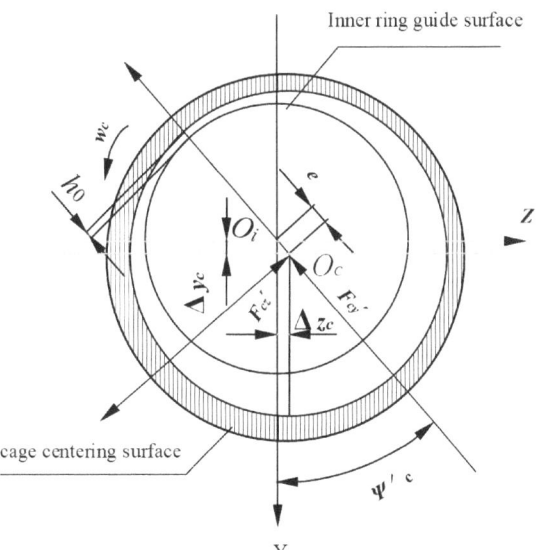

Figure 9. Contact geometry relationship between cage and ring.

The force F_c generated by solid-lubricating film pressure can be described by decomposition into F'_{cy} and F'_{cz} [33]. The forces in the normal and tangent directions are denoted, respectively. h_0 is where the minimum oil film thickness is located, rotate ψ_c is the angle relative to the inertial axis y_c, and w_c is the angular velocity of the cage around the x-axis of the inertial coordinate system. $\Delta y_c, \Delta z_c, e$ is the displacement deviation between the center of mass of the cage and the center of mass of the inner ring.

3.3. Differential Equation of Ball

Figure 10 shows the stress situation of the ball, and the equilibrium equation of the ball can be obtained as follows:

$$\begin{cases} m_b \ddot{x}_b = Q_{ej} \sin \alpha_{ej} - Q_{ij} \sin \alpha_{ij} + T_{\eta ej} \cos \alpha_{ej} - T_{\eta ij} \cos \alpha_{ij} - F_{R\eta ej} \cos \alpha_{ej} \\ \qquad + F_{R\eta ij} \cos \alpha_{ij} + F_{H\eta ej} \cos \alpha_{ej} - F_{H\eta ij} \cos \alpha_{ij} + P_{S\xi j} + P_{R\xi j} \\ m_b \ddot{y}_b = Q_{ej} \cos \alpha_{ej} - Q_{ij} \cos \alpha_{ij} - T_{\eta ej} \sin \alpha_{ej} + T_{\eta ij} \sin \alpha_{ij} + F_{R\eta ej} \sin \alpha_{ej} \\ \qquad - F_{R\eta ij} \sin \alpha_{ij} - F_{H\eta ej} \sin \alpha_{ej} + F_{H\eta ij} \sin \alpha_{ij} + F_{nj} - P_{S\eta j} - P_{R\eta j} \\ m_b \ddot{z}_b = T_{\xi ij} - T_{\xi ej} - F_{R\xi ij} + F_{R\xi ej} + F_{H\xi ij} - F_{H\xi ej} - Q_{cj} + F_{Dj} + F_{\tau j} \\ I_b \dot{\omega}_{bx} = (T_{\xi ij} - T_{R\xi ij}) \cos \alpha_{ij} \frac{D_w}{2} + (T_{\xi ej} - F_{R\xi ej}) \cos \alpha_{ej} \frac{D_w}{2} \\ \qquad - (P_{S\eta j} + P_{R\eta j}) \frac{D_w}{2} - J_x \dot{\omega}_{xj} \\ I_b \dot{\omega}_{by} - I_b \omega_{bz} \dot{\theta}_b = (F_{R\xi ij} - T_{\xi ij}) \sin \alpha_{ij} \frac{D_w}{2} + (F_{R\xi ej} - T_{\xi ej}) \sin \alpha_{ej} \frac{D_w}{2} \\ \qquad - G_{yj} - (P_{S\xi j} + P_{R\xi j}) \frac{D_w}{2} - J_y \dot{\omega}_{yj} \\ I_b \dot{\omega}_{bz} + I_b \omega_{by} \dot{\theta}_b = (T_{\eta ij} - F_{R\eta ij}) \frac{D_w}{2} + (T_{\eta ej} - F_{R\eta ej}) \frac{D_w}{2} - G_{zj} - J_z \dot{\omega}_{zj} \end{cases} \quad (4)$$

Equation (4) represents the variables used in the analysis. Where m_b refers to the mass of the ball; $\ddot{x}_b, \ddot{y}_b, \ddot{z}_b$ represent the acceleration components for the ball; $\omega_{bx}, \omega_{by}, \omega_{bz}$ for ball angular velocity; $\dot{\omega}_{bx}, \dot{\omega}_{by}, \dot{\omega}_{bz}$ or ball angular acceleration; θ for the revolution of the ball speed; I_b represents the moment of inertia of the ball; J_x, J_y, J_z are the moment of inertia components of the ball in the inertial coordinate; G_{yj}, G_{zj} are the components of the moment of inertia during the motion of the steel ball in the inertial coordinate system; $F_{R\xi ij}, F_{R\eta ij}, F_{R\xi ej}, F_{R\eta ej}$ are the friction forces in the contact entrance area between the steel ball and the raceway; $T_{\xi ij}, T_{\eta ij}, T_{\xi ej}, T_{\eta ej}$ are the traction force in the direction of the ξ and η axis of the contact between the ball and the inner and outer raceways; $F_{H\xi ij}, F_{H\eta ij}, F_{H\xi ej}, F_{H\eta ej}$ are all horizontal components of force acting on the center of steel ball; $P_{S\xi j}, P_{S\eta j}$ are the sliding friction forces on the surface of steel balls; $P_{R\xi j}, P_{R\eta j}$ are the rolling friction forces on the steel ball surface; $F_{nj}, F_{\tau j}$ are the inertial force component during the movement of the steel ball; F_{Dj} is the aerodynamic resistance of the gas to a single steel ball; and Q_{cj} is the collision force between the steel ball and the cage.

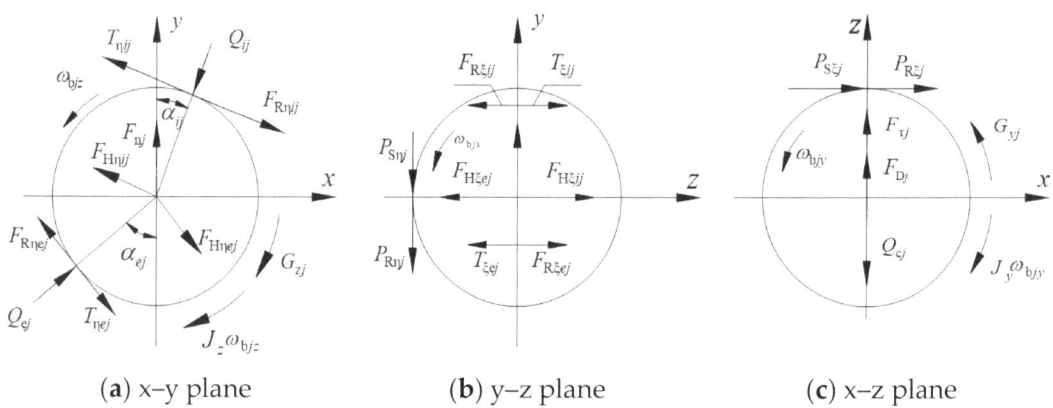

(a) x–y plane   (b) y–z plane   (c) x–z plane

Figure 10. Steel ball force diagram.

3.4. Differential Equation of Cage

In this study, the cage guidance method is inner ring guidance, and the force balance equation is as follows:

$$\begin{cases} m_c \ddot{x}_c = \sum_{j=1}^{Z} (P_{S\eta j} + P_{R\eta j}) \\ m_c \ddot{y}_c = \sum_{j=1}^{Z} \left[(P_{S\xi j} + P_{R\xi j}) \cos \varphi_j + Q_{cj} \sin \varphi_j \right] + F_{cy} \\ m_c \ddot{z}_c = \sum_{j=1}^{Z} \left[(P_{S\xi j} + P_{R\xi j}) \sin \varphi_j - Q_{cj} \cos \varphi_j \right] + F_{cz} \\ I_{cx} \dot{\omega}_{cx} - (I_{cy} - I_{cz}) \omega_{cy} \omega_{cz} = \sum_{j=1}^{Z} \left[(P_{S\xi j} + P_{R\xi j}) \frac{D_w}{2} - Q_{cj} \frac{d_m}{2} \right] + M_{cx} \\ I_{cy} \dot{\omega}_{cy} - (I_{cz} - I_{cx}) \omega_{cz} \omega_{cx} = \sum_{j=1}^{Z} (P_{S\eta j} + P_{R\eta j}) \frac{d_m}{2} \sin \varphi_j \\ I_{cz} \dot{\omega}_{cz} - (I_{cx} - I_{cy}) \omega_{cx} \omega_{cy} = \sum_{j=1}^{Z} (P_{S\eta j} + P_{R\eta j}) \frac{d_m}{2} \cos \varphi_j \end{cases} \quad (5)$$

m_c is cage quality; φ_j is the position angle of the j th s ball; d_m is the diameter of the bearing pitch; $\ddot{x}_c, \ddot{y}_c, \ddot{z}_c$ are cage acceleration; I_{cx}, I_{cy}, I_{cz} are cage three principal moments of inertia; $\omega_{cx}, \omega_{cy}, \omega_{cz}$ are cage angular velocity; $\dot{\omega}_{cx}, \dot{\omega}_{cy}, \dot{\omega}_{cz}$ are the cage angular acceleration; and Q_{cj} is the components of the collision force between the ball and the cage in the inertial coordinate system, respectively. For the calculation of dynamic differential equations of inner and outer rings, see reference [34].

4. Cage Stability Analysis Method

The movement of the cage is complex, and its stability is often assessed based on the shape of its centroid trajectory. A point trajectory indicates complete stability, while a single circle or periodic circle trajectory suggests a stable vortex state. When the cage centroid trajectory is polygonal, or even chaotic, it indicates that the centroid of the cage is divergent and in an unstable vortex state [35,36].

Figure 11 shows the variations in centroid vibration displacement of the cage in the Y and Z directions during time-domain analysis under constant load conditions obtained from dynamic simulation analysis. Curve 1 represents the displacement in the y-direction, while curve 2 represents the displacement in the Z-direction. The figure illustrates that as the cage vortex moves, the radial displacement of the centroid changes periodically over time, with the Z displacement leading to the Y-direction displacement. On this basis, the dynamic performance of a solid-lubricated ball-bearing cage is analyzed in the next section.

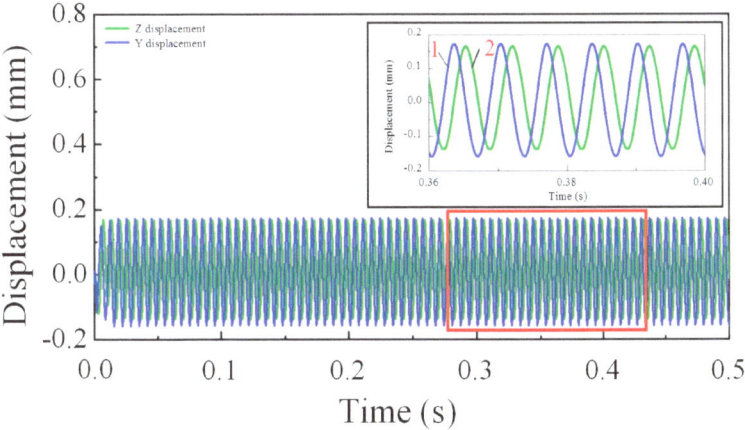

Figure 11. Cage centroid displacement.

To quantitatively analyze the cage when the centroid trajectory exhibits a vortex rather than a point, the change in centroid vortex speed can be combined to assess the stability of the cage. In the actual optimization design analysis, the instability of the cage movement is determined by calculating the ratio of the deviation in centroid vortex velocity.

The stability of the cage is determined by comparing the deviation of the centroid vortex velocity, as proposed by Ghaisas, et al. [23]. The ratio of the standard deviation of the centroid velocity to the average value can be calculated as follows:

$$\sigma_v = \frac{\sqrt{\sum_{i=1}^{n}(v_i - v_m)/(n-1)}}{v_m} \tag{6}$$

v_i is the vortex velocity of the centroid of the cage at different moments; v_m is the cage centroid average speed. The larger the ratio of the centroid vortex velocity deviation and the greater the change in the vortex velocity, the worse the stability of the cage is, and vice versa.

In the next section, the stability of the cage under inner ring guidance is analyzed by changing operating parameters such as axial load and radial load, as well as structural parameters such as the radial internal clearance, cage pocket gap, and guide gap, and the dynamic performance of the cage is discussed from the aspects of the cage centroid trajectory and the traction characteristics between the ball and the cage.

5. Analysis of Factors Affecting Cage Stability

5.1. Validation of the Cage Analysis Model

The model of the test bearing is 7204AC. The initial basic parameters and working conditions of the bearing are shown in Table 2. The bearing ring and ball are made of G95Cr18 material, and the surface is coated with MoS_2 film, while the cage is made of nylon 66. The material characteristics of nylon 66 are listed in Table 3.

Table 2. Basic bearing parameters.

Parameter		Value
Outside diameter (mm)	D	47
Bore diameter (mm)	d	20
Width (mm)	B	14
Nominal contact angle (°)	α	25
Diameter of sphere (mm)	D_w	7.938
Number of balls	Z	10
Initial radial internal clearance (mm)	u_0	0.091
Rotate speed (r/min)	n	20,000
MoS$_2$ film thickness (μm)	h	8

Table 3. Bearing material parameters.

Materials	Nylon 66	G95Cr18
Elasticity modulus E (GPa)	2.83	200
Poisson's ratio μ	0.4	0.28
Density ρ (g/cm^3)	1.15	7.8
Coefficient of linear expansion c_l (10^{-6} °C^{-1})	1.2	11.5
Heat conductivity coefficient λ (W/(m·°C))	0.25	29.3

In this paper, the classical example of Gupta [37,38] is used to verify the reliability of the proposed model. Gupta analyzes the stability of solid-lubricated high-speed angular contact ball bearing cages. Although the results of the cage stability obtained by the model are slightly different from those calculated by Gupta, the overall trend is consistent, which is caused by the different parameters of cage material and lubricant.

As can be seen from [38], the cage mass center orbit shape derived from the Gupta analysis is consistent with the results of the later sections of this paper. As the cage mass center whirl velocity at the center of mass of the cage increases, the force between the cage and the guide surface also increases. The comparison and analysis of the above results prove that the results calculated in this paper have certain accuracy and reliability. On the basis of the Gupta analysis model, a variety of structural parameters and working condition parameters are added to analyze the stability of the cage more comprehensively.

5.2. Influence of Axial Load on Cage Stability

Assuming the angular contact ball bearing used in the rotor system of axial load working condition had a rotating speed of n_i = 20,000 r/min, a radial load of F_r = 0 N, and axial loads of 50 N, 75 N, 100 N, 200 N, 300 N, and 400 N, respectively, the relationship between the centroid trajectory of the cage and the axial load is shown in Figure 12. The calculated speed deviation ratio of the bearing cage and the maximum force between the cage and the guide surface of the ring vary with the axial load, as shown in Figure 13.

The maximum force between the cage and ring guide surface decreases with the increase in axial load. This is mainly because the traction coefficient of a solid lubricant decreases with the increase in load. In Figure 5 of Section 2, it can be observed that the traction effect of the raceway on the ball weakens, leading to a reduction in the collision between the ball and the cage. The cage is guided by the inner ring, and the guiding force of the ring on the cage tends to remain stable. Therefore, within the range of 50 N to

400 N axial load, the cage remains stable, and the vortex velocity deviation ratio also tends to stabilize.

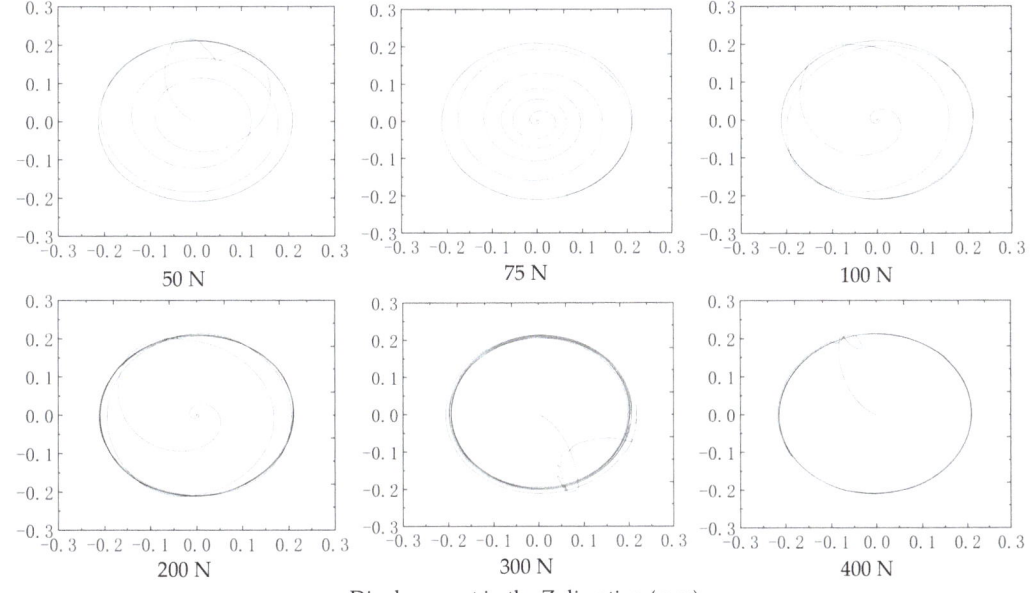

Figure 12. Centroid trajectories under different axial loads.

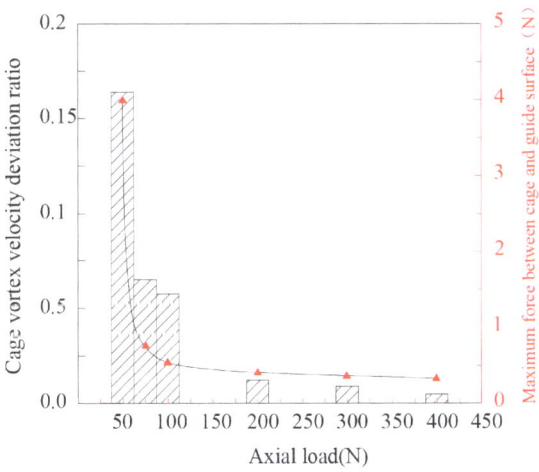

Figure 13. Cage stability and force under different axial loads.

5.3. Influence of Radial Load on Cage Stability

Angular contact ball bearings are under combined loading, and this section assumes that the bearing ring speed for n_i = 20,000 r/min and F_r = 2000 N for radial load, respectively, 0 N, 400 N, 800 N, 1200 N, 1600 N, and 2000 N. The centroid trajectories under different load ratios are shown in Figure 14. The calculated cage vortex velocity deviation ratio and the maximum force between the cage and the guide surface of the ring vary with the radial load, as shown in Figure 15.

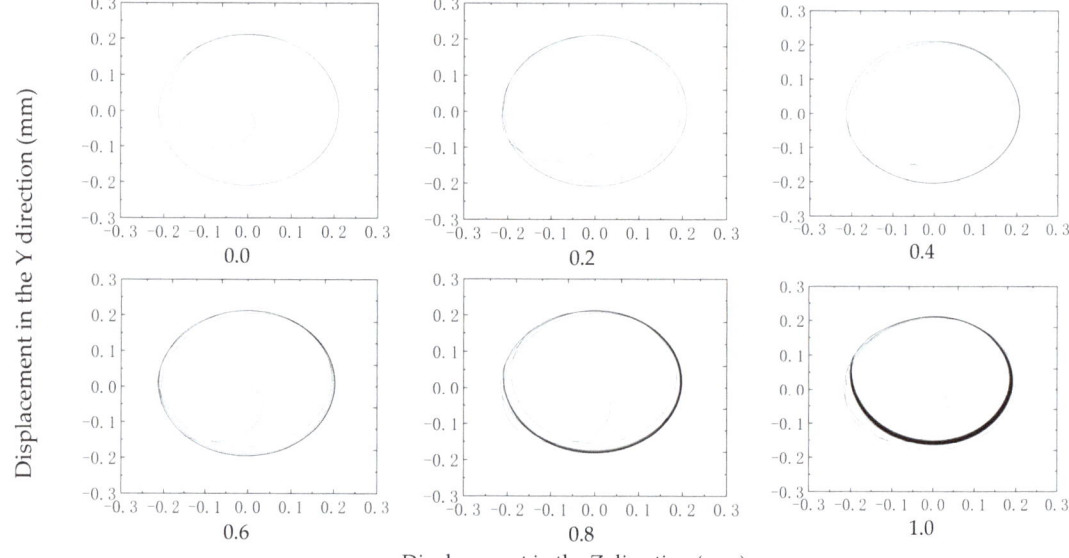

Figure 14. Centroid trajectories under different load ratios.

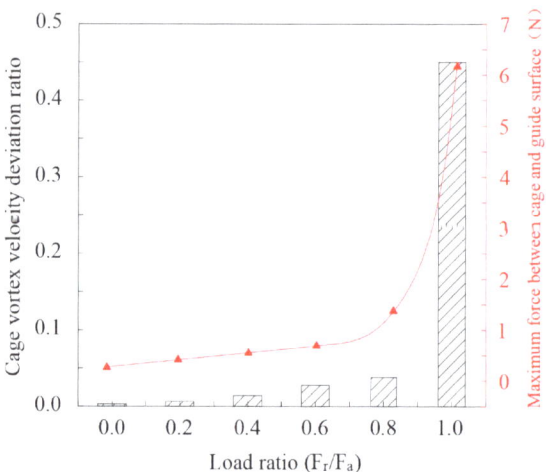

Figure 15. Cage stability and force under different load ratios.

The main reason for the cage movement characteristics shown in the figure is that when the bearing is subjected to a combined load, the load distribution of the balls becomes increasingly uneven with an increase in radial load. This uneven load distribution leads to a significant difference in the generation of traction force on the ring, resulting in a substantial change in the collision force between the ball and the cage pocket at different azimuth angles, which will cause the collision between the ring guide surface and the cage, and, ultimately, reduce the stability of the cage.

5.4. Influence of Radial Internal Clearance on Cage Stability

The influence of the initial radial internal clearance on the dynamic performance of the cage was studied when the cage pocket gap was 0.24 mm and the guide gap was 0.40 mm. With a bearing speed of n_i = 20,000 r/min, F_a = 2000 N and F_r = 800 N conditions, as shown in Figures 16 and 17, the velocity deviation ratio of the cage centroid and the maximum force between the cage and the guide surface of the ring change with the radial internal clearance.

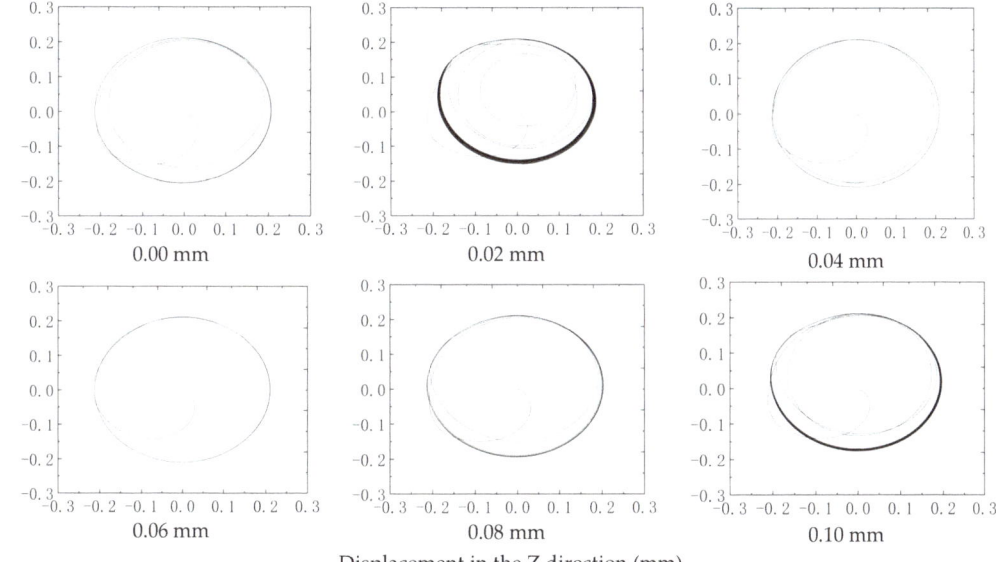

Figure 16. Centroid trajectories under different radial internal clearance.

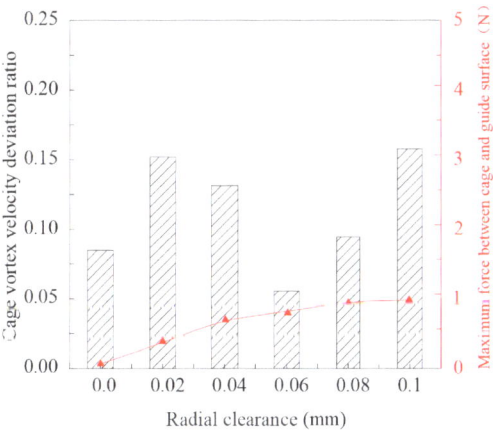

Figure 17. Cage stability and force under different radial internal clearance.

The results presented in Figure 17 demonstrate that the deviation ratio of the centroid vortex velocity initially increases, then decreases, and, eventually, increases with the increase of radial internal clearance. Although there is no clear linear relationship between the stability of the cage and the change in clearance, within the selected range of clearances, the lowest deviation ratio of the vortex velocity of the centroid and the smallest range of the centroid's trajectory circle when the radial internal clearance is 0.06 mm.

The force between the cage and the ring guide surface remains relatively constant as the radial internal clearance changes, indicating that the clearance has minimal influence on this force.

5.5. Influence of Guide Gap Change on Cage Stability

When the cage pocket gap is 0.24 mm and the gap ratio (the ratio of the cage pocket gap to the guide gap) ranges from 0.2 to 1.2, the bearing condition for speed is n_i = 20,000 r/min, F_a = 2000 N and F_r = 800 N, the influence of the guide gap on the cage stability is studied, as shown in Figures 18 and 19. Extract the cage of the centroid velocity deviation ratio, cage and ring guide surface maximum force changing with the guide gap.

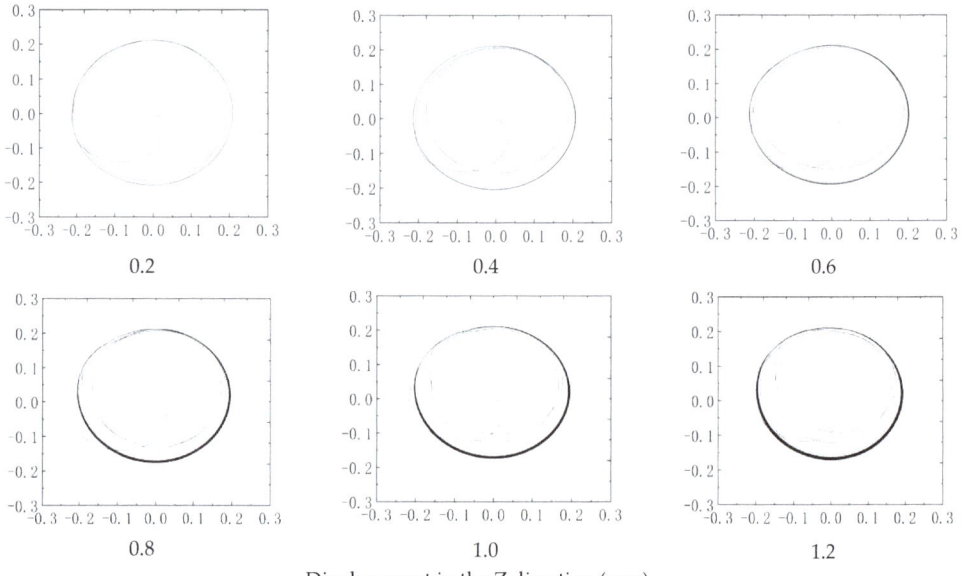

Figure 18. Centroid trajectories under different gap ratios (pocket gap is 0.24 mm).

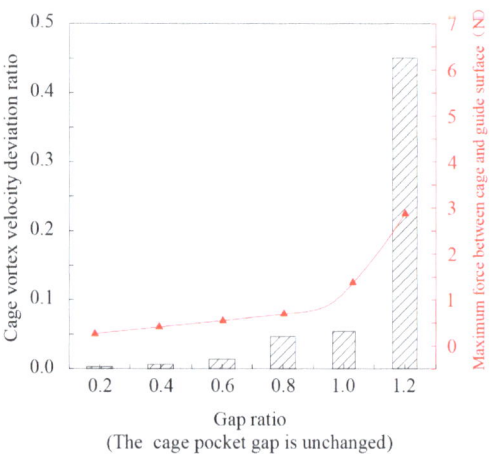

Figure 19. Cage stability and force under different gap ratios.

The primary cause of the cage's unstable movement is attributed to the gradual decrease in the guide gap and the subsequent reduction in the movement range of the centroid of the cage as the gap ratio increases. As a result, the collision between the ring and the cage obviously increases vibration, and the motion stability of the cage gradually deteriorates.

5.6. Influence of Cage Pocket Gap Change on Cage Stability

When the guide gap is 0.40 mm and the gap ratio (the ratio of the pocket gap to the guide gap) varies from 0.2 to 1.2, the influence of the change of pocket gap on the stability of the cage is studied. Under the conditions of an axial force of 2000 N, radial force of 800 N, and rotation speed of 20,000 r/min, extract the velocity deviation ratio of the cage centroid, cage, and ring guide surface maximum force changing with guide gap, as shown in Figures 20 and 21.

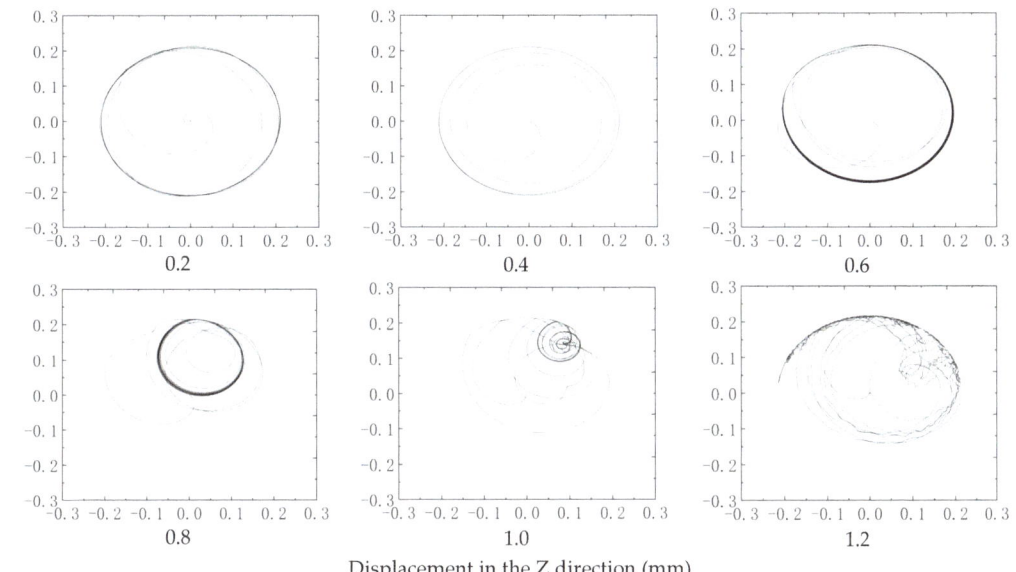

Figure 20. Centroid trajectories under different gap ratios (guide gap is 0.24 mm).

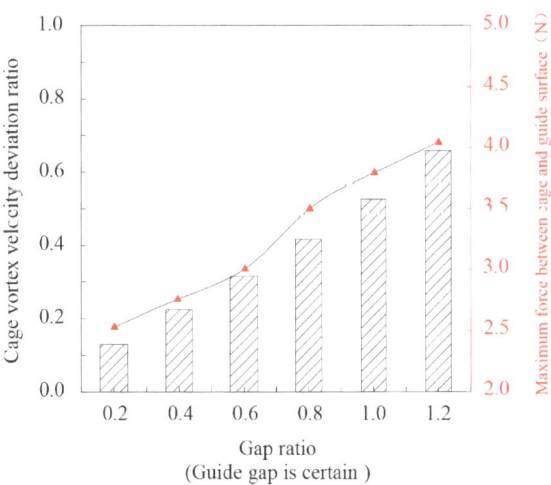

Figure 21. Cage stability and force under different gap ratios.

The main reason for the unstable movement of the cage is the decrease in the guide gap as the gap ratio increases. This leads to a decrease in the movement range of the cage's centroid and causes obvious collisions between the ring and the cage, resulting in increased vibration and weakened motion stability of the cage. Therefore, it is important to pay attention to the matching between the cage pocket gap and the guide gap in the design of the cage, ensuring that the cage remains in a stable vortex state to improve the reliability of bearing applications.

6. Conclusions

This study focuses on solid-lubricated angular contact ball bearings operating in an ultra-low temperature environment. The traction coefficient of the MoS_2 solid lubricating materials was first obtained through experiments, and the Gupta fitting model was constructed to derive the traction equation. Then, the bearing dynamic model was established, and the bearing cage stability analysis was carried out. The conclusions are drawn as follows:

(1) When the load is constant, the solid lubricating material's traction coefficient first rises and then slightly declines as the sliding speed rises. The traction coefficient falls with an increase in load while the sliding speed is constant. Less than 10% of errors exist between the experimental data and the traction curve fitted by the Gupta model, indicating high accuracy. MoS_2 has better friction performance.

(2) When the bearing speed n_i is 20,000 r/min and radial load F_r is 0 N, the cage stability increases with increasing axial force. When the bearing speed n_i is 20,000 r/min and F_a is 2000 N, the cage stability declines as the radial load increases.

(3) When the cage pocket gap was 0.24 mm, the guide gap was 0.40 mm, the bearing speed n_i is 20,000 r/min, F_a is 2000 N, and F_r is 800 N conditions. The cage stability is optimal when the radial internal clearance of the bearing is approximately 0.06 mm.

(4) The ratio of the cage pocket gap to the cage guide gap was studied, ranging from 0.2 to 1.2. When the ratio is 0.2, the stability of the cage is the best, which can improve the application reliability of the bearing.

Author Contributions: Conceptualization, B.S., H.L. and G.Z.; Data Curation, H.L. and G.Z.; Formal Analysis, H.L. and F.L.; Funding Acquisition, Y.C.; Investigation, G.Z., B.S. and H.L.; Methodology, B.S. and H.L.; Project Administration, B.S. and Y.C.; Resources, Y.C.; Software, H.L. and F.L.; Supervision, Y.C. and B.S.; Validation, F.L., B.S. and G.Z.; Writing—Original Draft, B.S. and H.L.; Writing—Review and Editing, B.S. and F.L. All authors have read and agreed to the published version of the manuscript.

Funding: This research was funded by the National Natural Science Foundation of China (52005158).

Data Availability Statement: The data used to support the findings of this study are available from the corresponding author upon request.

Conflicts of Interest: Authors Guangtao Zhang and Fengbo Liu were employed by the company China Aviation Optical-Electrical Technology Co., Ltd. The remaining authors declare that the research was conducted in the absence of any commercial or financial relationships that could be construed as a potential conflict of interest.

References

1. Antonov, S.A.; Bartko, R.; Nikul, P.A.; Kilyakova, A.Y.; Tonkonogov, B.P.; Danilov, A.M. The Current State of Development of Greases. *Chem. Technol. Fuels Oils* **2021**, *57*, 279–288. [CrossRef]
2. Porfiryev, Y.; Shuvalov, S.; Popov, P.; Kolybelsky, D.; Petrova, D.; Ivanov, E.; Tonkonogov, B.; Vinokurov, V. Effect of Base Oil Nature on the Operational Properties of Low-Temperature Greases. *ACS Omega* **2020**, *5*, 11946–11954. [CrossRef]
3. Zhang, Y.; Li, P.; Ji, L.; Liu, X.; Wan, H.; Chen, L.; Li, H.; Jin, Z. Tribological properties of MoS_2 coating for ultra-long wear-life and low coefficient of friction combined with additive g-C_3N_4 in air. *Friction* **2021**, *9*, 789–801. [CrossRef]
4. Ren, Y.; Zhang, L.; Xie, G.; Li, Z.; Chen, H.; Gong, H.; Xu, W.; Guo, D.; Luo, J. A review on tribology of polymer composite coatings. *Friction* **2021**, *9*, 429–470. [CrossRef]
5. John, M.; Menezes, P.L. Self-Lubricating Materials for Extreme Condition Applications. *Materials* **2021**, *14*, 5588. [CrossRef]
6. Kumar, R.; Antonov, M. Self-lubricating materials for extreme temperature tribo-applications. *Mater. Today Proc.* **2021**, *44*, 4583–4589. [CrossRef]
7. Wong, K.C.; Lu, X.; Cotter, J.; Eadie, D.T.; Wong, P.C.; Mitchell, K.A.R. Surface and friction characterization of MoS_2 and WS_2 third body thin films under simulated wheel/rail rolling–sliding contact. *Wear* **2008**, *264*, 526–534. [CrossRef]

8. Dunckle, C.G.; Aggleton, M.; Glassman, J.; Taborek, P. Friction of molybdenum disulfide–titanium films under cryogenic vacuum conditions. *Tribol. Int.* **2011**, *44*, 1819–1826. [CrossRef]
9. Kwak, W.; Lee, J.; Lee, Y.-B. Theoretical and experimental approach to ball bearing frictional characteristics compared with cryogenic friction model and dry friction model. *Mech. Syst. Signal Process.* **2019**, *124*, 424–438. [CrossRef]
10. Yang, X.; Li, R.; Wang, Y.; Zhang, J.; Yao, X. Improvement of mechanical and tribological performances of carbon nanostructure films by cryogenic treatment. *Tribol. Int.* **2021**, *156*, 106819. [CrossRef]
11. Hongtao, L.; Hongmin, J.; Xuemei, W. Tribological properties of ultra-high molecular weight polyethylene at ultra-low temperature. *Cryogenics* **2013**, *58*, 1–4.
12. Zhang, G.; Su, B.; Liu, P.; Wang, J.; Zhang, W. Experimental Research on Frictional Torque of Solid Lubricated Bearings in Liquid Nitrogen Environment. *Bearing* **2023**, *135*, 68–73.
13. Theiler, G.; Gradt, T. Friction and wear behaviour of polymers in liquid hydrogen. *Cryogenics* **2018**, *93*, 1–6. [CrossRef]
14. La, S.; Wang, J.; Zhang, X.; Zhou, Y. Frictional Behavior of a Micro-sized Superconducting Fiber in a Low-Temperature Medium: Experimental and Computational Analysis. *Acta Mech. Solida Sin.* **2018**, *31*, 405–415. [CrossRef]
15. La, S.; Liu, C.; Zhang, X. The mechanism of stick-slip phenomenon during friction process at low temperature environment. *AIP Adv.* **2019**, *9*, 065019. [CrossRef]
16. Gradt, T.; Schneider, T.; Hübner, W.; Börner, H. Friction and wear at low temperatures. *Int. J. Hydrog. Energy* **1998**, *23*, 397–403. [CrossRef]
17. Zhang, Z.; Klein, P.; Theiler, G.; Hübner, W. Sliding performance of polymer composites in liquid hydrogen and liquid nitrogen. *J. Mater. Sci.* **2004**, *39*, 2989–2995. [CrossRef]
18. Sápi, Z.; Butler, R. Properties of cryogenic and low temperature composite materials—A review. *Cryogenics* **2020**, *111*, 103190. [CrossRef]
19. Cui, L. A new fatigue damage accumulation rating life model of ball bearings under vibration load. *Ind. Lubr. Tribol.* **2020**, *29*, 1205–1215. [CrossRef]
20. Li, H.; Liu, H.; Li, H.; Qi, S.; Liu, Y.; Wang, F. Effect of cage-pocket wear on the dynamic characteristics of ball bearing. *Ind. Lubr. Tribol.* **2020**, *72*, 905–912. [CrossRef]
21. Choe, B.; Kwak, W.; Jeon, D.; Lee, Y. Experimental study on dynamic behavior of ball bearing cage in cryogenic environments, Part II: Effects of cage mass imbalance. *Mech. Syst. Signal Process.* **2019**, *116*, 25–39. [CrossRef]
22. Gao, S.; Han, Q.; Zhou, N.; Zhang, F.; Yang, Z.; Chatterton, S. Pennacchi, Dynamic and wear characteristics of self-lubricating bearing cage: Effects of cage pocket shape. *Nonlinear Dyn.* **2022**, *110*, 177–200. [CrossRef]
23. Ghaisas, N.; Wassgren, C.R. Sadeghi Cage Instabilities in Cylindrical Roller Bearings. *J. Tribol.* **2004**, *126*, 681–689. [CrossRef]
24. Pederson, B.M.; Sadeghi, F.; Wassgren, C.R. The Effects of Cage Flexibility on Ball-to-Cage Pocket Contact Forces and Cage Instability in Deep Groove Ball Bearings. *SAE Trans.* **2006**, *1*, 260–271.
25. Chen, S.; Chen, X.; Zhang, T.; Li, Q.; Gu, J. Cage motion analysis in coupling influences of ring guidance mode and rotation mode. *J. Adv. Mech. Des. Syst. Manuf.* **2019**, *13*, JAMDSM0054. [CrossRef]
26. Nogi, T.; Maniwa, K.; Matsuoka, N. A Dynamic Analysis of Cage Instability in Ball Bearings. *J. Tribol.* **2018**, *140*, 011101. [CrossRef]
27. Ryu, S.; Choe, B.; Lee, J.; Lee, Y. Correlation Between Friction Coefficient and Sound Characteristics for Cage Instability of Cryogenic Deep Groove Ball Bearings. In *Proceedings of the 9th IFToMM International Conference on Rotor Dynamics, v.2*; Springer: Cham, Switzerland, 2015; pp. 1921–1931.
28. Wen, B.; Wang, M.; Yun, X.; Zhang, X.; Han, Q. Analysis of cage slip in angular contact ball bearing considering non-Newtonian behavior of elastohydrodynamic lubrication. *IOP Conf. Ser. Mater. Sci. Eng.* **2021**, *1081*, 012013. [CrossRef]
29. Ma, S. A Study on Bearing Dynamic Features under the Condition of Multiball–Cage Collision. *Lubricants* **2022**, *10*, 9. [CrossRef]
30. Zhang, J.; Qiu, M.; Dong, Y.; Pang, X.; Li, J. Yang, Investigation on the Cage Whirl State of Cylindrical Roller Bearings under High Speed and Light Load. *Machines* **2022**, *10*, 768. [CrossRef]
31. Gupta, P.K. Traction coefficients for some solid lubricants for rolling bearing dynamics modeling. *Tribol. Trans.* **2000**, *43*, 647–652. [CrossRef]
32. Sugimura, J.; Okumura, T.; Yamamoto, Y.; Spikes, H.A. Simple equation for elastohydrodynamic film thickness under acceleration. *Tribol. Int.* **1999**, *32*, 117–123. [CrossRef]
33. Deng, S.; Jia, Q.; Xue, J. *Design Principle of Rolling Bearings*, 2nd ed.; China Standard Press: Beijing, China, 2014.
34. Lei, J.; Su, B.; Zhang, S.; Yang, H.; Cui, Y. Dynamics-Based Thermal Analysis of High-Speed Angular Contact Ball Bearings with Under-Race Lubrication. *Machines* **2023**, *11*, 691. [CrossRef]
35. Kingsbury, E.P.; Walker, R. Motions of an Unstable Retainer in an Instrument Ball Bearing. *J. Tribol. Trans. ASME* **1994**, *116*, 202–208. [CrossRef]
36. Kingsbury, E.P. Torque Variations in Instrument Ball Bearings. *ASLE Trans.* **1965**, *8*, 435–441. [CrossRef]
37. Gupta, P.K. Modeling of Instabilities Induced by Cage Clearances in Ball Bearings. *Tribol. Trans.* **1991**, *34*, 93–99. [CrossRef]
38. Gupta, P.K.; Forster, N.H. Modeling of Wear in a Solid-Lubricated Ball Bearing. *ASLE Trans.* **1987**, *30*, 55–62. [CrossRef]

Disclaimer/Publisher's Note: The statements, opinions and data contained in all publications are solely those of the individual author(s) and contributor(s) and not of MDPI and/or the editor(s). MDPI and/or the editor(s) disclaim responsibility for any injury to people or property resulting from any ideas, methods, instructions or products referred to in the content.

Article

Calibration of Oil Film Thickness Acoustic Reflection Coefficient of Bearing under Multiple Temperature Conditions

Fei Shang [1,2], Bo Sun [1,2,*], Shaofeng Wang [1,2], Yongquan Han [3], Wenjing Liu [1,2], Ning Kong [4], Yuwu Ba [1,2], Fengchun Miao [5] and Zhendong Liu [6]

[1] School of Mechanical Engineering, Inner Mongolia University of Science and Technology, Baotou 014010, China
[2] Inner Mongolia Key Laboratory of Intelligent Diagnosis and Control of Mechatronic Systems, Inner Mongolia University of Science and Technology, Baotou 014010, China
[3] School of Material and Metallurgy, Inner Mongolia University of Science and Technology, Baotou 014010, China
[4] School of Mechanical Engineering, University of Science and Technology Beijing, Beijing 100083, China
[5] Inner Mongolia North Heavy Industry Group Co., Ltd., Baotou 014000, China
[6] Baotou Special Equipment Inspection Institute, Inner Mongolia Institute of Special Equipment Inspection and Research, Baotou 014000, China
* Correspondence: sunbo2117@stu.imust.edu.cn

Citation: Shang, F.; Sun, B.; Wang, S.; Han, Y.; Liu, W.; Kong, N.; Ba, Y.; Miao, F.; Liu, Z. Calibration of Oil Film Thickness Acoustic Reflection Coefficient of Bearing under Multiple Temperature Conditions. *Lubricants* 2024, 12, 125. https://doi.org/10.3390/lubricants12040125

Received: 5 March 2024
Revised: 20 March 2024
Accepted: 4 April 2024
Published: 7 April 2024

Copyright: © 2024 by the authors. Licensee MDPI, Basel, Switzerland. This article is an open access article distributed under the terms and conditions of the Creative Commons Attribution (CC BY) license (https://creativecommons.org/licenses/by/4.0/).

Abstract: Rolling mill bearings are prone to wear, erosion, and other damage characteristics due to prolonged exposure to rolling forces. Therefore, regular inspection of rolling mill bearings is necessary. Ultrasonic technology, due to its non-destructive nature, allows for measuring the oil film thickness distribution within the bearing during disassembly. However, during the process of using ultrasonic reflection coefficients to determine the oil film thickness and distribution state of rolling mill bearings, changes in bearing temperature due to prolonged operation can occur. Ultrasonic waves are susceptible to temperature variations, and different temperatures of the measured structure can lead to changes in measurement results, ultimately distorting the results. This paper proposes using density and sound speed compensation methods to address this issue. It simulates and analyzes the oil film reflection coefficients at different temperatures, ultimately confirming the feasibility and effectiveness of this approach. The paper establishes a functional relationship between bearing pressure and reflection coefficients, oil film thickness, and reflection coefficients. This allows for the compensation of reflection coefficients under any pressure conditions, enhancing the accuracy of oil film thickness detection. The proposed method provides technical support for the maintenance of plate rolling processes in the steel industry.

Keywords: bearings; oil film thickness; acoustic reflection coefficient; temperature compensation; mathematical model

1. Introduction

Bearings, as ubiquitous mechanical components, wield significant influence over the operational conditions of machinery. The oil film, a critical parameter of bearings, serves as a reflection of their active status. Therefore, the state of the oil film can be used to assess the working condition of most machinery [1,2]. Under normal circumstances, the thickness of the oil film in bearings undergoes variations within a specific range, ensuring the normal functioning of the bearings. A fragile oil film can lead to bearing wear, impacting both the lifespan and operational precision of the bearings. On the other hand, an excessively thick oil film can result in viscous dissipative losses, affecting the overall performance of the machinery [3,4]. Hence, monitoring the oil film thickness in bearings to keep it within an appropriate range is crucial. In industrial production, methods commonly used to measure the thickness of bearing oil films include optical measurement [5–7], electrical measurement [8,9], and ultrasonic measurement [10,11]. Optical measurement requires ensuring the smoothness of the optical path and installing sensors in the bearings, making it impractical for use in rolling mill bearings. Electrical measurement relies too much on the conductivity of the bearing structure, leading to measurement failure in complex electromagnetic environments and susceptibility to external interference. Ultrasonic measurement, being easy to operate with strong adaptability, is widely used in various bearing oil film thickness measurements. Presently, leveraging the advantages of non-destructive ultrasonic testing, various methods for measuring the oil film thickness in bearings have been proposed, playing a pivotal role in industrial production.

Ultrasonic waves exhibit exceptional sensitivity to variations in the propagation medium, allowing for precise reflection of the oil film conditions at different states and thicknesses. Zhang et al. [12] employed piezoelectric elements as substitutes for traditional commercial ultrasound probes, addressing issues related to the large volume of conventional probes that hinder comprehensive component detection. Dou et al. [13] proposed using the reflection coefficient phase spectrum method, enabling accurate measurement of oil film thickness across a wide range. This demonstrates that ultrasound can precisely measure the thickness of bearing oil films at different levels. Beamish et al. [14,15] measured the radial bearing circumferential oil film thickness, utilizing ultrasound amplitude, phase, and resonance tilt technologies to collectively reflect the bearing oil film's thickness. Through finite element analysis, Dou et al. [16] assisted in examining the oil film thickness of bearing races. Combining finite element simulation with experimental methods enhanced the resolution of oil film thickness and identified uneven oil film distribution as the leading cause of measurement errors. Jia et al. [17] employed self-calibration and pre-calibration methods to measure the oil film thickness in heavy-duty hydroelectric generators, demonstrating high accuracy. Wei et al. [18] monitored bearing oil film thickness and state in aviation fuel pumps online. The maximum relative error met measurement requirements, highlighting ultrasound technology's superior online monitoring capability. Gray et al. [19], based on the different physical properties of lubricating oils, explored the replacement of different types of lubricating oils under ultrasound monitoring. This suggests that ultrasound can study lubrication mechanisms without altering components.

The studies above robustly demonstrate the significant advantages of ultrasound in the long-term monitoring of oil film thickness and its state. However, due to the sensitivity of ultrasound, measurement precision can be compromised by various interfering factors. In engineering detection, factors such as the coating thickness of bearings [20,21], ultrasound incident angle [22], and the temperature of the structure being measured [17,23] all have specific impacts on detection accuracy. When precise measurements are required, it is crucial to eliminate the errors above. During plate rolling processes, it is necessary to adjust the value of rolling force according to the rolling requirements. Different rolling forces will affect the operating temperature of bearings, thereby influencing the measurement accuracy of bearing oil film thickness. Some scholars have already measured bearing oil film thickness under thermal expansion conditions and achieved high-precision results [24,25]. However, most detection methods do not account for the influence of temperature on measurement accuracy. Particularly in environments with significant temperature variations, there is often a certain degree of error between most oil film measurements and actual results. During temperature changes, the density of the medium and its internal sound speed are affected. Recently, some scholars have proposed using density and proper speed compensation strategies to enhance oil film thickness measurement accuracy under various temperature conditions [17,23]. Therefore, this paper considers multiple aspects, including pressure, temperature, and sound velocity. Based on these considerations, by adjusting different bearing loads, the density and sound velocity corresponding to the temperature conditions of the load are calculated. This aims to verify the application of density and sound velocity compensation methods in oil film thickness measurement.

2. Basic Theory of Acoustic Models

In the process of ultrasound propagation through multiple layers of media, reflections, and transmissions occur at acoustic boundaries due to the varying acoustic impedance of each layer. Thickness is one of the primary factors influencing the propagation of ultrasound. Different thicknesses of the medium result in varying reflection effects on ultrasound. Exploiting this phenomenon, the thickness of the oil film in bearings can be calculated through reflection coefficient analysis. Given the relatively thin nature of the oil film, the intermediate oil film layer is simplified as a lightweight spring [26], as illustrated in Figure 1. I and III are the media on both sides of the oil film, and II is the oil film layer. Yellow and green only represent the medium on both sides of the oil film and have no actual meaning. The material parameters are shown in Table 1.

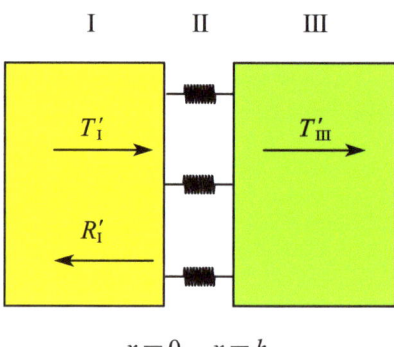

Figure 1. Ultrasound spring model.

Table 1. Material parameters.

Material	Density (kg/m³)	Speed of Sound (m/s)
Iron	7850	5900
Oil	910	1740

According to the propagation characteristics of elastic waves in a medium, the displacement and stress of longitudinal waves are given by [27]:

$$u(x) = Ae^{i\omega x/c} + Be^{-i\omega x/c}$$
$$\sigma(x) = i\omega z(Ae^{i\omega x/c} - Be^{-i\omega x/c}) \quad (1)$$

The transmission of sound waves between different media requires continuity of displacement and stress at the interface between these media, considering $x = 0$ and $x = h$ as the same boundary. It is understood that the continuous propagation of elastic waves necessitates equal displacement and stress on both sides of $x = 0$ and $x = h$. Neglecting the inertial factors of the oil film and assuming the amplitude of the incident sound wave is 1, the calculation formula for the acoustic reflection coefficient in the spring model is obtained [28].

$$R = \frac{z_1 - z_3 + i\omega(z_1 z_3)/K}{z_1 + z_3 + i\omega(z_1 z_3)/K} \quad (2)$$

This leads to the relationship between the reflection coefficient in the spring model and the oil film thickness [29].

$$R = \frac{(z_1 - z_3) + i\omega h z_1 z_3/(\rho_2 c_2^2)}{(z_1 + z_3) + i\omega h z_1 z_3/(\rho_2 c_2^2)} \quad (3)$$

3. Establishment of Finite Element Model

The properties of a solid are governed by the differential equations, which control phenomena such as the propagation of sound waves and deformation, along with other factors influencing substantial transformations. The equation is represented as follows:

$$\rho \frac{\partial^2 \mathbf{u}}{\partial t^2} = \nabla \cdot s + \mathbf{F}_v \qquad (4)$$

In the bearing structure, a coupled interface exists between solid and liquid. It requires simultaneous consideration heat-transfer equation and the fluid heat-transfer equation to control the transfer of heat within the bearing. The heat-transfer equation in the solid is represented as follows:

$$\begin{cases} \rho c_p \frac{\partial T}{\partial t} + \rho c_p \mathbf{u} \cdot \nabla T + \nabla \cdot \mathbf{q} = Q + Q_{ted} \\ \mathbf{q} = -k \nabla T \end{cases} \qquad (5)$$

The heat-transfer equation for the fluid is represented as follows in Equation (7), with its basic form being similar to the heat-transfer equation for solids.

$$\begin{cases} \rho c_p \frac{\partial T}{\partial t} + \rho c_p \mathbf{u} \cdot \nabla T + \nabla \cdot \mathbf{q} = Q + Q_p + Q_{vd} \\ \mathbf{q} = -k \nabla T \end{cases} \qquad (6)$$

By jointly controlling the aforementioned differential equations, the influence of changes in the oil film thickness of bearings on the ultrasound reflection coefficient under different temperature conditions is determined. This information establishes a formula for calculating the reflection coefficient under temperature compensation.

Ultrasound excitation signals typically use single-level or multi-level digital pulse signals. However, when propagating through multi-layered media, the output form of such pulse signals exhibits lower parameter accuracy. It contains numerous high-order harmonics, impacting the simulation accuracy of sound wave transmission and failing to meet simulation requirements. Therefore, a modulated sine function is chosen to exclude substantial interference from clutter in the detection signal, as shown in Equation (8).

$$v(t) = \exp\left[-\left(\frac{t - 2T_0}{T_0/2}\right)^2\right] \sin(2\pi f_0 t) \qquad (7)$$

Due to the significant computational time required for three-dimensional models in COMSOL, this model only considers the propagation of sound waves on a plane. A two-dimensional finite element model is established based on bearing data, and the material and boundary conditions of the finite element model are set as shown in Figure 2.

The model only analyzes the characteristics of the first echo; thus, it is necessary to eliminate the interference of multiple echoes on the measurement results. The Low-Reflecting Boundary (LRB) is commonly used to remove the interference of multiple reflected waves on the normal propagation of acoustic waves within the model. When the acoustic wave propagates to the boundary, the LRB ensures that the wave does not reflect in the time-domain or frequency-domain analysis. It acquires material data from the domain to create a perfect impedance match for pressure waves and shear waves.

When acoustic waves propagate in solids and fluids, their governing equations differ. When the acoustic wave propagates to the coupled boundary between solid and fluid, it is necessary to establish a coupling relationship to calculate the propagation of the acoustic wave at the coupled boundary. The Acoustic-Structure Boundary is used to couple pressure acoustic models with any structural component. This functionality couples with solid mechanics, elastic waves, shells, layered shells, membranes, and multibody dynamics interfaces. The coupling includes fluid loads on the structure and structural accelerations experienced by the fluid. For thin internal structures with fluid on both sides, such as shells or membranes, a slot is added in the pressure variable, and attention is paid to coupling the upper and lower sides.

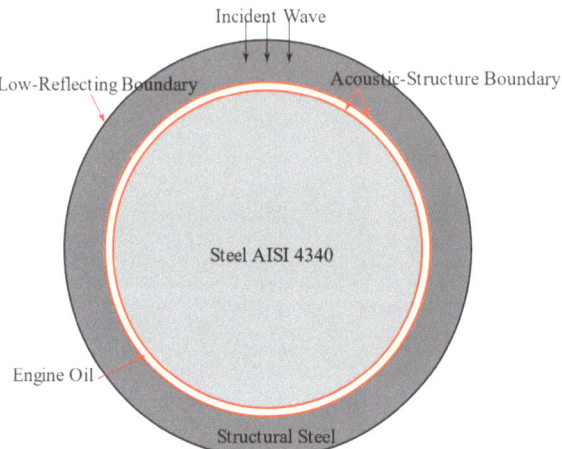

Figure 2. Finite element model: Boundary conditions and material settings.

Considering the need for efficient calculations while ensuring convergence during finite element computations, it is crucial to optimize computational efficiency. In finite element mesh generation, the mesh should be refined at the contact interfaces between different media based on the speed of sound in different media and the contact positions between media. Therefore, reasonable refinement of the mesh is required at various contact interface locations. When the mesh size is set between 1/6 and 1/5 of the wavelength, the relative error of the computed results is relatively small, proving that a mesh size between 1/6 and 1/5 of the wavelength is the optimal mesh density.

However, for the grids between the contact areas of different media layers, the mesh needs to be appropriately reduced, as reflections and transmissions of sound waves on the contact surface need to be calculated with high precision. The relative error approaches zero when the mesh size is less than 1/10 of the wavelength. Considering the computational time, a mesh size of 1/10 of the wavelength is chosen for these contact areas [22,30].

The model employs fully coupled transient methods, determining the wavelength of incident acoustic waves based on the mesh partition of the finite element model, ensuring convergence of the model within a certain range. By adjusting the boundary conditions of the model, convergence of the acoustic wave propagation at the boundary is ensured. Within the bearing model, there are multiple physics field couplings. The generalized α method within the implicit solver is employed for time stepping. The solver utilizes a free time step while determining the maximum step size to be 3×10^{-8}. The backward Euler method is set with a safety factor of 20 for linear elements. Residuals of each physics field tend to stabilize during computation, and the convergence plot in COMSOL exhibits a smooth trend, indicating good convergence of the model and correct convergence condition settings.

4. Simulation Results Analysis

4.1. Analysis of Reflected Ultrasonic Signals at Different Temperatures

After performing Fourier transform on the reflected ultrasonic waves, the amplitude and phase of the reflected signal can be obtained. The reflection coefficient of the oil film is typically calculated using the reference signal and the reflected signal after Fourier transformation. Considering the temperature factor, the reflected ultrasonic wave at 20 °C is chosen as the reference signal, while the reflected ultrasonic waves under other temperature conditions are used as the reflected signals [31].

$$|R(f)| = \frac{A(f)}{A_0(f)}|R_0(f)| \tag{8}$$

$$\Phi(f) = \theta(f) - \theta_0(f) + \Phi_0(f) \tag{9}$$

In ultrasonic measurement of bearing oil film thickness, there are three models involved: the spring model, the phase shift model, and the resonance model. Depending on the thickness of the oil film, the appropriate model is selected to calculate the oil film thickness. Since the effect of temperature on the signal characteristics corresponding to different oil film thicknesses is unknown, oil films with thicknesses of 20 µm and 100 µm are selected for analysis to assess the impact of temperature on the reflected acoustic wave signal. The 20 µm thick oil film exhibits characteristics of both the spring model and the phase shift model, while the 100 µm thick oil film presents characteristics of the resonance model. Therefore, the oil film with a thickness of 100 µm is chosen for the analysis and calculation of temperature interference.

Setting the oil film thickness to 20 µm, the model is configured with temperatures of 30 °C, 50 °C, 70 °C, 90 °C, and 110 °C in the solid and fluid heat transfer modules, and the computation of reflected ultrasonic waves is performed. The resulting time-domain signals are shown in Figure 3.

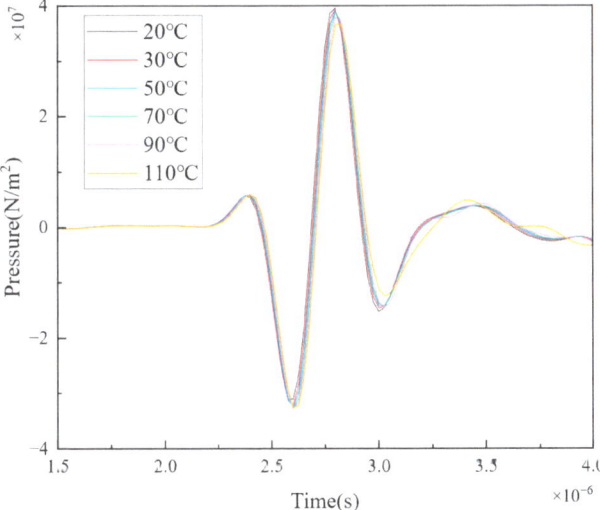

Figure 3. The influence of temperature on acoustic signals under the condition of a 20 µm thick oil film.

According to Figure 3, it can be observed that the amplitude of the reflected ultrasonic wave remains relatively unchanged as the temperature increases. However, the phase gradually shifts to the right. Performing Fourier transform on the data yields the results shown in Figure 4a,b. The change in the amplitude of the reflected ultrasonic wave in the frequency domain follows a similar pattern to its time-domain counterpart, with minimal and localized variations that make it challenging to discern the impact of temperature on the amplitude. The phase of the reflected ultrasonic wave exhibits a regular variation in the frequency range, with the phase trend gradually shifting to the left as the temperature increases within the 6 MHz frequency range. The phase signal change is more pronounced than the amplitude signal.

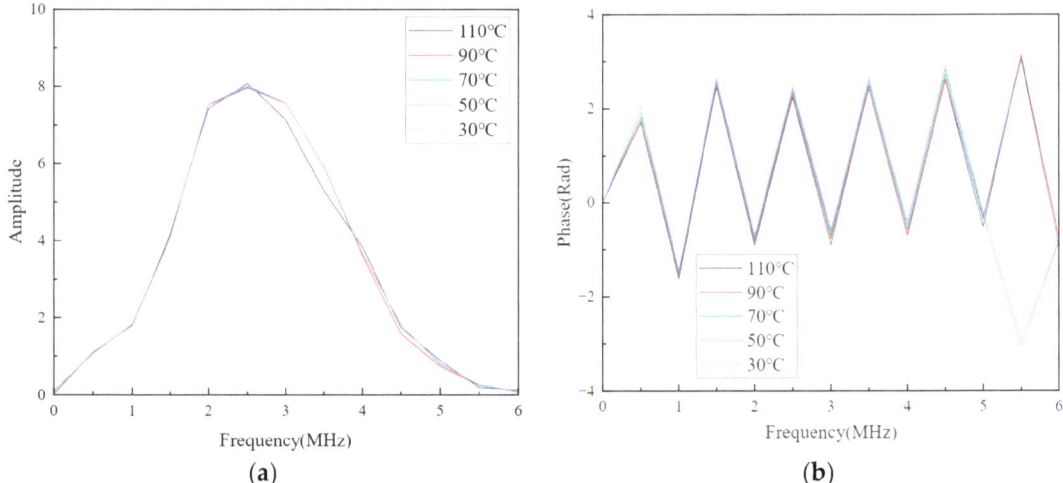

Figure 4. The Fourier transform data: (**a**) amplitude, (**b**) phase.

By substituting the signal information from Figure 4 into Equations (8) and (9), the amplitude and phase of the reflection coefficient in the frequency domain under the influence of temperature on the reflected ultrasonic wave for a 20 μm oil film thickness can be obtained, as shown in Figure 5. Figure 5a represents the amplitude of the reflection coefficient, where most of the reflection coefficients are entwined within the frequency range, making it challenging to analyze the impact of temperature on the reflected ultrasonic wave signal from the amplitude of the reflection coefficient. Figure 5b represents the phase of the reflection coefficient, and it can be observed that the phase of the reflection coefficient varies significantly under different temperature conditions. Moreover, the phases of the reflection coefficients are independent of each other, without entanglement or intertwining phenomena. The impact of temperature on the reflected ultrasonic wave can be more clearly reflected through the variation in the reflection coefficient phase.

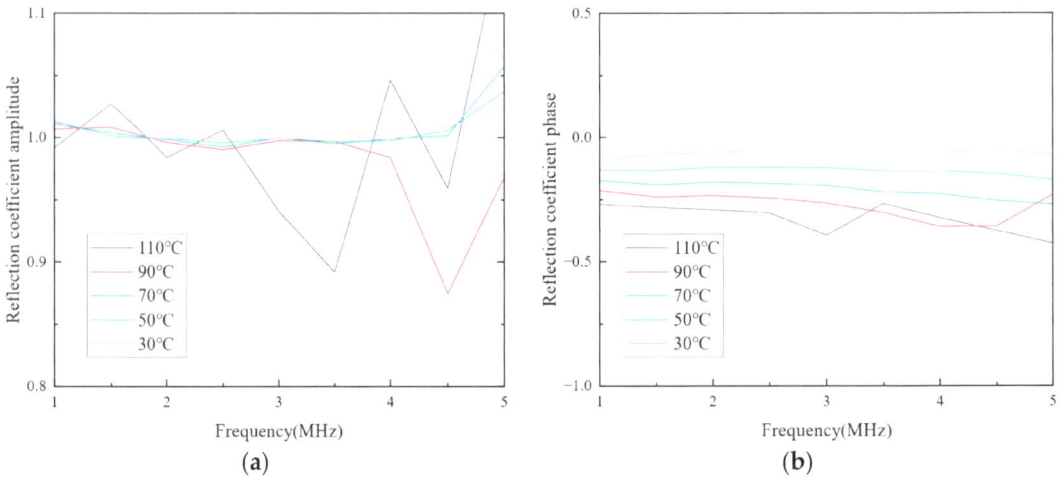

Figure 5. Frequency domain reflection coefficients: (**a**) amplitude, (**b**) phase.

Setting the oil film thickness in the finite element model to 100 μm, and keeping other conditions unchanged, the impact of temperature on the reflected ultrasonic wave is recalculated. The changes in the time-domain-reflected ultrasonic wave signals obtained from the finite element calculation are shown in Figure 6. Similar to the pattern observed with a 20 μm thick oil film, the amplitude of the reflected ultrasonic wave remains relatively unchanged when the temperature increases. However, the phase of the ultrasonic wave signal gradually shifts to the right as the temperature increases.

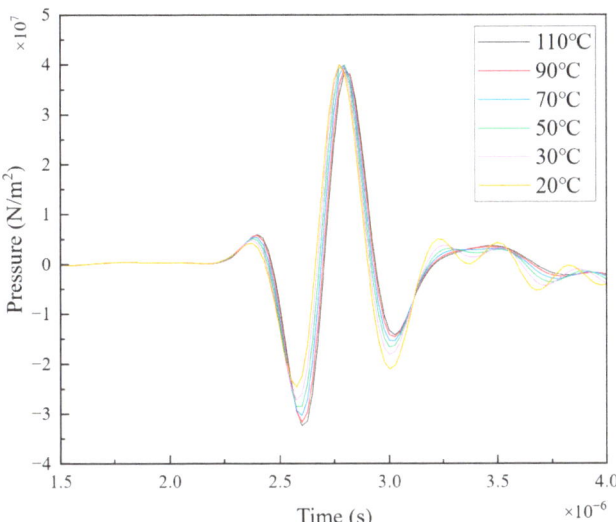

Figure 6. The influence of temperature on acoustic signals under the condition of a 100 μm thick oil film.

After performing Fourier transform on the time-domain signals, frequency-domain signal images are obtained, as shown in Figure 7. Figure 7a represents the amplitude image of the reflected ultrasonic wave, while Figure 7b represents the phase image of the reflected ultrasonic wave.

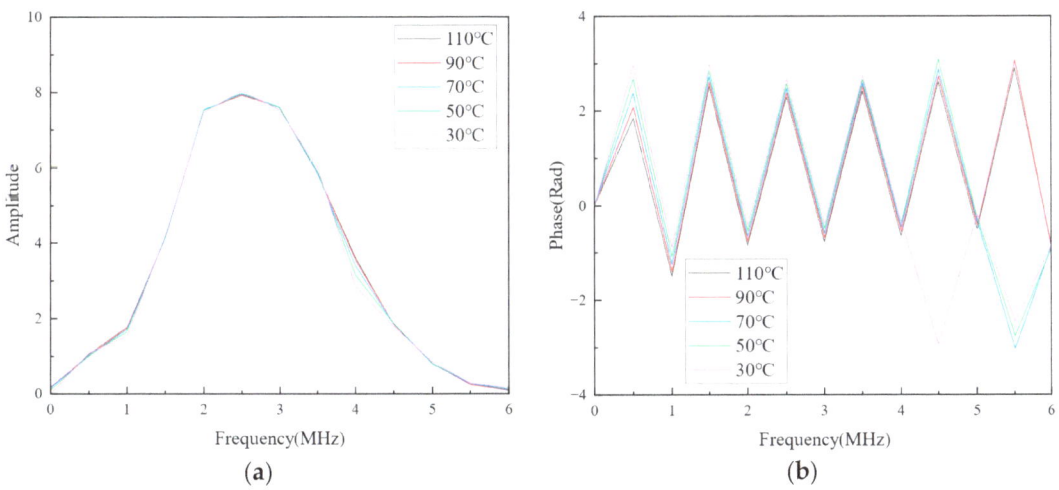

Figure 7. The Fourier transform data: (**a**) amplitude, (**b**) phase.

According to Figure 7, the variation in the amplitude of the reflected ultrasonic wave in the frequency domain follows a similar pattern to its time-domain counterpart. The amplitude changes are less pronounced, and the degree of amplitude variation is much smaller than that under the condition of a 20 μm oil film thickness. It is challenging to discern the impact of temperature on the reflected ultrasonic wave from the amplitude of the reflection coefficient. The phase of the reflected ultrasonic wave exhibits a regular variation in the frequency range, with the phase signal gradually shifting to the left as the temperature increases within the given range. The phase signal shows a regular and systematic change within the frequency range. Compared to the amplitude signal, the phase signal provides more compelling information, and its acquisition is relatively more straightforward, requiring less computational effort.

To demonstrate the influence of temperature on the phase of the ultrasonic wave, the reflection coefficient is used to characterize the relationship between the temperature and the reflected ultrasonic wave. Substituting the frequency-domain signals obtained after Fourier transform from Figure 7 into Equations (8) and (9), the amplitude and phase of the reflection coefficient in the frequency domain under the influence of temperature on the reflected ultrasonic wave for a 100 μm oil film thickness can be obtained, as shown in Figure 8.

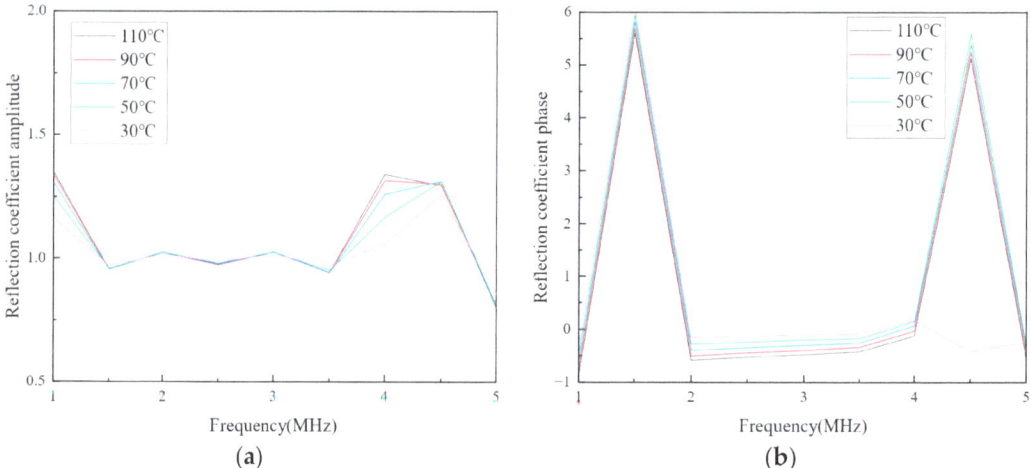

Figure 8. Frequency domain reflection coefficients: (**a**) amplitude, (**b**) phase.

Figure 8a represents the amplitude of the reflection coefficient in the frequency domain. Compared to Figure 5a, the entanglement of this reflection coefficient amplitude is lighter. However, due to the small degree of variation in the amplitude of the reflection coefficient within the frequency range, it remains challenging to judge the impact of temperature on the reflected ultrasonic wave from the amplitude. Figure 8b represents the phase of the reflection coefficient in the frequency domain. There is a noticeable difference in the phase of the reflection coefficient under different temperature conditions. The phase of the reflection coefficient within this range can be used to establish the relationship between the reflected ultrasonic wave and temperature. In the 2–4 MHz range, the variation in the phase of the reflection coefficient is relatively regular. With the increase in temperature, the phase of the reflection coefficient gradually decreases, consistent with the trend of signal changes in the time domain. However, it provides a more straightforward way to obtain relationship information.

Through simulations under the coupling conditions of reflected ultrasonic waves and temperature fields with oil film thicknesses of 20 μm and 100 μm, it can be observed that the temperature variation has minimal impact on the amplitude of the reflected ultrasonic wave. However, it does specifically affect the phase of the reflected ultrasonic wave. With the increase in temperature, there is a trend of overall rightward shift in the reflected ultrasonic wave signal, indicating that temperature influences the phase relationship of the reflected ultrasonic wave. For the 20 μm oil film thickness condition, the spring model and phase shift model control the characteristics of the reflected ultrasonic wave. In contrast, for the 100 μm oil film thickness condition, the resonance model holds the characteristics of the reflected ultrasonic wave. The simulation results also indicate that the effect of temperature on the reflected ultrasonic wave is independent of the model containing the reflected ultrasonic wave. In other words, when analyzing the impact of temperature on the reflected ultrasonic wave, there is no need to consider the influence of oil film thickness on the reflected ultrasonic wave.

Because temperature affects the phase and amplitude of acoustic signals but does not impact the overall form of the signals, the degree of temperature influence on the signals can be analyzed through amplitude attenuation and waveform expansion factors. The amplitude attenuation factors and waveform expansion factors under the conditions of 20 μm and 100 μm oil film thickness are shown in Figures 9 and 10, respectively.

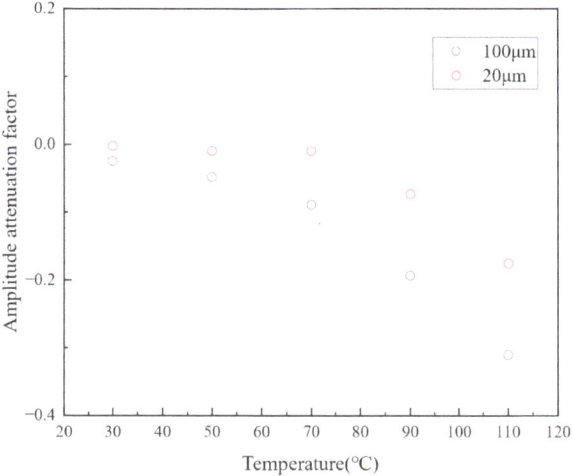

Figure 9. The trend of the amplitude attenuation factor.

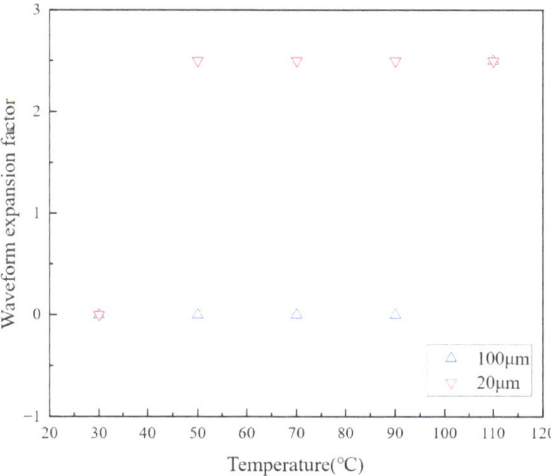

Figure 10. The trend of the waveform expansion factor.

The amplitude attenuation factor is employed to analyze the trend of amplitude changes between signals, which are calculated by comparing the amplitude of the original signal with that of the transformed signal. Figure 9 shows that with the increase in temperature, the amplitude attenuation factor gradually decreases, indicating a reduction in amplitude compared to the original signal for the transformed signal.

The waveform expansion factor is utilized to analyze the trend of time difference changes between signals, calculated by determining the time difference between the peak values of the original signal and the transformed signal. Figure 10 shows that the waveform expansion factor for the 100 µm oil film thickness increases with rising temperature, reaching a plateau when the temperature exceeds 30 °C. The 20 µm oil film thickness waveform expansion factor shows no significant change below 90 °C due to the giant time step in the finite element calculation. Further refinement of the time step can yield more accurate waveform expansion factors. By analyzing the existing waveform expansion factors, it can be inferred that the peak values of the reflected acoustic signals gradually shift to the right as the temperature increases.

In the above analysis of the frequency and time domain characteristics of the signals, it has been demonstrated that temperature changes affect both the amplitude and phase of the reflected acoustic signals. The next step will involve compensating for these temperature-induced effects on the reflected acoustic signals to improve the accuracy of oil film thickness measurements.

4.2. Compensation of Reflected Acoustic Signals

The temperature variation has a significant impact on the acoustic properties of the medium. The density and speed of sound of the material jointly control acoustic properties. Compensating for the reflected acoustic signals involves adjusting the medium's density and speed of sound. For lubricating oil compensation, Dowson and others proposed the following compensation formula.

$$\rho_T = \rho_0 \frac{1 + C_1 p}{1 + C_2 p} - C_3(T - T_0) \tag{10}$$

Table 2 shows the variation function of the sound velocity and density of the medium at different temperatures. The variation in the density and sound velocity of the medium at different temperatures in COMSOL follows this functional trend. When calculating the thickness of the bearing oil film based on the acoustic reflection coefficient, it is necessary to consider the density and sound velocity of the medium. Usually, when calculating the thickness of the oil film, the influence of the bearing temperature on the reflection coefficient is not considered, and the actual density and sound velocity do not match the theoretical density and sound velocity. Correcting the medium density and sound velocity in COMSOL can compensate for the reflected acoustic signal. In practice, adjusting the incident acoustic signal can compensate for the reflected acoustic signal at different temperatures [23].

Table 2. The density and speed of sound of the medium at different temperatures [23]. Reproduced with permission from Yaping Jia, Tonghai Wu, Pan Dou and Min Yu, Wear; published by Elsevier, 2024.

Medium	Density	Speed of Sound
Oil	$\rho = 910/(1 + 0.0007(T - 20))$	$c = 0.0039T^2 - 3.39T + 1740$
Iron	$\rho = 7850/(1 + 3\alpha(T - 273))$	$c = 5900 \times 10^3(1 - 11\alpha(T - 273))$

In COMSOL, the parameters of each material were adjusted according to the functions in Table 2, and finite element calculations were conducted again. The time-domain reflection acoustic signals are shown in Figure 11. In the figure, the reflected acoustic signals at various temperatures, after compensation, are nearly overlapping, demonstrating that compensating for the density and speed of sound is a practical approach to enhance measurement accuracy.

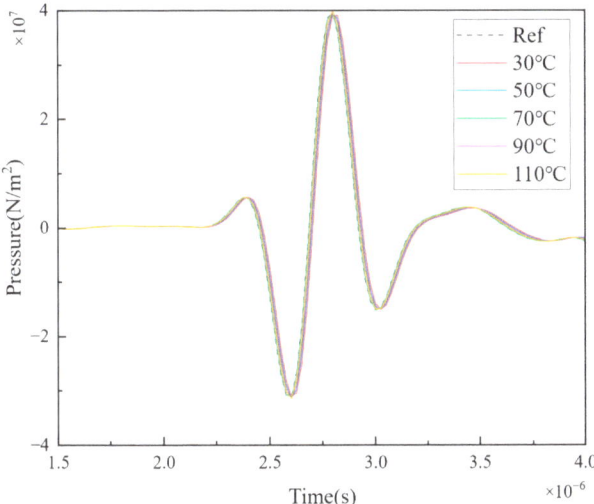

Figure 11. The time-domain signal waveforms after compensation.

The Fourier-transformed signals after compensation are presented in Figure 12, illustrating their amplitude and phase spectra.

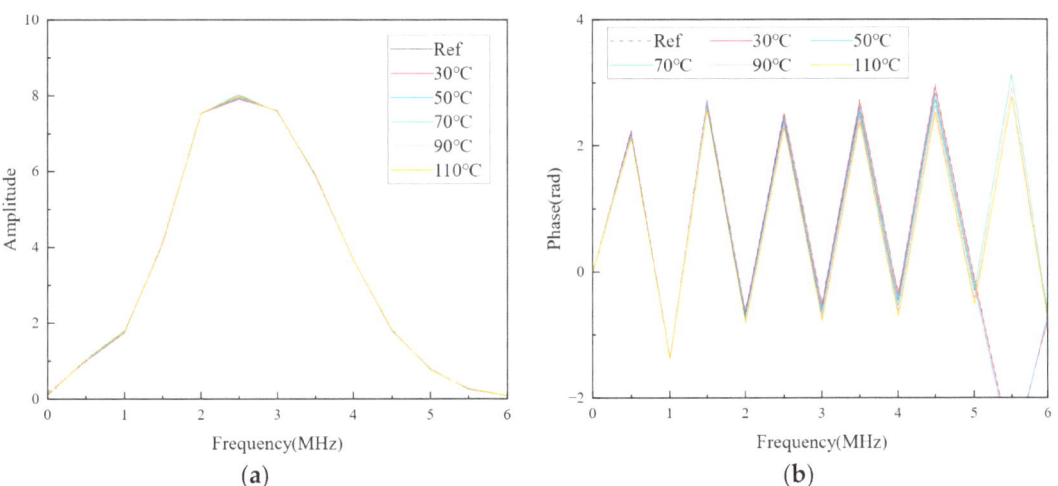

Figure 12. The Fourier-transformed signals after compensation: (**a**) amplitude, (**b**) phase.

The calculated reflection coefficient amplitude and phase after compensation are shown in Figure 13. The amplitude curves of the reflection coefficient are closely aligned and indistinguishable between different temperatures. Compared to uncompensated signals, the compensated curves are numerically closer, making it nearly impossible to discern the impact of temperature on the reflection ultrasonic signals based on the reflection coefficient amplitude alone. In the compensated reflection coefficient phase, it is evident that the temperature-induced phase lag effect persists, but its magnitude is reduced. The compensated reflection coefficient phase is less than -0.25, half of the uncompensated reflection coefficient phase. This demonstrates the effectiveness of paying for density and speed of sound in the frequency domain. For practical oil film thickness measurements that do not require extremely high precision, temperature compensation for the medium's density and speed of sound is sufficient to meet basic measurement requirements.

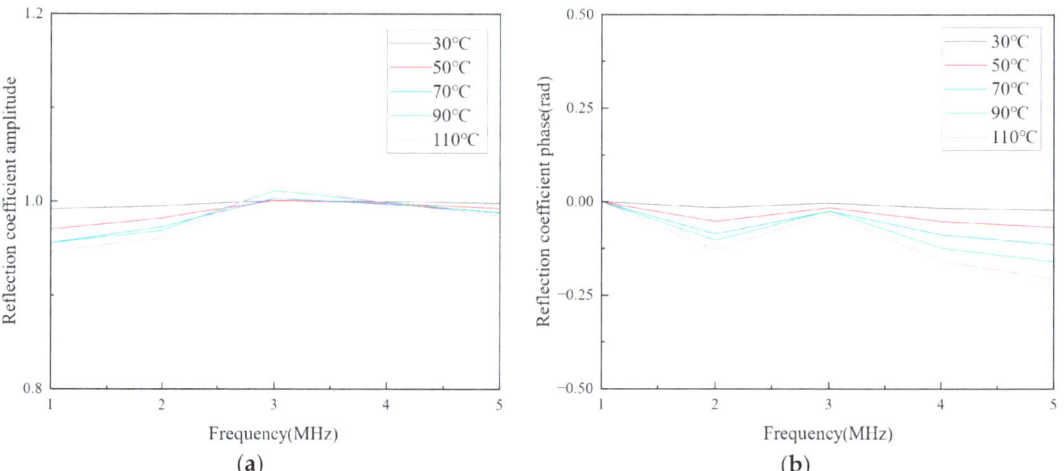

Figure 13. The frequency domain reflection coefficient of the compensated signals: (**a**) amplitude, (**b**) phase.

In the time domain, the signals' amplitude decay factor and waveform expansion factor are depicted in Figure 14. The maximum difference between amplitude decay factors is 0.0012, indicating that the amplitudes of the compensated signals between different temperatures remain almost unchanged. Due to the enormous time step, the waveform expansion factor of the compensated signals does not change, making it impossible to assess the correlation between waveform signals using the waveform expansion factor.

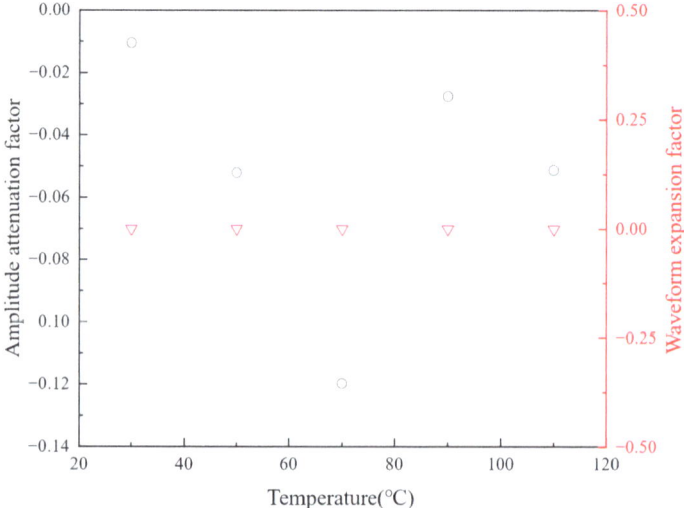

Figure 14. The trends of amplitude decay factors and waveform expansion factors for the compensated signals.

The feasibility of improving the accuracy of oil film thickness measurement by compensating for density and speed of sound under temperature variations has been demonstrated through the characteristics of signals in both time and frequency domains. Depending on temperature changes, the relationship function between the density and speed of sound for solid media is represented by Equation (11).

$$\begin{aligned} \rho_t &= 7850/(1 + 3\alpha(T - 273)) \\ c_t &= 5900 \times 10^3(1 - 11\alpha(T - 273)) \end{aligned} \tag{11}$$

The relationship function between density and speed of sound in liquid media is given by Equation (12).

$$\begin{aligned} \rho_t &= 910/(1 + 0.0007(T - 20)) \\ c_t &= 0.0039T^2 - 3.39T + 1740 \end{aligned} \tag{12}$$

Equations (11) and (12) represent functions describing the relationship between temperature and density, and temperature and speed of sound, respectively. The reflection coefficient is related to the medium's density and speed of sound. A relationship function between the reflection coefficient and temperature is established, considering temperature as the independent variable. As the temperature changes, the medium's density and speed of sound change synchronously in the equation to avoid computational errors arising from discrepancies between actual and theoretical densities and speeds of sound. The formula for calculating the compensated reflection coefficient is given by Equation (13).

$$R(T) = \frac{z_2(T) - z_1(T)}{z_2(T) + z_1(T)} \tag{13}$$

In the time domain, compared to the reflected acoustic wave signal without temperature compensation, the compensated signal exhibits similar amplitudes, and the trend of phase shift is less noticeable. By calculating the amplitude attenuation factor and waveform expansion factor of different temperature waveforms, it is observed that the factors of the compensated signal tend to approach 0, indicating a convergence of different signals. In the frequency domain, judging the effect of temperature on the reflected acoustic wave based on the trend of amplitude change in the reflection coefficient is difficult. However, the phase of the reflection coefficient varies significantly with the temperature, with an upward shift in temperature causing a downward shift in the phase of the reflection coefficient. Through signal analysis in both the time and frequency domains, the effectiveness of temperature compensation for the reflected acoustic wave signal is demonstrated. This technique can be effectively applied to the measurement of bearing oil film thickness under various temperature conditions, thereby improving the accuracy of bearing oil film thickness measurement.

5. Compensation of Reflection Coefficients in Sliding Bearings under Multiple Temperature Conditions

During the sheet rolling process, variations in rolling force lead to different shear forces on the lubricating oil in the rolling mill bearings, resulting in varying temperatures of the bearing oil film. Under normal rolling loads, bearing temperatures typically range from 45 to 55 °C. If the bearing temperature exceeds this range, wear and scuffing may occur, and so temperatures must be maintained within the normal range to extend the working life of the bearings. A three-dimensional finite element model is established using COMSOL, incorporating modules such as Heat Transfer in Films Interface, Hydrodynamic Bearing Interface, Solid Mechanics Interface, Heat Transfer in Solids Interface, and coupling through the Multiphysics Branch with the inclusion of the Thermal Expansion module. This model simulates the bearings' temperature and oil film thickness variations under different pressure conditions.

Different journal loads are set in the hydrodynamic bearing based on different rotational speed conditions to calculate the oil film thickness and temperature under various loads. The oil film thickness and temperature of the bearing under different loads are shown in Figure 15. It can be observed that there is a linear relationship between the bearing oil film thickness and the bearing load, with the oil film thickness decreasing as the load increases. The bearing oil film temperature shows a quadratic function relationship with the load, with the oil film temperature increasing as the load increases.

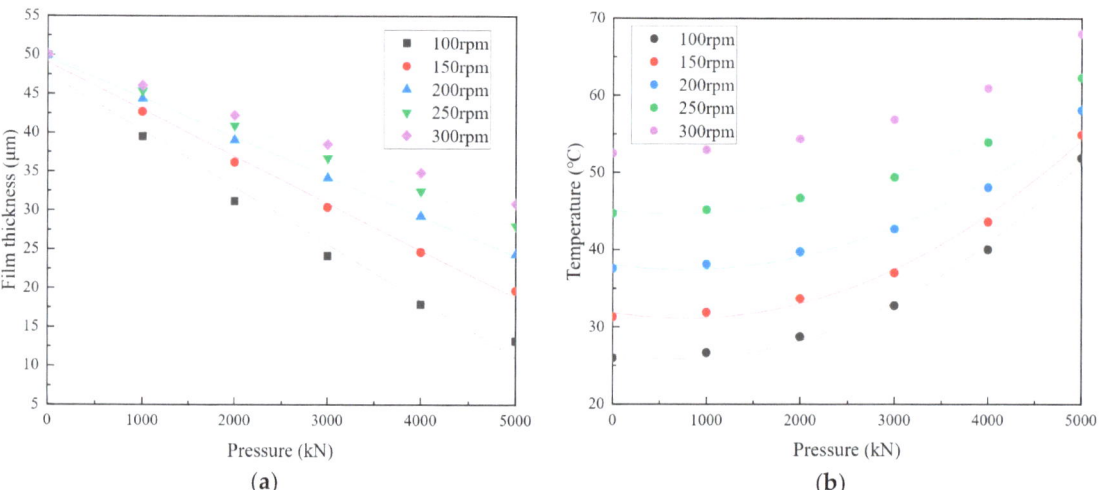

Figure 15. Variations in oil film thickness and temperature under different rotational speeds: (**a**) pressure–film thickness, (**b**) pressure–temperature.

The fitting functions for pressure–temperature and pressure–film thickness are given by Equation (14). Based on these appropriate functions, temperatures and film thicknesses under different pressure conditions can be calculated, avoiding the inefficiency caused by multiple simulations. This demonstrates the functional relationship between pressure, temperature, and film thickness in actual bearings. When measuring bearings in a factory setting, this fitting function can be utilized to calculate and analyze oil film thickness under different conditions.

$$f = kP + b$$
$$t = AP^2 \times 10^{-6} + BP + C \qquad (14)$$

The coefficients in Equation (14) are listed in Table 3.

Table 3. Coefficients of fitting functions for temperature and film thickness equations.

Parameter Items	Numeric Value				
Rotation speed (rpm)	100	150	200	250	300
Slope k	−0.0066	−0.0057	−0.0049	−0.0043	−0.0038
Coefficient b	46.5105	48.4431	49.3195	49.7299	49.9161
Coefficient A	1.4117	1.2507	1.2317	1.1163	0.9667
Coefficient B	−0.00206	−0.00181	−0.00220	−0.00208	−0.00172
Coefficient C	26.5625	31.7585	38.3206	45.4576	53.1523

Considering the variations in density and sound speed induced by temperature, the reflection coefficients of the oil film under different pressure conditions are calculated based on the equation. Furthermore, the reflection coefficients after compensation for density and temperature are shown in Figure 16.

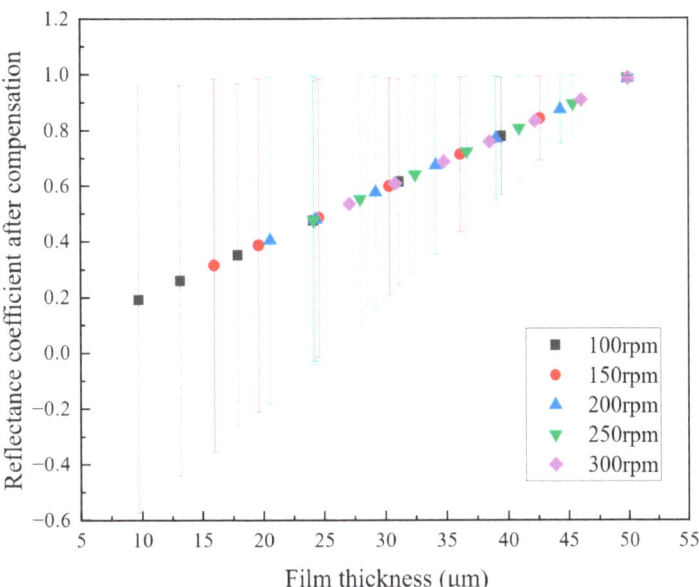

Figure 16. Reflection coefficients of the bearing oil film.

The error values between uncompensated reflection coefficients and compensated reflection coefficients are shown in Figure 17. The errors exhibit a linear trend between different oil film thicknesses, and the error gradually decreases with the increase in oil film thickness.

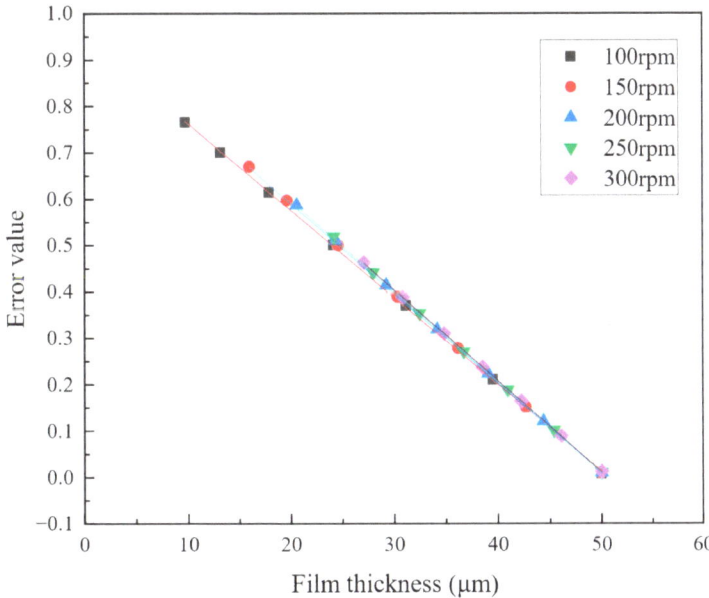

Figure 17. Error values of the reflection coefficient for bearing oil film thickness.

When the oil film thickness is less than 35 μm, there is a significant error between the reflection coefficients, making it impractical to calculate the oil film thickness changes using uncompensated reflection coefficients. However, when the oil film thickness exceeds 35 μm, the error between the reflection coefficients is relatively tiny. When measurement precision is not critical, uncompensated reflection coefficients can be used to calculate the oil film thickness. In cases of thin oil film, most of the acoustic waves transmit through the oil film without significant reflection at the interface with steel. Therefore, the reflection coefficient for a thin oil film is small. As the temperature increases, the density and speed of sound for the oil film and steel are significantly affected, resulting in increased reflection coefficients. This effect impacts the measurement of oil film thickness. Conversely, in the case of a thick oil film, most acoustic waves reflect at the oil film–steel interface, with only a tiny portion transmitting through. Consequently, the influence of temperature on the reflection coefficients is minimal in this scenario, resulting in a slight difference between the reflection coefficients under different temperatures. The fitted function for the reflection coefficient error in Figure 17 is given by Equation (15).

$$e = k_e h + a_e \tag{15}$$

The coefficients of the fitted function are provided in Table 4.

Table 4. Coefficients of the error–oil film thickness fitting function.

Parameter Items	Numeric Value				
Rotation speed (rpm)	100	150	200	250	300
Coefficient k_e	−0.018	−0.019	−0.019	−0.019	−0.019
Coefficient a_e	0.95	0.95	0.97	0.98	0.99

The fitting function of the reflection coefficient error indicates that the temperature's influence on the reflection coefficient exhibits an exponential change, i.e., the bearing pressure shows an exponential change in its effect on the reflection coefficient. Figure 18 shows the variation between bearing pressure and reflection coefficient error.

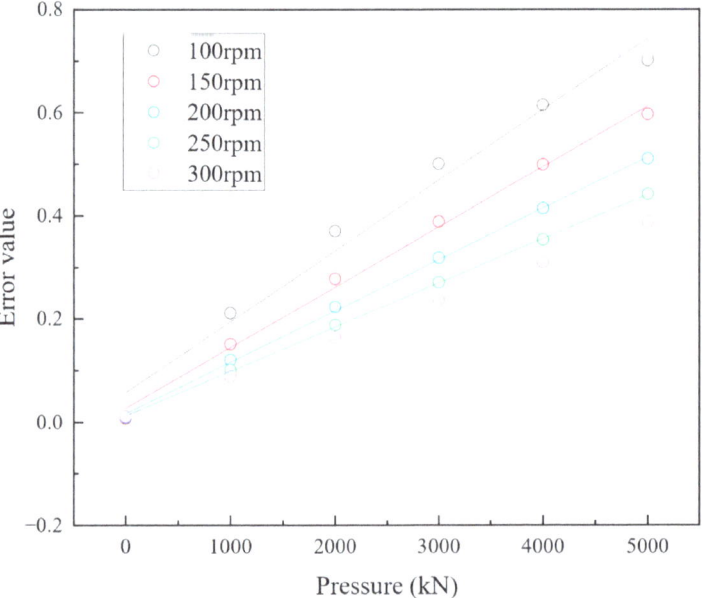

Figure 18. Error values of the reflection coefficient for bearing oil film pressure.

The relationship between bearing pressure and error values of the reflection coefficient is also an exponential function. Under the known conditions of bearing pressure, after measuring the reflection coefficient of the bearing oil film, the error in the reflection coefficient can be compensated using Equation (16), avoiding the calculation errors caused by unknown oil film thickness and distribution.

$$E = k_E P \times 10^{-5} + a_E \tag{16}$$

The coefficients of the fitting function are shown in Table 5.

Table 5. Coefficients of the error–pressure fitting function.

Parameter Items	Numeric Value				
Rotation speed (rpm)	100	150	200	250	300
Coefficient k_E	12.49	11.05	9.65	8.47	7.49
Coefficient a_E	0.077	0.038	0.022	0.015	0.012

According to Equation (16), the reflection coefficient error of the oil film thickness can be calculated under any bearing pressure conditions. When the bearing pressure exceeds a certain level, using this formula to calculate the reflection coefficient error becomes less significant.

The mathematical model's goodness of fit is analyzed using the coefficient of determination R^2, where a value approaching 1 indicates a good fit. However, R^2 is greatly influenced by the independent variables, and having a large number of independent variables may render R^2 ineffective. Therefore, the adjusted coefficient of determination, which considers the degrees of freedom, is used to characterize the fit of the mathematical model. The R^2 and adjusted R^2 of the aforementioned mathematical model are presented in Table 6. As the coefficients in Table 6 approach one, this indicates an excellent fit of the mathematical model, suggesting that it can be used to compensate for errors in the reflection coefficients under multiple temperature conditions, thereby achieving compensation in the bearing model measurement.

Table 6. Coefficient of determination and adjusted coefficient of determination.

	R^2					Adjusted R^2				
Figure 15a	0.9717	0.99086	0.99754	0.99953	0.99992	0.96604	0.98903	0.99705	0.99944	0.99999
Figure 15b	0.99544	0.99311	0.99159	0.99324	0.9956	0.99241	0.98852	0.98598	0.98873	0.99266
Figure 17	0.99985	0.99985	0.99999	0.99998	0.99999	0.99982	0.99982	0.99999	0.99998	0.99998
Figure 18	0.97878	0.99573	0.99916	0.99976	0.99987	0.97348	0.99467	0.99895	0.99970	0.99983

6. Conclusions

This paper introduces an error compensation method for measuring oil film thickness using ultrasonic waves at different temperatures. The process is applied to measure the oil film thickness of bearings at various temperatures. Conclusions are drawn through the analysis of finite element models under different temperature conditions, specifically focusing on the calculation results of the bearing oil film acoustic reflection coefficients.

(1) Temperature is crucial to the measurement of the acoustic reflection coefficient in bearing oil film thickness. As the temperature increases, the phase of the reflected sound wave lags and the amplitude gradually decreases. This phenomenon is verified through analysis in the time and frequency domains. Therefore, when accurately measuring the thickness of bearing oil film, it is necessary to consider the impact of temperature on the density and sound speed of the medium and compensate for the reflection coefficient.

(2) By adjusting the density and sound speed, the measurement model of the oil film reflection coefficient under multiple temperature conditions was compensated. The time-domain simulation results showed that in the compensated model, the reflected sound waves almost overlapped, with small changes in amplitude attenuation factors, but there was still a lag in phase. In the frequency domain, the reflected sound wave amplitudes under different temperatures were similar and remained stable. This presented a challenge for solving the phase lag problem.

(3) Based on signal compensation under multiple temperature conditions, a model of liquid dynamic pressure bearing was established, and the changes in the oil film of the bearing under different load conditions were analyzed. As the load increases, the temperature of the oil film increases and the thickness of the oil film decreases. The simulation results show that there is an error between the thickness of the oil film and the reflection coefficient. By establishing an error model, the calculation problem of reflection coefficient error is solved, which lays a foundation for analyzing the influence of temperature on reflection coefficient error.

Based on the findings from the research above, a method for detecting the thickness of hydrodynamic-bearing oil films under multiple temperature conditions has been established. This method addresses the errors introduced by temperature variations in oil film thickness measurements, ensuring the feasibility of measuring hydrodynamic bearing oil film thickness under different temperature conditions. It improves the accuracy of oil film thickness detection that may be affected by temperature issues, providing theoretical support for industrial applications.

Author Contributions: Conceptualization, F.S. and B.S.; methodology, B.S.; software, Y.B.; validation, B.S., S.W. and Y.H.; formal analysis, W.L.; investigation, N.K.; resources, Y.B.; data curation, F.M.; writing—original draft preparation, F.S. and B.S.; writing—review and editing, Y.H. and Z.L.; visualization, W.L.; supervision, S.W.; project administration, F.S.; funding acquisition, S.W., N.K. and Y.B. All authors have read and agreed to the published version of the manuscript.

Funding: This research was funded by the Science and Technology Planning Project of Inner Mongolia Autonomous Region, grant number 2022YFSH0126 and by the Fundamental Research Funds for Inner Mongolia University of Science & Technology, grant number 2023QNJS072, and by the Natural Science Foundation of Inner Mongolia Autonomous Region, grant number 2023ZD12, and by theBasic scientific research business expenses project of universities directly under the Inner Mongolia Autonomous Region, grant number 2024YXXS058.

Data Availability Statement: Please contact the corresponding author for data.

Acknowledgments: We would like to thank the Science and Technology Planning Project of Inner Mongolia and the Fundamental Research Funds for Inner Mongolia University of Science & Technology for sponsoring this research. We also appreciate the support of the School of Mechanical Engineering at Inner Mongolia University of Science & Technology for this study. Additionally, we thank the editors and reviewers for their assistance with this paper. In this paper, a large language model was used to optimize the grammatical content of the article.

Conflicts of Interest: Author Fengchun Miao was employed by the company Inner Mongolia North Heavy Industry Group Co., Ltd. The remaining authors declare that the research was conducted in the absence of any commercial or financial relationships that could be construed as a potential conflict of interest.

Nomenclature

Symbol	Unit	
u	m	Displacement of elastic waves.
ω	Hz	Angular frequency of incident sound wave.
A	-	Amplitudes of the incident waves.
B	-	Amplitudes of the reflected waves.
c	m/s	Speed of sound.
z	kg/(m²·s)	Acoustic impedance. $z = \rho c$.
R	-	Acoustic reflection coefficient.
ρ	kg/m³	Density.
K	N/m	$K = -dp/dh$.
s	Pa	Stress tensor.
$\mathbf{u}$	m/s	Velocity field defined at the velocity grid nodes.
c_p	J/kg·°C	Solid heat capacity under constant pressure.
$\mathbf{F}_v$	N	Load vector.
Q	W	Heat source.
k	W/(m·K)	Reliable thermal conductivity.
Q_p	W	Pressure work.
Q_{ted}	W	Thermoelastic damping.
$v(t)$	m/s	Specified velocity on the boundary.
Q_{vd}	W	Viscous dissipation.
$\|R(f)\|$	-	Amplitude of the reflection coefficient.
T_0	s	Signal period.
$A_0(f)$	-	Amplitude of the reference signal in the frequency domain.
$A(f)$	-	Amplitude of the reflected ultrasonic wave in the frequency domain.
$\Phi(f)$	rad	Phase of the reflection coefficient.
$\|R_0(f)\|$	-	Reflection coefficient of the reference signal in the frequency domain, typically taken as one.
$\theta_0(f)$	rad	Phase of the reference signal in the frequency domain.
$\theta(f)$	rad	Phase of the reflected ultrasonic wave in the frequency domain.
ρ_T	kg/m³	Density at temperature T.
$\Phi_0(f)$	rad	Phase of the reflection coefficient of the reference signal in the frequency domain, typically taken as 0.
C_1	Pa^{-1}	Density–pressure coefficient.
ρ_0	kg/m³	Density at temperature T_0.
C_3	K^{-1}	Density–temperature coefficient.
C_2	Pa^{-1}	Density–pressure coefficient.
α	m/m/K	Thermal expansion coefficient.
ρ_t	kg/m³	Density varying with temperature.
c_t	m/s	Speed of sound running with temperature.
T	K	Temperature.
$Z_i(T)$	kg/(m²·s)	Temperature-compensated acoustic impedance.
$R(T)$	-	Temperature-compensated reflection coefficient.
P	N	Pressure applied to the bearing.
f	m	Bearing oil film thickness.
e	-	Reflection coefficient error.
h	m	Temperature of the bearing oil film.

References

1. Peng, H.; Zhang, H.; Shangguan, L.; Fan, Y. Review of Tribological Failure Analysis and Lubrication Technology Research of Wind Power Bearings. *Polymers* **2022**, *14*, 3041. [CrossRef]
2. Marko, M.D. The Impact of Lubricant Film Thickness and Ball Bearings Failures. *Lubricants* **2019**, *7*, 48. [CrossRef]
3. Xu, F.; Ding, N.; Li, N.; Liu, L.; Hou, N.; Xu, N.; Guo, W.; Tian, L.; Xu, H.; Lawrence Wu, C.; et al. A review of bearing failure Modes, mechanisms and causes. *Eng. Fail. Anal.* **2023**, *152*, 107518. [CrossRef]
4. Zhu, S.; Yuan, W.; Cong, J.; Guo, Q.; Chi, B.; Yu, J. Analysis of regional wear failure of crankshaft pair of heavy duty engine. *Eng. Fail. Anal.* **2023**, *154*, 107635. [CrossRef]
5. Zhang, Y.; Wang, W.; Zhang, S.; Zhao, Z. Optical analysis of ball-on-ring mode test rig for oil film thickness measurement. *Friction* **2016**, *4*, 324–334. [CrossRef]

6. Mu, B.; Qu, R.; Tan, T.; Tian, Y.; Chai, Q.; Zhao, X.; Wang, S.; Liu, Y.; Zhang, J. Fiber Bragg Grating-Based Oil-Film Pressure Measurement in Journal Bearings. *IEEE Trans. Instrum. Meas.* **2019**, *68*, 1575–1581. [CrossRef]
7. Deng, Y.; Zhong, S.; Lin, J.; Zhang, Q.; Nsengiyumva, W.; Cheng, S.; Huang, Y.; Chen, Z. Thickness Measurement of Self-Lubricating Fabric Liner of Inner Ring of Sliding Bearings Using Spectral-Domain Optical Coherence Tomography. *Coatings* **2023**, *13*, 708. [CrossRef]
8. Xie, K.; Liu, L.; Li, X.; Zhang, H. Non-contact resistance and capacitance on-line measurement of lubrication oil film in rolling element bearing employing an electric field coupling method. *Measurement* **2016**, *91*, 606–612. [CrossRef]
9. Cheng, M.H.; Chiu, G.T.C.; Franchek, M.A. Real-Time Measurement of Eccentric Motion with Low-Cost Capacitive Sensor. *IEEE/ASME Trans. Mechatron.* **2013**, *18*, 990–997. [CrossRef]
10. He, Y.; Wang, J.; Gu, L.; Zhang, C.; Yu, H.; Wang, L.; Li, Z.; Mao, Y. Ultrasonic measurement for lubricant film thickness with consideration to the effect of the solid materials. *Appl. Acoust.* **2023**, *211*, 109563. [CrossRef]
11. Wang, J.; He, Y.; Shu, K.; Zhang, C.; Yu, H.; Gu, L.; Wang, T.; Li, Z.; Wang, L. A sensitivity-improved amplitude method for determining film thickness based on the partial reflection waves. *Tribol. Int.* **2023**, *189*, 109010. [CrossRef]
12. Zhang, K.; Wu, T.; Meng, Q.; Meng, Q. Ultrasonic measurement of oil film thickness using piezoelectric element. *Int. J. Adv. Manuf. Technol.* **2018**, *94*, 3209–3215. [CrossRef]
13. Dou, P.; Wu, T.; Luo, Z. Wide Range Measurement of Lubricant Film Thickness Based on Ultrasonic Reflection Coefficient Phase Spectrum. *J. Tribol.* **2019**, *141*, 031702. [CrossRef]
14. Beamish, S.; Li, X.; Brunskill, H.; Hunter, A.; Dwyer-Joyce, R. Circumferential film thickness measurement in journal bearings via the ultrasonic technique. *Tribol. Int.* **2020**, *148*, 106295. [CrossRef]
15. Beamish, S.; Dwyer-Joyce, R.S. Experimental Measurements of Oil Films in a Dynamically Loaded Journal Bearing. *Tribol. Trans.* **2022**, *65*, 1022–1040. [CrossRef]
16. Dou, P.; Wu, T.; Luo, Z.; Yang, P.; Peng, Z.; Yu, M.; Reddyhoff, T. A finite-element-aided ultrasonic method for measuring central oil-film thickness in a roller-raceway tribo-pair. *Friction* **2022**, *10*, 944–962. [CrossRef]
17. Jia, Y.; Dou, P.; Zheng, P.; Wu, T.; Yang, P.; Yu, M.; Reddyhoff, T. High-accuracy ultrasonic method for in-situ monitoring of oil film thickness in a thrust bearing. *Mech. Syst. Signal Process.* **2022**, *180*, 109453. [CrossRef]
18. Wei, S.; Wang, J.; Cui, J.; Song, S.; Li, H.; Fu, J. Online monitoring of oil film thickness of journal bearing in aviation fuel gear pump. *Measurement* **2022**, *204*, 112050. [CrossRef]
19. Gray, W.A.; Dwyer-Joyce, R.S. In-situ measurement of the meniscus at the entry and exit of grease and oil lubricated rolling bearing contacts. *Front. Mech. Eng.* **2022**, *8*, 1056950. [CrossRef]
20. Dou, P.; Zou, L.; Wu, T.; Yu, M.; Reddyhoff, T.; Peng, Z. Simultaneous measurement of thickness and sound velocity of porous coatings based on the ultrasonic complex reflection coefficient. *NDT E Int.* **2022**, *131*, 102683. [CrossRef]
21. Dou, P.; Zheng, P.; Jia, Y.; Wu, T.; Yu, M.; Reddyhoff, T.; Liao, W.; Peng, Z. Ultrasonic measurement of oil film thickness in a four-layer structure for applications including sliding bearings with a thin coating. *NDT E Int.* **2022**, *131*, 102684. [CrossRef]
22. Shang, F.; Sun, B.; Zhang, H. Measurement of Air Layer Thickness under Multi-Angle Incidence Conditions Based on Ultrasonic Resonance Reflection Theory for Flange Fasteners. *Appl. Sci.* **2023**, *13*, 6057. [CrossRef]
23. Jia, Y.; Wu, T.; Dou, P.; Yu, M. Temperature compensation strategy for ultrasonic-based measurement of oil film thickness. *Wear* **2021**, *476*, 203640. [CrossRef]
24. Zhang, M.; Ma, X.; Guo, N.; Xue, Y.; Li, J. Calculation and Lubrication Characteristics of Cylindrical Roller Bearing Oil Film with Consideration of Thermal Effects. *Coatings* **2023**, *13*, 56. [CrossRef]
25. Mo, H.; Hu, Y.; Quan, S. Thermo-Hydrodynamic Lubrication Analysis of Slipper Pair Considering Wear Profile. *Lubricants* **2023**, *11*, 190. [CrossRef]
26. Tattersall, H.G. The ultrasonic pulse-echo technique as applied to adhesion testing. *J. Phys. D Appl. Phys.* **1973**, *6*, 819–832. [CrossRef]
27. Achenbach, J.D. *Wave Propagation in Elastic Solids*; North-Holland Pub. Co.: Amsterdam, The Netherlands, 1973; p. 425.
28. Reddyhoff, T.; Dwyer-Joyce, R.; Harper, P. Ultrasonic measurement of film thickness in mechanical seals. *Seal. Technol.* **2006**, *2006*, 7–11. [CrossRef]
29. Zhu, J.; Li, X.; Beamish, S.; Dwyer-Joyce, R.S. An ultrasonic method for measurement of oil films in reciprocating rubber O-ring seals. *Tribol. Int.* **2022**, *167*, 107407. [CrossRef]
30. Shang, F.; Sun, B.; Li, H.; Zhang, H.; Liu, Z.; Meng, X.; Jiang, L.; Xv, G. Detection Method for Bolt Loosening and Washer Damage in Flange Assembly Structures Based on Phased Array Ultrasonics. *Res. Nondestruct. Eval.* **2024**, 1–27. [CrossRef]
31. He, Y.; Gao, T.; Guo, A.; Qiao, T.; He, C.; Wang, G.; Liu, X.; Yang, X. Lubricant Film Thickness Measurement Based on Ultrasonic Reflection Coefficient Phase Shift. *Tribology* **2021**, *41*, 1–8.

Disclaimer/Publisher's Note: The statements, opinions and data contained in all publications are solely those of the individual author(s) and contributor(s) and not of MDPI and/or the editor(s). MDPI and/or the editor(s) disclaim responsibility for any injury to people or property resulting from any ideas, methods, instructions or products referred to in the content.

Article

Simulation Analysis and Experimental Study on the Fluid–Solid–Thermal Coupling of Traction Motor Bearings

Hengdi Wang [1], Han Li [1], Zheming Jin [2], Jiang Lin [2], Yongcun Cui [1,*], Chang Li [3], Heng Tian [1] and Zhiwei Wang [1,2,*]

1. School of Mechatronics Engineering, Henan University of Science and Technology, Luoyang 471003, China; 9901432@haust.edu.cn (H.W.); li_han0013@163.com (H.L.); tianheng_1988@163.com (H.T.)
2. BH Technology Group Co., Ltd., Taizhou 318050, China; jzm_2001@163.com (Z.J.); jj008gx@163.com (J.L.)
3. Shandong Chaoyang Bearing Co., Ltd., Dezhou 253200, China; tyc698@163.com
* Correspondence: 9906172@haust.edu.cn (Y.C.); wangzw87@yeah.net (Z.W.); Tel.: +86-0379-64231479 (Y.C.& Z.W.)

Abstract: The traction motor is a crucial component of high-speed electric multiple units, and its operational reliability is directly impacted by the temperature increase in the bearings. To accurately predict and simulate the temperature change process of traction motor bearings during operation, a fluid–solid–thermal simulation analysis model of grease-lubricated deep groove ball bearings was constructed. This model aimed to simulate the temperature rise of the bearing and the grease flow process, which was validated through experiments. The results from the simulation analysis and tests indicate that the temperature in the contact zone between the bearing rolling element and the raceway, as well as the ring temperature, initially increases to a peak and then gradually decreases, eventually stabilizing once the bearing's heat generation power and heat transfer power reach equilibrium. Furthermore, the established fluid–solid–thermal coupling simulation analysis model can accurately predict the amount of grease required for effective lubrication in the bearing cavity, which stabilizes along with the bearing temperature. The findings of this research can serve as a theoretical foundation and technical support for monitoring the health status of high-speed EMU traction motor bearings.

Keywords: bearing; grease lubrication; fluid–solid–thermal coupling; flow behavior; temperature rise

Citation: Wang, H.; Li, H.; Jin, Z.; Lin, J.; Cui, Y.; Li, C.; Tian, H.; Wang, Z. Simulation Analysis and Experimental Study on the Fluid–Solid–Thermal Coupling of Traction Motor Bearings. *Lubricants* **2024**, *12*, 144. https://doi.org/10.3390/lubricants12050144

Received: 27 March 2024
Revised: 22 April 2024
Accepted: 22 April 2024
Published: 25 April 2024

Copyright: © 2024 by the authors. Licensee MDPI, Basel, Switzerland. This article is an open access article distributed under the terms and conditions of the Creative Commons Attribution (CC BY) license (https://creativecommons.org/licenses/by/4.0/).

1. Introduction

In rail transit vehicles such as high-speed electric multiple units, traction motors play a crucial role in converting electrical energy to kinetic energy. The performance of these motors directly impacts the driving performance, range, and energy efficiency of the train. Bearings, as integral components of traction motors, significantly influence the motor's performance and lifespan. Selecting appropriate lubricants is essential for optimal bearing performance [1,2]. Grease lubrication is the primary mode for high-speed EMU bearings [3], with the lubricant having the shortest service life in bearing parts. Analyzing the grease condition of the bearing is thus highly significant [4]. The operating conditions of traction motors are intricate [5], with varying speed, load, and external factors leading to intense friction and heat generation within the bearing. This can result in lubrication failure, posing a risk to the safe operation of rolling bearings and potentially causing mechanical failures [6–8]. Therefore, understanding the temperature rise mechanism of the bearing and the flow behavior of the grease is crucial to assess bearing performance under complex conditions like high speed and heavy load. This analysis is essential for predicting and preventing bearing failure, ultimately enhancing the reliability and stability of traction motor and high-speed train operations.

Lugt, Velickov, and Tripp [9] conducted a study on the erratic behavior of grease in bearings, often characterized by a highly variable temperature signal. The temperature typically increases at the onset of bearing operation, a phenomenon commonly associated with grease agitation. The study revealed that the initial filling condition of the bearing plays a crucial role in determining the lifespan of the grease within the bearing. Xu, Zhang, Huang, and Wang [10] developed a heat transfer model specific to high-speed railway bearings, taking into account the unique working conditions and structural characteristics of double-row tapered roller bearing assemblies. The researchers calculated the heat source and external heat dissipation of the bearing, established reasonable lubrication boundary conditions, and created a finite element model using ANSYS 15.0. Chai, Ding, and Wang [11] utilized the computational fluid dynamics software FLUENT 15.0 to simulate the flow field in a herringbone grooved bearing, accounting for factors such as viscosity, temperature, and heat transfer.

Ma, Li, Qiu, Wu, and An [12] proposed a precise calculation model for the heating rate of grease-lubricated spherical roller bearings using the local heat source analysis method. They utilized the thermal network method to develop a transient thermal model of the bearing seat–shaft system. Xiu, Xiu, and Gao [13] focused on the mathematical model of the oil film temperature field of ultra-high-speed dynamic and static pressure sliding bearings. They conducted numerical simulations of the oil film temperature field of the bearing using FLUENT 11.0 software. These simulations not only allow for the prediction of the flow state and thermal performance of the bearing oil but also facilitate the identification of design flaws in the bearing, providing valuable insights for improving bearing design. Wang, Sun, Chang, and Liu [14] utilized the computational fluid dynamics method to establish a flow analysis model of grease lubrication in the bearing cavity, considering the flow characteristics of grease during bearing operation. They conducted a detailed analysis of the flow characteristics of grease lubrication in the bearing cavity. Furthermore, Zhang, Lin, Fan, Su, and Lu [15] employed ANSYS FLUENT 19.2 software to investigate and analyze the temperature fields of two oil chambers with the same speed, thickness, and bearing area of the oil cavity. They summarized the temperature distribution patterns of the two cavity oil films.

Since the mid-20th century, Palmgren [16] utilized the bearing friction torque test method to derive an empirical formula for calculating the frictional moment of bearings. Deng, Xie, Liao, Zhou, and Deng [17–19] further developed a formula for calculating the friction moment of high-speed angular contact ball bearings through bearing dynamic analysis. They also investigated the correlation between axial preload, cage structural parameters, and bearing friction moment. Harris and Kotzalas [20] introduced a method for calculating local heat generation in ball bearings and analyzing bearing temperature using a thermal grid method. Ai, Wang, Wang, and Zhao [21] established a thermal mesh model for a double-row tapered roller bearing based on the generalized Ohm's law. They explored the impact of rotational speed, filling grease ratio, and roller diameter on bearing temperature. Additionally, Takabi and Khonsari [22,23] developed a bearing thermal analysis model to examine the relationship between lubricating oil viscosity, convection coefficient, temperature gradient, and thermally induced preload in bearings. De-xing, Weifang, and Miaomiao [24–26] considered the influence of lubricating oil and thermal expansion to establish a thermal grid analysis model of bearings and analyzed the factors that cause the temperature rise of bearings. Local temperature rises in bearings can affect fluid density and viscosity as a result of a variety of phenomena, including elastic flow lubrication and rolling/sliding friction [27,28]. While these are important phenomena to consider when modeling lubricated elliptical contacts, they are beyond the scope of current work and CFD capabilities, and the fluid is considered isothermal throughout the simulation [29]. Nevertheless, this approach is not applicable for analyzing grease lubricated bearings, as grease viscosity varies significantly with temperature and heat generation is dynamic, precluding isothermal considerations. Some researchers have conducted simulations and analyses of the temperature distribution in grease-lubricated bearings to determine the optimal grease filling amount under operational conditions. However, there is limited research on the changes in grease flow behavior within the bearing cavity.

During the test, the temperature of the grease-lubricated deep groove ball bearing in the traction motor increases significantly after the initial grease filling, leading to an alarm from the temperature detector. To better understand the mechanism behind this temperature rise, it is crucial to accurately simulate the entire process. Existing research primarily focuses on heat generation and transfer calculations to determine the final stable temperature of the bearing. However, these studies often overlook the impact of lubricant viscosity changes on heat generation and do not consider heat transfer to the bearing seat or air convection. This dynamic process, from the start of operation to the bearing temperature reaching equilibrium, is critical as the maximum temperature of the outer ring influences the temperature monitoring element's threshold setting. This paper introduces a novel approach that considers the effects of grease viscosity and temperature on bearing friction power consumption. By treating temperature as a variable and grease viscosity as an intermediary, the frictional power consumption at different contact zone temperatures is calculated. An iterative method is employed to analyze the variable friction power consumption conditions, providing insights into the overall temperature rise process of the bearing. A fluid–solid thermal analysis model is utilized to investigate the temperature rise behavior mechanism and grease fluidity in grease-lubricated deep groove ball bearings of traction motors, with experimental validation confirming the simulation's accuracy.

2. Establishment of a Fluid–Solid–Thermal Coupling Model

To comprehensively study the temperature rise behavior mechanism and grease flow behavior of grease-lubricated bearings, it is essential to consider the heat generation and heat transfer characteristics of the bearings, as well as the dynamic characteristics of the fluid–structure interaction model. The process involves establishing a bearing dynamics model to determine the friction source and calculate the frictional heat, followed by the establishment of thermal conductivity, convection, and thermal radiation models for the bearings. The calculated results serve as thermal boundary conditions for the fluid–solid–thermal coupling simulation model, which is used to analyze the temperature rise characteristics of the bearing and the fluidity of the grease.

2.1. Heat Generation Model

The power consumption of the traction motor bearing in the service state results from various factors. This includes friction power consumption between the steel ball and the raceway due to material elastic hysteresis and frictional power dissipation between the steel ball and the raceway due to differential sliding. Spin sliding friction power dissipation due to the steel ball moves relative to the raceway. The rotation of the steel ball is influenced by fluid flow resistance and stirring resistance, leading to viscous loss in the steel ball oil film. Additionally, frictional power dissipation arises from hydrodynamic lubrication and relative sliding between the bearing ring guide surface and the cage [30].

$$H_{total} = H_E + H_D + H_S + H_{cb} + H_L \tag{1}$$

where H_{total} is the total frictional power consumption of the bearing, H_E is the elastic hysteresis friction power consumption between the ball and the raceway, H_D is the friction power consumption caused by the differential sliding between the ball and the raceway, H_S is the frictional power consumption caused by the spin of the ball, H_L is the frictional power consumption of viscous stirring loss, and H_{cb} is the friction power consumption between the cage and the ball.

This friction power consumption is further divided into four parts for thermal boundary setting in fluid–structure interaction simulation models: friction between the ball and the inner ring, friction between the ball and the outer ring, friction caused by sliding between the ball and the cage, and friction from viscous stirring loss in the cage. According to the calculation of the bearing dynamics model, with the change in lubricant temperature, the change trend of the friction power consumption value of each part is shown in Figure 1.

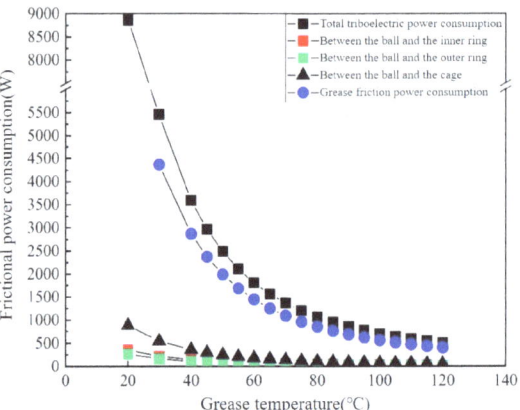

Figure 1. Effect of grease temperature on frictional power consumption.

With the increase in grease temperature in the contact zone, the total friction power consumption of the bearing shows a gradual decreasing trend, and the decreasing rate becomes smaller and smaller. The total frictional power consumption follows the same trend as the viscosity of the grease as a function of temperature, with viscous stirring loss contributing the most, followed by friction between the ball and the cage, and the least coming from friction between the ball and the inner and outer rings.

2.2. Heat Transfer Model

There are three forms of heat transfer, heat conduction between solids, heat convection between solids and fluids, and two thermal radiations separated from each other by space. According to the real working conditions of the traction motor bearing during the test operation, the heat transfer diagram of the bearing is shown in Figure 2, where Q_1 and Q_2 are the parts of the ball and the inner and outer raceways in contact with the heat conduction, Q_3 and Q_4 are the parts of the heat convection between the ball and the grease, Q_5 and Q_6 are the parts of the heat conduction between the outer ring of the bearing and the inner ring of the bearing and the bearing seat, respectively, Q_7 and Q_8 are the parts of heat convection between the grease and the bearing seat of the grease storage chamber on the left and right sides, and Q_9 is the part of the heat convection between the bearing seat and the air.

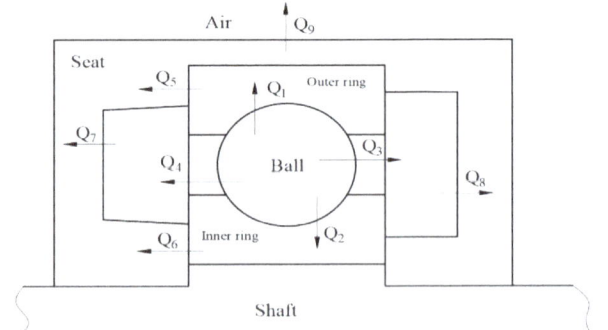

Figure 2. Schematic diagram of bearing heat transfer.

Heat conduction refers to the heat transfer from an object with a high temperature to an object with a low temperature due to the movement between molecules when two objects are in contact. It occurs in the form of heat flux density, and the energy transfer formula is as follows, which can be used to calculate Q_1, Q_2, Q_5, and Q_6:

$$Q_0 = q_n \cdot S = -\lambda \frac{\partial T}{\partial n} \cdot S \qquad (2)$$

where q_n—heat flux, W/m^2; S—thermal conduction contact area, m^2; λ—thermal conductivity, W/(m·K); $\frac{\partial T}{\partial n}$—rate of change in temperature.

Convective heat transfer refers to the heat exchange that occurs when a fluid flows through a solid wall. Convective heat transfer can be divided into natural convection and forced convection, and the convective heat transfer formula is as follows, which can be used to calculate Q_3, Q_4, Q_7, Q_8, and Q_9:

$$Q_n = h_v \cdot S(T_w - T_f) \qquad (3)$$

where T_w—wall temperature, °C; T_f—fluid temperature, °C; S—convection area, m^2; h_v—convective heat transfer coefficient.

In the oil–air lubrication system, the heat dissipation in the bearing cavity is mainly completed by the forced convection of compressed air. Therefore, in this paper, the convection coefficient calculation model of the ball bearing model proposed by Crecelius et al. [31] is selected to calculate the convective heat transfer coefficient of each surface of the ball bearing.

$$h_v = 0.0986 \left[\frac{n}{v}(1 \pm \frac{D_w \cos \alpha}{d_m})\right]^{1/2} \lambda_f P_r^{1/3} \qquad (4)$$

where n—bearing working speed, r/min; v—the kinematic viscosity of the lubricant, m^2/s; d_m—bearing pitch diameter, mm; λ_f—thermal conductivity of fluids; P_r—Plante number.

2.3. The Bearing Simulation Model Was Established

2.3.1. Establishment of Coupling Analysis Model

The fluid–solid thermal coupling analysis model is established to show the solution idea more intuitively. The algorithm flow chart is shown in Figure 3.

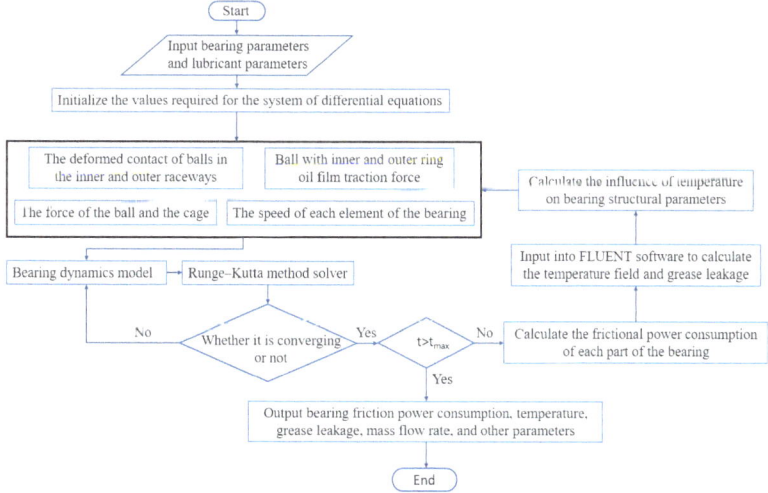

Figure 3. Algorithm flow chart.

Firstly, the working condition parameters and structure parameters of the bearing are input, and the differential equations of bearing dynamics are established [30]. The Runge–Kutta method is used to solve. When the results converge, six kinds of friction power consumption in the bearing are calculated. Then, the friction power consumption is input into Fluent for simulation analysis, and the temperature, grease leakage, and grease flow rate are solved. Then, the influence of the temperature on bearing parameters and friction power consumption is calculated to achieve the coupling analysis.

The fourth-order Runge–Kutta method is an iterative method commonly used to solve systems of nonlinear differential equations. The solution principle is as follows:

$$\begin{cases} y' = f(t, y) \\ y(t_0) = y_0 \\ y_{n+1} = y_n + \frac{h}{6}(k_1 + 2k_2 + 2k_3 + k_4) \end{cases} \quad (5)$$

$$\begin{cases} k_1 = f(t_n, y_n) \\ k_2 = f(t_n + \frac{h}{2}, y_n + \frac{h}{2}k_1) \\ k_3 = f(t_n + \frac{h}{2}, y_n + \frac{h}{2}k_2) \\ k_4 = f(t_n + h, y_n + k_3) \end{cases} \quad (6)$$

Thus, the next value (y_{n+1}) is determined by the initial value (y_n), the product of iteration step h, and the slope(k) of an estimate. The slope is the weighted average of the following slopes:

$$\text{slope}\frac{h}{6}(k_1 + 2k_2 + 2k_3 + k_4) \quad (7)$$

2.3.2. Establishment of Geometric Model

In this paper, a deep groove ball bearing for the nontransmission end of a high-speed rail traction motor is used as the object to construct a temperature field simulation and analysis model. The bearing model is 6215, the specific structural parameters are shown in Table 1, and the physical-property-related parameters are shown in Table 2.

Table 1. Structural parameters of bearings.

Parameter	Meaning	Value	Parameter	Meaning	Value
d	Inside diameter	75 mm	D	Outside diameter	130 mm
B	Width	25 mm	d_m	Bearing pitch diameter	102.5 mm
f_i	Coefficient of curvature radius of inner raceway groove	0.5097	f_e	Coefficient of curvature radius of outer raceway groove	0.5268
D_W	Steel ball diameter	17.462 mm	N	Number of steel balls	11
λ	Radial clearance	46~71 μm	E_0	The elastic modulus of steel	2.08×10^{11} Pa
C_r	Basic dynamic radial load rating	66,000 N	C_{or}	Basic static radial load rating	49,500 N

Table 2. Physical parameters of the fluid–solid model.

	Rings and Ball (Bearing Steel)	Cage (Brass)	Bearing Seat (Grey Cast Iron)	Grease
Density (kg/m^3)	7850	8300	7200	880
Specific heat (j/kg·k)	475	385	447	1845
Thermal conductivity (w/m·k)	44.5	118	39.2	0.145

Mesh was drawn and calculated by using FLUENT 2021 R1 software. In order to improve the efficiency of data processing and ensure the accuracy of simulation data, the following simplifications are made in the establishment of geometric models without affecting the accuracy of simulation analysis results:

(1) The cage rivet has little influence on the temperature field analysis of the whole bearing, so the cage structure is simplified and drawn as an ordinary cage in the pre-treatment modeling.

(2) Ignore the chamfer design of the bearing in the actual test operation, reduce the number of meshes and the difficulty of drawing, and improve the accuracy of meshes.

(3) The diameter of the rolling element is taken as the minimum size within the tolerance range so as to increase the gap with the inner and outer ring raceways and reduce the calculation difficulty of the overall model.

(4) During the operation of the bearing, the ball is not in direct contact with the inner and outer rings, and a lubricating film will be formed in the contact area. The empirical formula of oil film thickness (8) is fully considered in flow field modeling, and mesh refinement is carried out in this area.

$$h_0 = 2.69 \frac{\alpha^{0.53}(\eta_0 u)^{0.67} R_x^{0.466}}{E_0^{0.073} Q_{max}^{0.067}} (1 - 0.61 e^{-0.72K}) \quad (8)$$

where R_x is the equivalent radius of curvature of the ball along the rolling direction; Q_{max} is the maximum load on the ball; K is the ellipticity of the contact ellipse; E_0 is the equivalent elastic modulus; α is the pressure index of viscosity; η_0 is the dynamic viscosity at atmospheric pressure; u is the average surface velocity.

The result of the volume mesh division of the simulation model is shown in Figure 4. The radial section of the bearing is taken to show the mesh division.

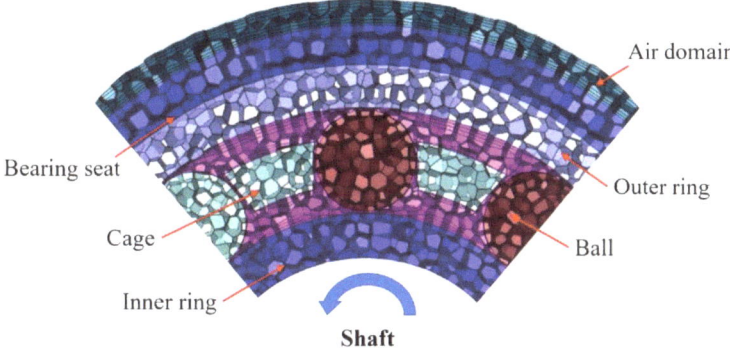

Figure 4. Draw the model volume mesh. The air domain and bearing seat are selected with a mesh size of 0.5 mm. The inner and outer rings and cages are selected with a mesh size of 0.3 mm. The balls and bearing cavity areas are divided into tetrahedral meshes of 0.2 mm. At least three layers of boundary mesh are added to the fluid–structure coupling boundary, and the mesh accuracy meets the calculation requirements.

2.3.3. Boundary Condition Settings

Figure 5 shows a simplified simulation model of fluid–structure interaction, which consists of the air–fluid domain, the simplified model of the bearing seat, the inner and outer rings of the bearing, and the grease fluid domain (red wireframe selection area). Among them, the outermost layer is the air layer, the outermost air wall is set to be room temperature, the air and the bearing seat are convection heat exchange, the bearing seat wraps the bearing and the grease storage chamber, the inner ring is in contact with the shaft, the heat dissipation performance of the bearing seat is better than that of the shaft, and the heat dissipation of the shaft is not considered. The grease outlet is the red ring shown in the diagram, and the outlet is convection air by default.

According to the bearing heat generation analysis mentioned above, the value of friction power consumption corresponding to different operating temperatures of the grease is calculated. The frictional power consumption varies with the viscosity of the lubricant and the viscosity of the grease varies with the temperature of the contact zone, as shown in Figure 6. Therefore, it is difficult to set the thermal boundary of the bearing temperature field simulation, and this paper proposes an iterative method of bearing grease temperature and bearing friction power consumption to realize the setting of variable friction power consumption boundary conditions. A simplified, iterative schematic is shown in Figure 7.

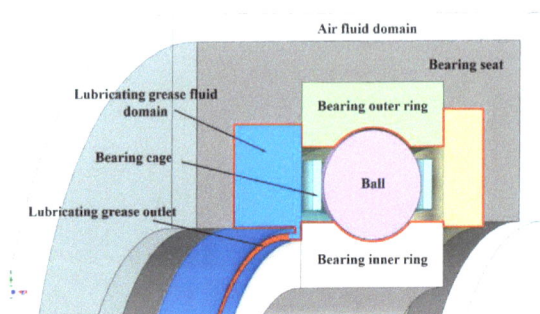

Figure 5. Simplified model diagram.

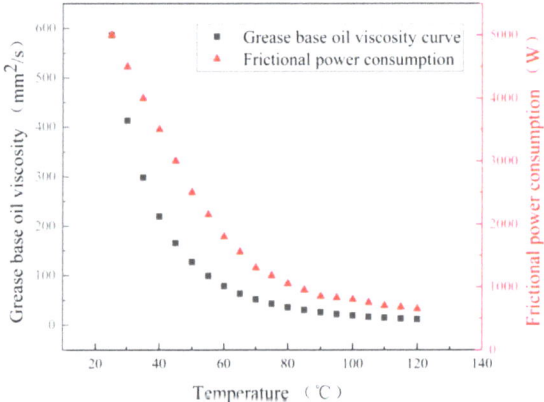

Figure 6. Frictional power consumption and grease viscosity vary with temperature.

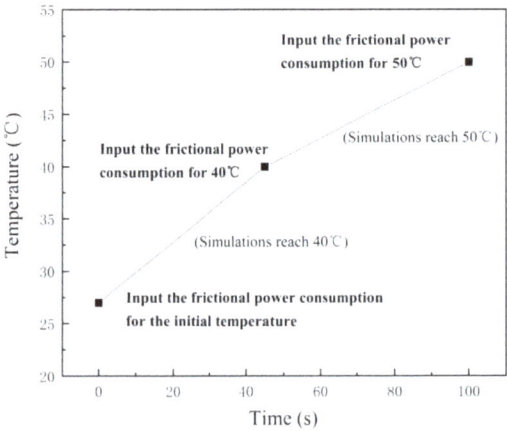

Figure 7. Diagram of the iteration.

The working condition of Figure 6 is 6.9 KN radial load, no axial load, and the rotational speed is 4800 r/min. A line plot of the total frictional power consumption of the bearing as a function of contact zone temperature and grease viscosity is calculated.

The viscosity temperature curve of the grease exhibited a similar trend to the friction power consumption curve of the bearing. It was observed that higher grease viscosity led to increased heat generation from grease stirring, resulting in greater friction power consumption of the bearing in this specific working condition, and vice versa.

The iterative principle of variable frictional power consumption as the thermal boundary condition is that the monitoring point of the grease temperature change in the bearing cavity is set firstly, and the initial frictional power consumption at room temperature of 25 °C is taken as the initial thermal boundary. After starting the simulation, when the grease temperature in the contact zone rises to 40 °C, the frictional power consumption in the grease operating environment of 40 °C is set. Then, the simulation continues, and the above operations are repeated to complete the approximate iteration and realize the input of thermal boundary conditions under variable operating conditions. Smaller temperature spans led to higher iterative accuracy and a more realistic depiction of the bearing's temperature rise process.

3. Transient Simulation and Analysis of Grease-Lubricated Bearings

The speed of the inner ring is 4800 r/min, the outer ring is fixed, and the rotational speeds of the cage and the balls are calculated according to the bearing dynamics. The rolling parameter is taken as the boundary condition of the motion wall. With the bearing cavity filled with grease, and the grease occupying 50% of both the left and right grease storage chambers, transient simulation analysis was conducted to analyze the two-phase distribution of oil and gas and the temperature rise of the bearing. The study observed the temperature rise and distribution in a bearing as the grease temperature detection point reached different temperatures over time. Three grease monitoring points near the raceway were selected to reflect the grease temperature by averaging their values.

3.1. Temperature Field Analysis of Grease-Lubricated Bearings

Before the simulation began, three grease monitoring points near the raceway were selected to reflect the grease temperature in the bearing cavity by averaging the average value. As the temperature of the monitoring point rises, the variable frictional power consumption is used as the boundary condition. The temperature distribution of the outer and inner ring temperatures of the bearing as the grease monitoring points reach different temperatures over the simulation time is analyzed, as shown in Figures 8 and 9. In order to facilitate the observation of the temperature rise of the bearing, the maximum temperature and minimum temperature of the inner and outer rings of the bearing are drawn as curves, as shown in Figure 10. The monitoring points record the change in the average temperature of the inner and outer rings of the bearing with the number of iteration steps. The temperature rise curve of the inner and outer rings of the bearing is shown in Figure 11.

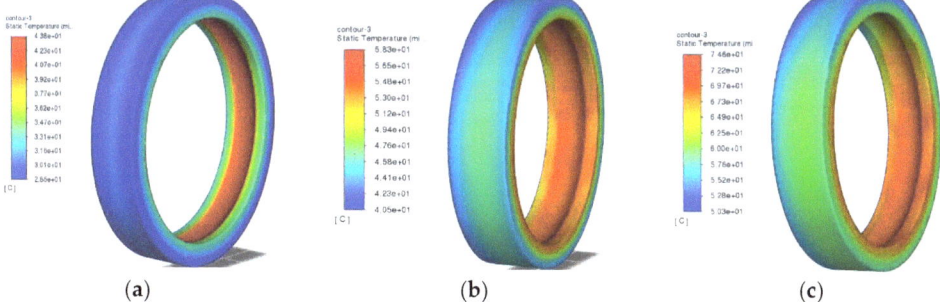

Figure 8. Diagram of the temperature distribution of the outer ring of the bearing. (**a**) The monitoring point reaches 40 °C. (**b**) The monitoring point reaches 60 °C. (**c**) The monitoring point reaches 100 °C.

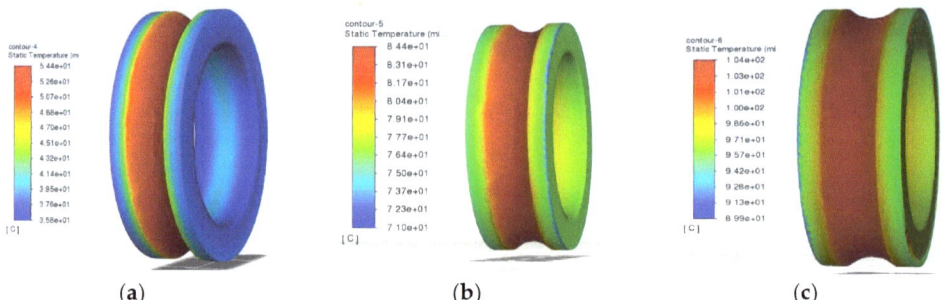

Figure 9. Diagram of the temperature distribution of the inner ring of the bearing. (**a**) The monitoring point reaches 40 °C. (**b**) The monitoring point reaches 60 °C. (**c**) The monitoring point reaches 100 °C.

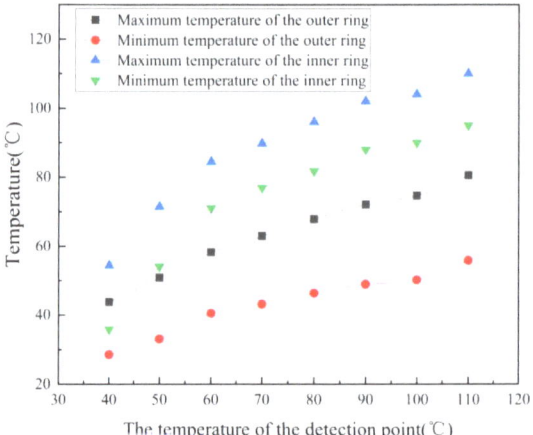

Figure 10. Temperature rise diagram of the inner and outer rings of the bearing.

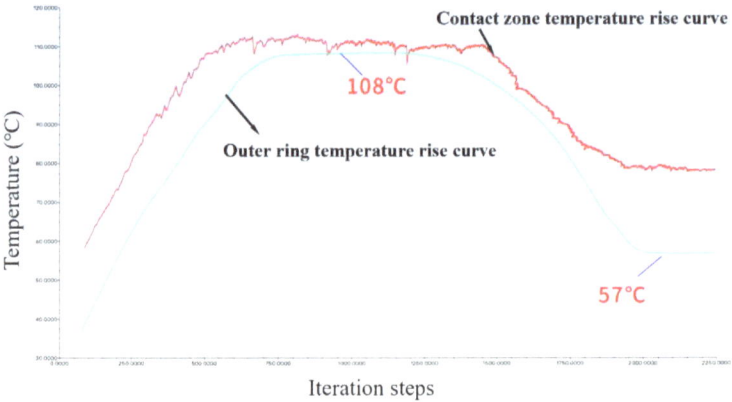

Figure 11. Temperature rise curve of the inner and outer rings of the bearing.

Observe the temperature distribution of the outer and inner rings of the bearings in Figures 8 and 9. As the simulation time increases, the overall temperature of the bearing shows an upward trend. The temperature distribution in the outer ring in Figure 8c is obviously not as uniform as in Figure 8a. The reason is that the inner ring of the bearing is mated to the spindle, and the outer ring is fixed to the bearing housing. The grease is deposited in the lower part of the bearing due to gravity, and the heat generated by the grease stirring is high. There is heat conduction between the bearing and the bearing seat in the upper part, and the convection heat dissipation of the air occurs. As a result, the temperature distribution of the outer ring of the bearing shows that the temperature of the upper half of the ring is less than that of the lower half of the ring. Since the inner ring rotates with the shaft, the overall temperature distribution uniformity of the inner ring is better than that of the outer ring.

As can be seen from Figure 10, the temperature of the inner ring of the bearing is higher than that of the outer ring of the bearing. With the increase in running time, the temperature of the inner and outer rings shows an upward trend, but the temperature growth rate gradually decreases. From the graph of the interaction between viscosity, temperature, and frictional power consumption of the grease in Figure 6, it can be seen that the lower the viscosity, the smaller the frictional power consumption, so the temperature growth rate is reduced. The trend of the temperature rise pattern of the inner and outer rings of the bearing is consistent with that of Figure 6.

The red line in Figure 11 represents the temperature rise curve of the average temperature in the contact zone of the inner ring. The blue line represents the temperature rise curve of the average temperature of the bearing outer ring. During the temperature rise process, the heat generation of the bearing is more than the heat dissipation. When the bearing is operated for a period of time, the temperature of the outer ring of the bearing gradually reaches a maximum of 108 °C. The cavity temperature is high and the grease is sheared and thinned. This process discharges excess grease out of the bearing cavity. The heat generated by the grease stirring in the cavity is reduced, and the bearing temperature is reduced. When there is only a moderate amount of grease left in the bearing cavity and it is effectively lubricated, the heat generation and heat dissipation reach an equilibrium state. At the end of the run-in phase, the temperature of the outer ring of the bearing reaches an equilibrium of 57 °C.

3.2. Simulation Analysis of Grease Flow Behavior

The flow behavior of the grease in the bearing cavity was analyzed under the condition that the bearing cavity was full of grease and the grease of the left and right grease chambers accounted for 50%, and the initial grease filling amount is shown in Figure 12. The red area represents the grease area, the blue area represents the gas cavity area, and the blank area is the inner and outer rings of the bearing, the bearing seat, and the shaft. Figure 13 shows the radial and axial cross-sections of the bearing cavity and the flux of grease at the outlet at different temperatures at the monitoring point.

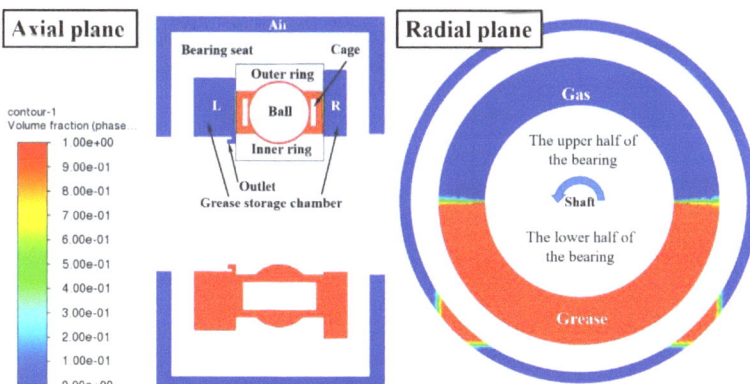

Figure 12. Initial filling grease status diagram.

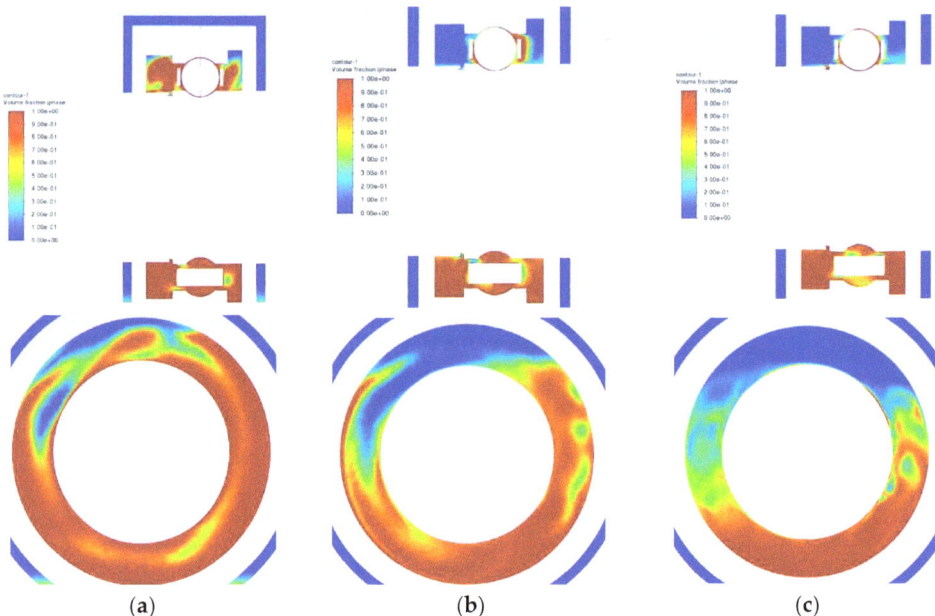

Figure 13. Two-phase distribution diagram of oil and gas in the bearing. (**a**) The monitored temperature reaches 40 °C. (**b**) The monitored temperature reaches 60 °C. (**c**) The monitored temperature reaches 100 °C.

The grease chamber is initially filled with 50% grease, and under the effect of gravity, the grease collects in the lower half ring of the grease chamber. As the bearing rotates, the grease in the bearing cavity is squeezed to both sides of the grease chamber. Due to the presence of inertial centrifugal force, the grease is thrown to the upper half ring of the grease chamber as shown in Figure 13a. The temperature of the grease at the monitoring point increases as the running time increases. The grease flows out of the outlet and the amount of grease in the bearing cavity and grease chamber decreases. The reason for this is that the temperature increases the viscosity of the grease and the phenomenon of shear thinning of the grease occurs.

In order to study the temperature rise mechanism and grease flow behavior during bearing operation, it is important to investigate the amount of grease involved in effective lubrication in the contact zone. Therefore, it is necessary to analyze the amount of residual grease in the bearing cavity. The simulation results are shown in Figure 14, and the formula for calculating the amount of residual grease in the bearing cavity is as follows.

$$M = M_0 + \int_{t_0}^{t} w_i - w_o dt \tag{9}$$

where M_0 is the initial grease filling mass, kg; w_i is the amount of grease flowing into the bearing per unit time, kg/s; w_o is the amount of grease flowing out of the bearing per unit time, kg/s.

With the rotation of the inner ring, cage, and ball bearing, a large amount of grease in the bearing cavity rushes into the grease chambers on both sides. This leads to a sudden drop in the amount of grease in the bearing cavity and an increase in the amount of grease in the grease chambers on both sides, at which time the rate of flow out of the bearing cavity reaches a maximum. The flow state of the grease is gradually stabilized with the increase in the running time. The amount of grease in the bearing cavity and the grease chambers fluctuates and tends to be in dynamic equilibrium. The rate of grease inflow and outflow in the bearing cavity fluctuates above and below the 0 scale. As shown in the partially enlarged Figure 14b, the amount of grease in the bearing cavity after reaching the dynamic equilibrium fluctuates above and below 10.45 g.

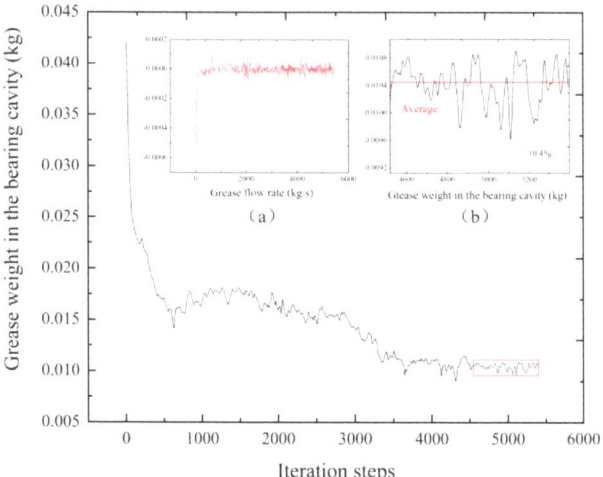

Figure 14. Flow characteristics of grease in the bearing cavity.

The mass flow rate at the bearing outlet is shown in Figure 15. The amount of grease in the bearing cavity and in the grease chamber is much greater than the amount required for bearing lubrication. At the beginning of the bearing operation, the grease outflow rate at the outlet is high. Because of the large amount of grease in the initial period, the heat generated by stirring is large. When the temperature inside the bearing cavity rises, the viscosity of the grease decreases, aggravating the grease outflow. When the bearing completes the break-in stage, the grease content in the bearing cavity reaches the optimum, the temperature reaches the equilibrium state, and the grease discharge behavior at the outlet stops.

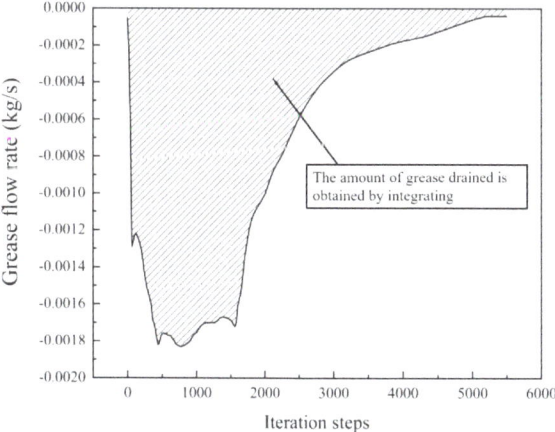

Figure 15. Grease flow rate at outlet.

4. Test Verification of the Temperature Field of Grease-Lubricated Bearings

To verify the accuracy of the above simulation model and simulation results, a current bearing testing machine is used to verify the temperature rise change of the outer ring of the bearing as well as verify the residual amount of grease in the bearing cavity once the bearing is running stably. The testing machine mainly includes the main body of the testing machine, power system, axial and radial loading system, temperature sensor and data acquisition system and other components; the testing machine is shown in Figure 16.

Figure 16. Bearing temperature rise testing machine.

The test condition is radial load of 6.9 KN and no axial load, the rotational speed is 4800 r/min, and the length of the test is 600 min. The temperature of the outer ring of the bearing measured by the temperature sensor of the testing machine is shown in Figure 17.

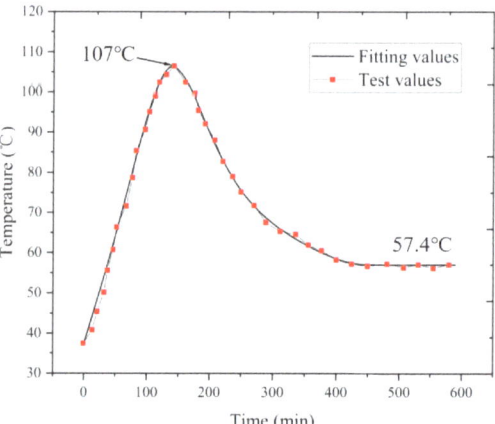

Figure 17. Temperature rise curve of the outer ring of the bearing.

As can be seen in the figure, during the time period of 0–140 min of the test, the temperature rises sharply due to the large friction power consumption in the bearing cavity, where heat generation is much higher than heat dissipation. The maximum temperature of 107 °C was reached in 140 min. In the time period of 140–400 min, the temperature in the bearing cavity is high, and the viscosity of the grease decreases. At this time, the grease discharge rate at the outlet reaches the maximum, the stirring heat is greatly reduced, and the bearing outer ring temperature is reduced. When the test time reaches 400 min, the heat generation and dissipation reach equilibrium. At this time, the amount of grease in the bearing cavity reaches the optimum, and the temperature is finally stabilized at

57.4 °C. Compared with the previous simulation results in Figure 11, the error of the peak temperature of the outer ring of the bearing is 0.93%, and the error of the final stabilized temperature is 0.69%, so the simulation results are consistent with the test results.

Before the test begins, the net weight of the bearing element in the unfilled state is weighed. After the testing machine runs for 600 min, the bearings are unloaded and weighed again. The weight of the bearing is shown in Figure 18.

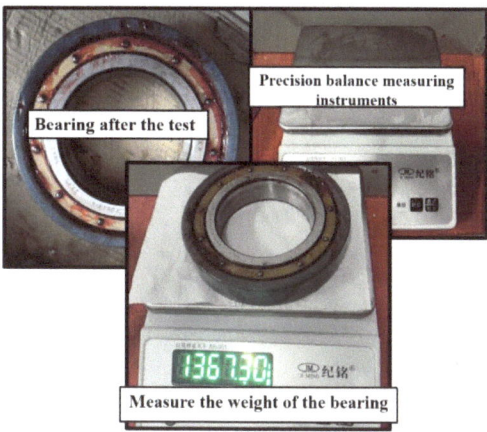

Figure 18. Bearing weighing test.

The total weight of the bearing and the grease is 1367.3 g, of which the net weight of the bearing is 1351 g, and the weight of the grease in the bearing cavity is 11.3 g. Compared with the remaining grease in the bearing cavity simulated above, the simulation error is 7.52%. The amount of residual grease in the bearing cavity accounts for 25.57% of the initial filling amount. The simulation results are highly accurate, and the margin of grease in the bearing cavity can be simulated and predicted more accurately.

There are several reasons for the error between the simulation value and the experimental value. Simulation cannot completely fit the actual modeling, there is a certain simplification. Some parts of the grid drawing of the model are not fine enough. And the test is affected by indoor humidity, air flow, and other environmental factors, some of which cannot be fully considered in the simulation analysis. The above factors cause some deviation between simulation and test values.

5. Conclusions

This study focuses on accurately predicting temperature rise in high-speed EMU traction motor bearings. A fluid–solid–thermal simulation analysis model was constructed for grease-lubricated deep groove ball bearings. The simulation of temperature rise and grease flow behavior was validated through experiments. The findings offer a theoretical foundation and technical assistance for monitoring the health of traction motor bearings in high-speed EMUs. The main conclusions are as follows:

(1) A transient temperature field simulation method considering multi-factor heat dissipation and variable friction power consumption is proposed, which accurately predicts the change process of temperature between ball and raceway, as well as outer ring temperature during grease homogenization. The temperature in the bearing contact zone and ring initially increases to the maximum and then gradually decreases. Finally, it tends to stabilize once heat generation and dissipation reach equilibrium.

(2) The fluid–solid–thermal coupling simulation analysis model can accurately predict the content of grease involved in effective lubrication in the bearing cavity, which tends to stabilize with the stability of the bearing temperature. The simulation analysis results

and test results show that the grease content in the 6215 bearing cavity stabilizes at 25.57% of the initial filling amount.
(3) In a comparison of simulation and experimental results, the error of the peak temperature of the outer ring is 0.93%, the error of the final stable temperature is 0.69%, and the error of the residual amount of grease in the bearing cavity is 7.52%, which prove the accuracy and effectiveness of the proposed simulation model.

Author Contributions: Conceptualization, H.W., H.L. and Z.J.; Data curation, H.L., J.L. and Z.J.; Formal analysis, H.L., Y.C. and Z.W.; Funding acquisition, Y.C.; Investigation, H.W., Z.J. and C.L.; Methodology, H.W. and H.L.; Project administration, H.W., Z.J. and Y.C.; Resources, Y.C.; Software, H.L., J.L. and C.L.; Supervision, Y.C. and H.W.; Validation, Z.J., J.L. and H.T.; Writing—original draft, H.L. and Z.W.; Writing—review and editing, H.W., H.T. and J.L. All authors have read and agreed to the published version of the manuscript.

Funding: This research was funded by the National Natural Science Foundation of China (52105182, 52005158), Zhejiang Province Key R&D Plan (2021C01095) and Key R&D Plan of Ningbo City (2023Z006).

Data Availability Statement: The data used to support the findings of this study are available from the corresponding author upon request.

Conflicts of Interest: Authors Zheming Jin, Jiang Lin and Zhiwei Wang were employed by the company BH Technology Group Co., Ltd. Author Chang Li was employed by the company Shandong Chaoyang Bearing Co., Ltd. The remaining authors declare that the research was conducted in the absence of any commercial or financial relationships that could be construed as a potential conflict of interest.

References

1. Ye, J.; Yang, L. Developing of Bearings for Rail Transit Vehicles. *Bearing* **2013**, *12*, 61–65.
2. Peng, F.; Zhang, Y. General Review of Rail Transit Bearing Fault Diagnosis and Life Prediction Technology. *Urban Mass Transit* **2020**, *23*, 162–168.
3. Yao, L.; Yang, H. The Trend of Greases for Railway Overseas. *Pet. Prod. Appl. Res.* **2006**, *3*, 50–53.
4. Li, Q.; Chen, L.; Xu, J.; Chen, G.; Yang, G. Failure Modes of Rolling Bearings for High-speed EMUs and Prospects for Countermeasures. *Bearing* **2024**, *3*, 1–8.
5. Wang, Y.; Cao, J.; Tong, Q.; An, G.; Liu, R.; Zhang, Y.; Yan, H. Study on the Thermal Performance and Temperature Distribution of Ball Bearings in the Traction Motor of a High-Speed EMU. *Appl. Sci.* **2020**, *10*, 4373. [CrossRef]
6. Kavathekar, S.; Upadhyay, N.; Kankar, P.K. Fault Classification of Ball Bearing by Rotation Forest Technique. *Procedia Technol.* **2016**, *23*, 187–192. [CrossRef]
7. Yu, G. Fault feature extraction using independent component analysis with reference and its application on fault diagnosis of rotating machinery. *Neural Comput. Appl.* **2015**, *26*, 187–198. [CrossRef]
8. Huang, W.; Sun, H.; Luo, J.; Wang, W. Periodic feature oriented adapted dictionary free OMP for rolling element bearing incipient fault diagnosis. *Mech. Syst. Signal Process.* **2019**, *126*, 137–160. [CrossRef]
9. Lugt, P.M.; Velickov, S.; Tripp, J.H. On the Chaotic Behavior of Grease Lubrication in Rolling Bearings. *Tribol. Trans.* **2009**, *52*, 581–590. [CrossRef]
10. Xu, J.; Zhang, J.; Huang, Z.; Wang, L. Calculation and finite element analysis of the temperature field for high-speed rail bearing based on vibrational characteristics. *J. Vibroeng.* **2015**, *17*, 720–732.
11. Chai, D.; Ding, Q.; Wang, B. Temperature field simulation of herringbone grooved bearing based on FLUENT software. In Proceedings of the 2016 IEEE Advanced Information Management, Communicates, Electronic and Automation Control Conference (IMCEC), Xi'an, China, 3–5 October 2016.
12. Ma, F.; Li, Z.; Qiu, S.; Wu, B.; An, Q. Transient thermal analysis of grease-lubricated spherical roller bearings. *Tribol. Int.* **2016**, *93*, 115–123. [CrossRef]
13. Xiu, S.; Xiu, P.; Gao, S. Simulation of Temperature Field of Oil Film in SuperHigh Speed Hybrid Journal Bearing Based on FLUENT. *Adv. Mater. Res.* **2009**, *69–70*, 296–300. [CrossRef]
14. Wang, B.M.; Sun, C.; Xiao, C.; Liu, H. Numerical Analysis of Flow Characteristics of Grease in High-speed Angular Contact Ball Bearing. *Lubr. Eng.* **2018**, *43*, 109–115+131.
15. Zhang, S.; Lin, X.; Fan, C.; Su, G.; Lu, Y. Study on Structural Optimization and Temperature Characteristics of Hydrostatic Bearing Based on Fluent. *Hydraul. Pneum. Seals* **2019**, *39*, 21–24.
16. Palmgren, A. *Ball and Roller Bearing Engineering*; SKF Industries: Philadelphia, PA, USA, 1945.

17. Deng, K.; Xie, P.; Liao, H.; Zhou, G.; Deng, S. Mechanism of Abnormal Fluctuation of Friction Torque of Control Moment Gyro Bearing Assembly. *J. Aerosp. Power* **2023**, *38*, 752–768.
18. Deng, S.; Li, X.; Wang, J.; Wang, Y.; Teng, H. Analysis on Friction Torque Fluctuation of Angular Contact Ball Bearings. *J. Mech. Eng.* **2011**, *47*, 104–112. [CrossRef]
19. Deng, S.; Hua, X.; Zhang, W. Analysis on Friction Torque Fluctuation of Angular Contact Ball Bearing in Gyro Motor. *J. Aerosp. Power* **2018**, *33*, 1713–1724.
20. Harris, T.A.; Kotzalas, M.N. *Advanced Concepts of Bearing Technology*; CRC Press: Boca Raton, FL, USA, 2007.
21. Ai, S.; Wang, W.; Wang, Y.; Zhao, Z. Temperature rise of double-row tapered roller bearings analyzed with the thermal network method. *Tribol. Int.* **2015**, *87*, 11–22. [CrossRef]
22. Takabi, J.; Khonsari, M.M. Experimental testing and thermal analysis of ball bearings. *Tribol. Int.* **2013**, *60*, 93–103. [CrossRef]
23. Takabi, J.; Khonsari, M.M. On the thermally-induced failure of rolling element bearings. *Tribol. Int.* **2016**, *94*, 661–674. [CrossRef]
24. Zheng, D.; Chen, W.; Li, M. An optimized thermal network model to estimate thermal performances on a pair of angular contact ball bearings under oil-air lubrication. *Appl. Therm. Eng.* **2018**, *131*, 328–339.
25. Neurouth, A.; Changenet, C.; Ville, F.; Arnaudon, A. Thermal modeling of a grease lubricated thrust ball bearing. *Proc. Inst. Mech. Eng. Part J J. Eng. Tribol.* **2014**, *228*, 1266–1275. [CrossRef]
26. Zheng, D.; Chen, W. Thermal performances on angular contact ball bearing of high-speed spindle considering structural constraints under oil-air lubrication. *Tribol. Int.* **2017**, *109*, 593–601. [CrossRef]
27. Pouly, F.; Changenet, C.; Ville, F.; Velex, P.; Damiens, B. Investigations on the power losses and thermal behaviour of rolling element bearings. *Proc. Inst. Mech. Eng. Part J J. Eng. Tribol.* **2009**, *224*, 925–933. [CrossRef]
28. Sadeghi, F.; Sui, P.C. Thermal Elastohydrodynamic Lubrication of Rolling/Sliding Contacts. *J. Tribol.* **1990**, *112*, 189–195. [CrossRef]
29. Peterson, W.; Russell, T.; Sadeghi, F.; Berhan, M.T.; Stacke, L.E.; Ståhl, J. A CFD investigation of lubricant flow in deep groove ball bearings. *Tribol. Int.* **2020**, *154*, 106735. [CrossRef]
30. Lei, J.; Su, B.; Zhang, S.; Yang, H.; Cui, Y. Dynamics-Based Thermal Analysis of High-Speed Angular Contact Ball Bearings with Under-Race Lubrication. *Machines* **2023**, *11*, 691. [CrossRef]
31. Crecelius, W.J.; Pirvics, J. *Computer Program Operation Manual on SHABERTH. A Computer Program for the Analysis of the Steady State and Transient Thermal Performance of Shaft-Bearing Systems*; US Air Force Technical Report AFAPL-TR-76-90; Defense Technical Information Center: Fort Belvoir, VA, USA, 1976.

Disclaimer/Publisher's Note: The statements, opinions and data contained in all publications are solely those of the individual author(s) and contributor(s) and not of MDPI and/or the editor(s). MDPI and/or the editor(s) disclaim responsibility for any injury to people or property resulting from any ideas, methods, instructions or products referred to in the content.

Article

WLI, XPS and SEM/FIB/EDS Surface Characterization of an Electrically Fluted Bearing Raceway

Omid Safdarzadeh [1,*], Alireza Farahi [2], Andreas Binder [1], Hikmet Sezen [3] and Jan Philipp Hofmann [3]

[1] Institute for Electrical Energy Conversion, Technical University of Darmstadt, 64283 Darmstadt, Germany; abinder@ew.tu-darmstadt.de
[2] Faculty of Electrical Engineering, K. N. Toosi University of Technology, Tehran 1631714191, Iran; a.farahi@email.kntu.ac.ir
[3] Surface Science Laboratory, Department of Materials and Earth Sciences, Technical University of Darmstadt, Otto-Berndt-Strasse 3, 64287 Darmstadt, Germany; hsezen@surface.tu-darmstadt.de (H.S.); hofmann@surface.tu-darmstadt.de (J.P.H.)
* Correspondence: osafdarzadeh@ew.tu-darmstadt.de

Abstract: Electrical bearing currents may disturb the performance of the bearings via electro-corrosion if they surpass a limit of ca. 0.1 to 0.3 A/mm^2. A continuous current flow, or, after a longer time span, an alternating current or a repeating impulse-like current, damages the raceway surface, leading in many cases to a fluting pattern on the raceway. Increased bearing vibration, audible noise, and decreased bearing lubrication as a result may demand a replacement of the bearings. Here, an electrically corroded axial ball bearing (type 51208) with fluting patterns is investigated. The bearing was lubricated with grease lubrication and was exposed to 4 A DC current flow. It is shown that the electric current flow causes higher concentrations of iron oxides and iron carbides on the bearing raceway surface together with increased surface roughness, leading to a mixed lubrication also at elevated bearing speeds up to 1500 rpm. The "electrically insulating" iron oxide layer and the "mechanically hard" iron carbide layer on the bearing steel are analysed by WLI, XPS, SEM, and EDS. White Light Interferometry (WLI) is used to provide an accurate measurement of the surface topography and roughness. X-ray Photoelectron Spectroscopy (XPS) measurements are conducted to analyze the chemical surface composition and oxidation states. Scanning Electron Microscopy (SEM) is applied for high-resolution imaging of the surface morphology, while the Focused Ion Beam (FIB) is used to cut a trench into the bearing surface to inspect the surface layers. With the Energy Dispersive X-ray spectrometry (EDS), the presence of composing elements is identified, determining their relative concentrations. The electrically-caused iron oxide and iron carbide may develop periodically along the raceway due to the perpendicular vibrations of the rolling ball on the raceway, leading gradually to the fluting pattern. Still, a simulation of this vibration-induced fluting-generation process from the start with the first surface craters—of the molten local contact spots—to the final fluting pattern is missing.

Keywords: bearing surface damage; electric bearing currents; rotor-to-ground current; bearing voltage; electrical wear; electrically damaged bearing; electro-corrosion; bearing oil degradation; bearing electrical failure

Citation: Safdarzadeh, O.; Farahi, A.; Binder, A.; Sezen, H.; Hofmann, J.P. WLI, XPS and SEM/FIB/EDS Surface Characterization of an Electrically Fluted Bearing Raceway. *Lubricants* **2024**, *12*, 148. https://doi.org/10.3390/lubricants12050148

Received: 20 February 2024
Revised: 15 April 2024
Accepted: 21 April 2024
Published: 27 April 2024

Copyright: © 2024 by the authors. Licensee MDPI, Basel, Switzerland. This article is an open access article distributed under the terms and conditions of the Creative Commons Attribution (CC BY) license (https://creativecommons.org/licenses/by/4.0/).

1. Introduction: Electrical Bearing Currents and Bearing Surface Damage

Inverter-driven variable-speed AC electrical machines, mainly three-phase synchronous or induction machines, are used in many industrial applications–in electric and hybrid cars, trains, street cars, ships, aircrafts, household appliances etc.–at different power and speed levels. Fast-switching power electronic devices, based on Silicon or Silicon-Carbide-semiconductor transistors, supply the torque-generating currents to the machine's three-phase stator windings with a fundamental frequency typically up to 1 kHz. In addition, a fluctuating electrical potential of the stator winding with respect to the electrically grounded base of the electrical machine occurs with the switching frequency, typically at several kHz, called common mode (CM) voltage [1]. This CM voltage causes parasitic high-frequency (HF) currents of small amplitude from the motor electrical terminals to the electric ground of the drive system [2], which flow through the conductive and capacitive motor components, e.g., motor bearings [3,4]. These currents may also indirectly induce a current flow in the bearings magnetically. These HF bearing currents may cause a corrugation of the bearing raceway surface, causing increased audible noise and motor vibrations, a deterioration of the bearing lubricant, increased bearing friction losses and bearing heating, and, in the worst-case, a bearing failure [5–7], e.g., via a cage mechanical break. Different bearing current effects are described in the literature, e.g., [8].

First, with fully lubricated bearings, apart from harmless capacitive HF currents via the electrically insulating lubrication film (e.g., film thickness h = 0.2 µm), considerable discharge currents may occur in the lubrication film [9] if the electrical voltage at the lubricant film surpasses its breakdown voltage (e.g., U = 6 V). The corresponding electrical field strength $E = U/h$ (e.g., 30 kV/mm) is ionizing the lubricant molecules [10–14], leading to short sparks of durations typically around 1 µs. The resulting EDM bearing current (EDM: Electric Discharge Machining [1]) oscillates due to the inductive and capacitive machine components in the MHz-range. These electric currents discharge the parasitic capacitances of the electrical machine [1,15]. The endangerment of the bearing currents is assessed with "apparent" bearing current densities $J = I/A_{Hz}$, where I is the amplitude of the current and A_{Hz} is the bearing *Hertz*'ian area [1]. If J < 0.3 A/mm², the corresponding craters, with diameters of typically around 1 µm on the bearing raceway surface, are sufficiently flattened by the roller elements [1,16]. If J > 0.3 A/mm², for EDM bearing currents, the exposed surface shows a grey trace (called "grey frosting" [2,15]), which may still allow a further bearing operation [17].

Second, the HF CM winding-stator currents excite an additional HF magnetic field inside the machine, which, due to *Faraday*'s law, induces a voltage on a loop path, composing both bearings, rotor shaft, and stator iron parts, resulting in so-called circular bearing currents [1]. These currents occur at both low and medium speeds in AC machines of frame sizes ca. above 200 mm [1,15]. Compared to the EDM bearing currents, the circular bearing currents happen more frequently, have longer durations, higher amplitudes, and lower frequencies, and are more dangerous for the bearings [15]. Similar to the circular bearing currents, rotor-to-ground bearing currents may endanger the bearings, even in machines with smaller frame sizes below 200 mm, when the rotor is electrically connected to the ground with a low HF impedance, e.g., via a gear box or a milling device [1]. Any bearing current between the ball and the raceway is limited through the so-called a-spots, due to the lubrication film and the covering oxide layer. The very high local current densities at the a-spots result in very high local temperatures. These temperatures degrade the lubricant by molecule dissociation, release carbon, thus blackening the oil and reducing the lubrication effect. The effect of the bearing current flow may gradually develop into the fluting pattern [1,18,19].

The cause-and-effect chain of the generation of the fluting pattern has been under investigation for a long time [3–8,18–21]. A theoretically established detailed physical model, proven by simulation, is currently missing. The following mechanism is obviously involved. Due the surface roughness, the real mechanical contact area between the ball and raceway is smaller than the calculated *Hertz*'ian area A [22]. The elastic and plastic deformations of the asperity peaks within the *Hertz*'ian area [23] provide the real mechanical contact area. The increased surface roughness due the current flow leads to micro-Elasto-Hydrodynamic Lubrication (micro-EHL), where the oil is pressed into the surface valleys [24]. Thus, for a rough surface, asperity peaks may still have electrical contacts (a-spots) [25], even when bearing parts are dominantly separated via a lubrication film [26]. At DC bearing currents, the bearing voltage drop between inner and outer bearing rings remains constant around 1 V, independent of the amplitude of bearing current [27]. This voltage is sufficient to cause very high local contact temperatures above 1000 °C on moving contact partners in a very short time (below 1 µs). This leads to melting of the micro-size volume of the *a*-spot with a typical radius of 1 µs [25], allowing the current flow. Similar local temperatures occur at inverter-fed motors at impulse bearing current flow. Catalytic effects of the metal-oxide debris promote the oxidation process in the lubricant [28] and lead to an increased lubricant conductivity. Due to this current flow, a build-up of an oxide layer on the bearing raceway surface occurs [20]. In electrical steel contacts, it is well established that the oxide surface layers are formed due to the sliding of the asperities, causing a significant amount of Fe^{2+} and Fe^{3+} compared to the bulk steel [29,30]. These oxides form an insulating layer between two mechanically touching bodies, constraining the current flow to the tiny *a*-spots areas [31], which are only very short in contact due to the bearing movement.

For fluting generation, the four following conditions are necessary: (1) a relative movement between metallic contact bodies, (2) a lubrication of the contact, (3) an electrical current flow, and (4) an electrically insulating surface (oxide) layer. With a single-contact rolling steel (ball-to-ball set-up, [32]) it was examined whether without lubrication, the current flow leads to a harsh mechanical wear; however, no fluting pattern occurs, so lubrication is necessary to produce fluting. At stand still, also without a lubrication, a considerable electrical current may flow without any harmful surface deterioration for a significant amount of time, so relative motion is necessary for the fluting [20]. Without an electrical current, no fluting occurs. With current density amplitudes ($J > 0.3$ A/mm^2) and with sufficient time of current flow, the fluting occurs. The fluting generation with DC currents is most effective. A fluting pattern were produced with the single-contact set-up in [32], so the number of balls in a bearing is not necessarily an explanation for the periodic fluting pattern. The oxide layer itself is at a long term operation in a certain "equilibrium" thickness, which we proved by experiment: (i) With acid-cleaned contact partners, the voltage drop at single-contact set-up in [32] was, at the beginning, close to 0 V, but after a few tens of seconds of rotation, this voltage raised to ca. 0.4 V, corresponding to the steel melting voltage, proving that the oxide layer has grown again to a stable equilibrium thickness. (ii) The steel balls of single-contact set-up in [32] were heated to 600 °C to grow a thick oxide layer. Then, at the beginning of rotation, the contact voltage drop was close to 2 V, but decreased after some tens of minutes rotation to ca. 0.6 V, indicating that the oxide layer thickness was decreased again.

We believe that the intact oxide layer is damaged by the *a*-spots at the beginning of the fluting generation due to the current flow. By the rims of these spots, the rolling ball masses obtain force impulses, perpendicular to the moving direction, that cause the masses to vibrate within the elastic lubrication film perpendicular to the raceway surface. These small vibrations vary the film thickness periodically. Thus, the electrical current flow may affect the raceway periodically, e.g., via surface oxidation and carburization. The purpose of this paper is to explore the effect of electrical current flow on topography and composition changes of the bearing raceway, comparing a damaged region to a non-damaged region, in one of the fluted bearings, reported in [19]. The findings give insights on effects of the electric current flow on the raceway surface and shall contribute to understand the fluting-generation mechanism [18,19].

2. Method, Setup and Sample

In the following section, after introducing the setup and the sample, White Light Interferometry (WLI) is used to measure the surface topography and roughness of the fluted bearing raceway. X-ray Photoelectron Spectroscopy (XPS) analyzes the chemical composition and oxidation states of the bearing surface. Scanning Electron Microscopy (SEM) is applied for high-resolution imaging of surface morphology, while Focused Ion Beam (FIB) is used to manipulate the bearing material and to inspect the surface layers. Finally, via Energy Dispersive X-ray spectrometry (EDS), the presence of composing elements in the surface layer is identified, determining their relative concentrations.

2.1. Experimental Setup

A setup composed of two axial ball bearings was produced (Figure 1a), where the bearing force F_b, the bearing speed n, and the momentary bearing current i_b can be adjusted. An axial ball bearing has been used in order to ensure the same force on each ball element and to dismantle the bearing easily for inspections. For transferring the electrical current from the rotating parts to the stationary parts without using mechanical brushes, two axial bearings are used in series (Figure 1b). Hence, no extra contact resistance, e.g., of a contacting sliding silver brush, is added to the path of electric measurements.

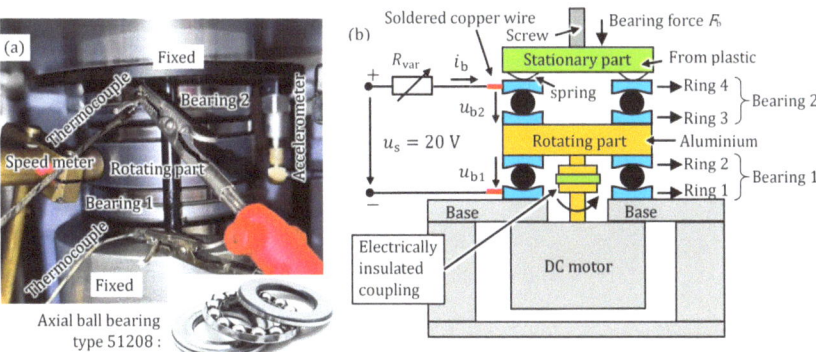

Figure 1. Test setup with two axial ball bearings in series. (**a**) Experimental test rig, (**b**) Schematic structure. The equipment is described in Table 1.

The feeding DC electric circuit is connected to the bearings via two copper wires, soldered to the stationary rings 1 and 4 of the two tested axial bearings (Figure 1b). This bearing setup is connected via R_{var} to a DC source voltage $u_s = 20$ V. The bearing voltage u_b between inner and outer bearing rings remains, as explained in Section 1, almost constant at around 1 V (e.g., in Figure 2), showing fluctuations due to the changing number of a-spots (e.g., in Figure 3). Based on (2), to set a constant current i_b via R_{var}, $u_s \gg 2 \cdot u_b$ must be observed. Hence $u_s = 20$ V is a proper choice. The bearing voltage drop u_b is measured directly via the copper wires, soldered to the rings 1 and 4. The resistivity of the bearing steel and of the aluminium in the rotating part is very small. As a result, the voltage drops at the stationary contact, i.e., between ring 3 and the rotating aluminium part and between the rotating aluminium part and ring 2, are negligible compared to the voltage drop at the rolling contacts, i.e., between the balls and the raceways. A screw and a nut adjust the bearing force F_b on a spring ring, placed mechanically in series with the bearings, which is measured via a load cell. Thus, a bearing force F_b up to a maximum of 1600 N is applied on both in series mounted bearings. The rotating rings of the bearings are placed on the rotating part (Figure 1b), which is connected to the shaft of a driving variable-speed DC motor via an electrically insulated coupling. Ring 4 is electrically insulated from the screw, so the applied current only passes through the bearings. All four rings 1–4 are centred mechanically to the rotating axis (not shown in Figure 1b for simplicity). The temperatures of ring 1 (ϑ_{r1}) and of ring 4 (ϑ_{r4}) are directly measured via thermocouples.

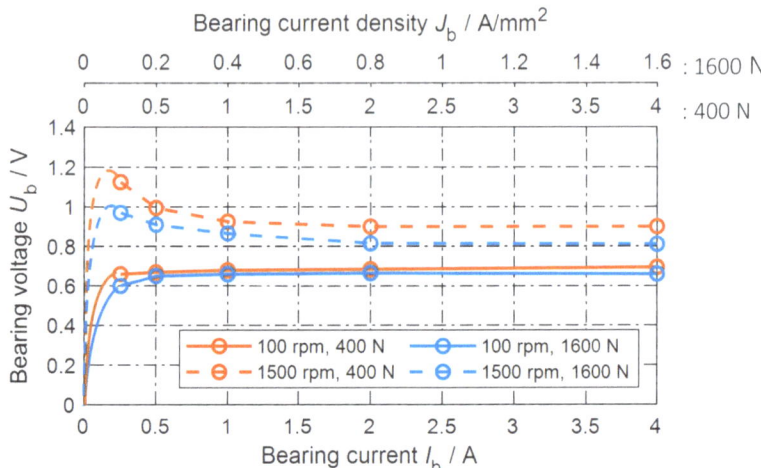

Figure 2. Measured bearing voltage U_b per bearing versus measured DC bearing current I_b and versus apparent bearing current density J_b, for two bearing speeds n = 100 rpm, 1500 rpm and two bearing forces F_b = 400 N, 1600 N.

Table 1. Bearing lubricant and instrumentation for the test-rig and surface measurements.

Term	Description
Bearing lubricant	Mineral-oil-based grease *Arcanol MUITI3* (company *Schaeffler*, Herzogenaurach, Germany). Thickener: Lithium soap. Base oil viscosity at 40 °C: 110 mm^2/s, at 100 °C: 12 mm^2/s. Operating temperature -20 °C to 120 °C.
Axial bearing	Type 51208, 15 balls, inner diameter 40 mm, outer diameter 68 mm, width 19 mm (company *Schaeffler*). Static load capacity 97 kN, dynamic load capacity 44 kN.
Digital oscilloscope	*Waverunner LT364L*, 500 MHz, accuracy 2 mV, vertical resolution 8 bits, 1 M samples capture memory.
Current clamp	*IWATSU SS-250*, 100 MHz, max. 30 A, accuracy $\pm 1.0\%$ or ± 10 mA.
Voltage probe	*TT-SI200*, 200 MHz, ± 60 V, accuracy $\pm 1.0\%$ or ± 20 mV.
WLI profile meter	*SmartWLI-compact* (company *gbs*) with *Nikon* objective lens 50X/0.55.
XPS instrument and *Argon* sputter gun	*Escalab 250* (company *Thermo Fisher*, East Grinstead, UK), with monochromic Al Kα X-ray source ($h\nu$ = 1486.7 eV), 650 µm spot size, 120 W excitation. Chamber pressure $\leq 5 \times 10^{-10}$ mbar. High-resolution pass energy 20 eV. Survey spectra pass energy 50 eV. 3 kV Ar$^+$ sputtering beam with raster area 1×1 mm^2.
SEM [1]/FIB [2]/EDS [3] instrument 1	*JEOL JIB 4600F*, with a dual-beam FIB, with EDX detector *INCA Energy 350* (company *Oxford instruments*, Abingdon, UK), at TU *Darmstadt*.
SEM/EDS instrument 2	Field emission scanning electron microscopy (FESEM), *Quanta 450 FEG* (company *FEI*, Lausanne, Switzerland), with *QUANTAX* EBSD/EDS (company *Bruker*, Billerica, MA, USA), at TU *Isfahan*.
Sandpaper	Emery paper, grit P400, grains of aluminum oxide (company *Smirdex*).
Polishing machine	*Saphir 330* (company *ATM*), polishing with SiC paper (company *Cloeren*) with the following grit numbers: P120, P220, P320, P500, P1200, P500, P4000, each for approx. 2 min.

[1] SEM: Scanning electron microscopy, [2] FIB: Focused ion beam, [3] EDS: Energy dispersive spectroscopy.

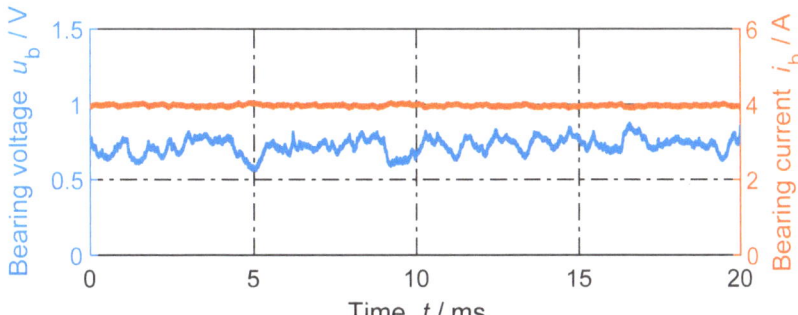

Figure 3. Example of measured DC bearing voltage u_b and DC bearing current i_b for n = 100 rpm, F_b = 400 N, and ϑ_b = 33 °C. Average bearing voltage U_b = 0.73 V. Grease lubricant *Arcanol MULTI3*.

The measured non-linear voltage-current characteristic at DC current supply, with grease lubricant *Arcanol MULTI3*, is given in Figure 2. Beforehand, u_{b1} and u_{b2} were measured separately in a similar set-up with a low-resistance mercury slip ring, showing $u_{b1} \approx u_{b2}$. Hence, we simplified the measured bearing voltage u_b as an average (1):

$$u_b = \frac{u_{b1} + u_{b2}}{2}, \qquad (1)$$

where u_{b1} and u_{b2} are the bearing voltage drops, as shown in Figure 1b. To determine the apparent bearing current density $J_b = I_b / A_{Hz}$, the calculated value of the *Hertz*'ian areas $A_{Hz,400N}$ = 1.0 mm² and $A_{Hz,1600N}$ = 2.5 mm² [33] are used. An external fan heater controls the average bearing temperatures 31 °C $\leq \vartheta_b \leq$ 35 °C. The measured bearing voltage U_b and current I_b are time-averaged over 2 s, e.g., $U_b = \left(\int_0^{2s} u_b(t) dt \right)/2s$. In the following, the attribute "time-averaged" will be skipped. The influence of the (time-averaged) DC bearing current I_b, of the apparent bearing current density J_b, of the bearing axial force F_b, and of the bearing rotational speed n on the (time-averaged) bearing voltage U_b is shown in Figure 2.

With an increasing bearing current I_b in Figure 2, the number of contact *a*-spots increases (A-fritting effect), and the existing contact *a*-spots enlarge (B-fritting effect) [27,31]. Therefore, the contact resistance decreases, resulting in a nearly constant bearing voltage U_b at higher DC bearing currents $I_b > 1$ A. With DC currents, an almost constant bearing voltage occurs, similar to the range 0.5 V $< U_b <$ 1.5 V, which has been reported previously [16,21,34,35]. At an increased speed, the greater lubrication film thickness h results in a sligthly increased bearing voltage U_b (Figure 2).

Since $u_b \approx 1V \approx$ const. $<< u_s = 20$ V DC = const., via R_{var} in Figure 1b, nearly a constant value for the momentary bearing current i_b is set:

$$i_b = \frac{u_s - 2 \cdot u_b}{R_{var}}. \qquad (2)$$

For instance, in Figure 3, the measured momentary bearing current i_b ranges within 3.88 A $< i_b <$ 4.09 A, which is considered practically a constant value at $I_b \approx 4$ A. The repeated electrical breakdown (fritting effect [27]) of the surface oxide layer and of the tribofilm covering the bearing steel surface results in fluctuations of the momentary bearing voltage u_b around the average bearing voltage U_b, while i_b flows via many contacting *a*-spots of a varying number.

2.2. Bearing Sample for Surface Investigations

A variety of different test conditions were examined for the fluting investigation [19]. Each test, with its individual test conditions, was performed with new bearings, giving each test an individual name, i.e., A14, see Table 1. A section of ring 1, see Figure 1b, from test A14 was used as test specimen for the investigations and results in Sections 3–5. Figure 4 shows the raceway of this ring 1. Table 2 lists the corresponding conditions of test A14. In the following discussions of Sections 3–5, when a damaged region is reported, an area inside the fluting valley of the fluting pattern of Figure 4 is meant, otherwise it is mentioned explicitly.

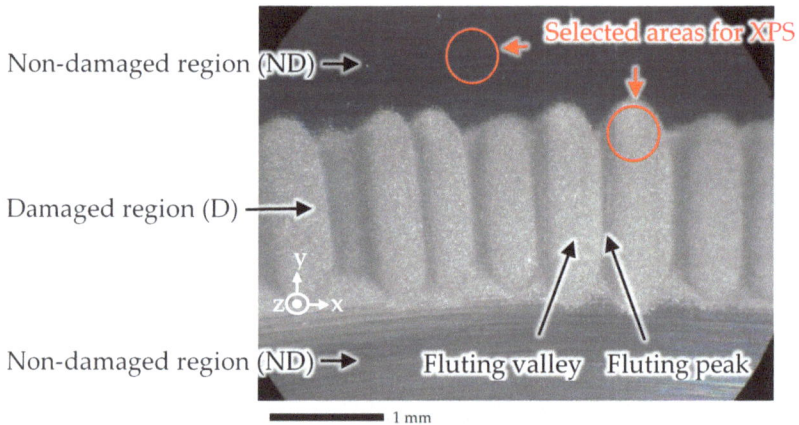

Figure 4. Photographed example of the raceway surface of ring 1 from test A14 via a light microscope. Conditions of test A14: See Table 2. "Damaged region (D)" represents the fluted raceway area. "Non-damaged region (ND)" represents the area outside the fluted raceway. Selected areas for the XPS measurements are shown with red circles.

Table 2. Conditions of test A14 with the test setup in Figure 1.

Bearing current I_b	4 A, DC
Ring 1: Electrical polarity	Negative
Axial bearing force F_b	400 N
Calculated bearing *Hertz*'ian area A_{Hz}	1.04 mm^2
Apparent bearing current density J_b	3.85 A/mm^2
Bearing speed n	1500 rpm
Grease lubricant	Arcanol MULTI3
Ring 1 temperature ϑ_{r1}	44.6 °C
Test duration	24 h

The investigated ring 1 was too big to be mounted on the sample holder of the used surface-inspecting instruments in Sections 3–5. Hence, the ring was cut into smaller pieces, with the longest dimension shorter than 18 mm, via an electric discharge machining (EDM) wire-cut process. The wire-cutting avoids a high temperature on the sample. It is one of the most precise and low-damaging techniques to cut hard steel in order to investigate the cross-section of the sample in Section 5. Prior to each of the following measurements, the sample is first cleaned with a cotton cloth, then with ethanol and deionized water, and finally blow-dried with an air flow.

3. WLI Results

White light interferometry (WLI) is a non-contact optical method for measuring surface heights, varying between tens of nanometers and a few centimetres. The damaged raceway surface of test A14 is measured via the WLI profilometer in Table 1 with a magnification of 240, a resolution of 0.19 μm in x- and y-direction, and less than 2 nm in z-direction.

Figure 5a shows a pseudo-colored view of the measured surface after removing the "fundamental" form deviation, described by 3rd order polynomial. Based on the rules in ISO 25178 [36], used to calculate the surface parameters of Figure 5a, the *Gaussian* L-filter (ISO 16610-61 [37]) with a cut-off wavelength of $\lambda_c = 25$ μm is applied to remove Large scale height components, thus eliminating a waviness with wavelengths $\lambda > \lambda_c$. Then, the *Gaussian* S-filter (ISO 16610-61 [37]), with a cut-off wavelength of $\lambda_s = 0.25$ μm, is applied to remove Small scale height components, thus eliminating the measurement noise. The selected cut-off wavelengths λ_c, λ_s are suitable for the shown roughness in Figure 5a, since λ_c and λ_s are, respectively, bigger and smaller than the typical a-spot diameters of 1 μm. The resulting surface parameters in given in Figure 5d. The histogram in Figure 5b shows the distribution of heights of the 1311 × 787 data points in Figure 5a with a bin width of 0.01 μm. The *Abbott-Firestone* curve [38] in Figure 5b gives the cumulated curve of the height distribution. It is used to calculate the functional parameters in Figure 5d. The functional parameters for volumes are calculated for the material ratios $S_{mr1} = 10\%$ and $S_{mr2} = 80\%$ [36]. The measured roughness has 10% of the measured heights above $c_1 = 0.28$ μm and 20% by $c_2 = -0.22$ μm below the mean plane. The highest roughness peak is 1.8 μm above, and the lowest about -3.3 μm below the mean plane. The measured rms surface roughness in Figure 5a is $S_q = 0.31$ μm.

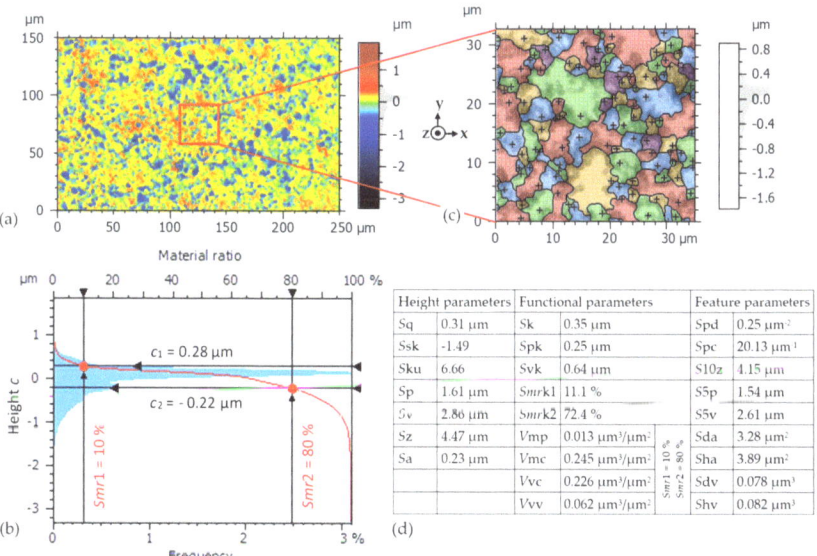

Figure 5. Test A14: Measured surface of a 250 μm × 150 μm area within a fluting valley of the fluted bearing raceway of Figure 4 (resolution in $x-$and $y-$direction 0.19 μm, in $z-$directions < 2 nm), after removing the third-degree polynomial form. (**a**) Pseudo-colored surface view. (**b**) Histogram of 1311 × 787 measured points and corresponding *Abbott-firestone* curve. (**c**) *Watershed* segmentation of motifs for a part of (**a**) with *Wolf* pruning (height criterium 5% S_z). (**d**) Determined surface parameters of (**a**), using a *Gaussian* L−filter ($\lambda_c = 25$ μm) and S-filter ($\lambda_s = 0.25$ μm) according to ISO 25178.

For these high surface peaks in Figure 5a, it is likely that the lubrication film is penetrated, causing local mechanical and electrical contacts within the encountered surface of the *Hertz'*ian area. The relatively deep valleys in Figure 5a may entrap the burnt lubricant, namely the carbon particles. The lubrication film thickness h = 0.62 μm is calculated based on the measured bearing force, bearing speed, and lubrication viscosity according to [33,39,40]. We assumed equal rms surface roughness values S_q = 0.31 μm for the raceway and the balls, resulting in a composite rms surface roughness $\sigma = \sqrt{0.31 \ \mu m^2 + 0.31 \ \mu m^2} = 0.44$ μm. Hence, the calculated lambda value $\Lambda = h/\sigma = 1.41 < 3$ indicates a "boundary lubrication state" even at the elevated speed n = 1500 rpm as a result of the big surface roughness. The reason for this big roughness is the exposition of the bearing to the high bearing current density 3.85 A/mm^2 (factor 10 above the admissible limit) during test A14, resulting in electrical asperity contacts even at the higher speed level 1500 rpm.

To characterize a surface, it is helpful to identify the repeating patterns, called "motifs" [41]. A part of the surface in Figure 5a is segmented into motifs in Figure 5c with the *watershed* algorithm [42], using *Wolf* pruning with a height criterium 5% S_z [42]. This height criterium of 5% S_z for the pruning is appropriately selected, since the resulting motif dimensions in Figure 5c are in the range of some micrometers, which is the same range of the roughness in lateral direction due to the *a*-spots in Figure 5a. Figure 5c does not show any repeating pattern because the *a*-spots caused an irregular roughness. So, the surface in Figure 5a is described better with the height and functional parameters of ISO 25178 in Figure 5d, introduced for a general surface, rather than with feature parameters describing the motifs.

4. XPS Results

In the following, the composition and variation of iron oxide phases of the raceway surface are studied by depth profiling with successive Ar$^+$-ion sputtering at 3 kV with an investigated raster area of 1×1 mm^2, using the XPS instrument from Table 1. The energy calibration of the XPS instrument was performed via measurements of the Cu 2p, Ag 3d, and Au 4f core levels and the Ag *Fermi* edge. XPS with Ar$^+$-sputtering depth profiling is a reliable tool for determining surface sensitive (<10 nm) elemental compositions. EDS and XRD are less suitable for the surface composition analysis, as EDS has a rather high penetration depth of the electron beam and of the photon emission (~10 μm), resulting in low sensitivity for thin surface layers such as of oxides, hydroxide, and carbon derivatives. XRD cannot distinguish between different iron oxide phases in the layer (e.g., Fe_2O_3 and Fe_3O_4) due to their similar lattice constants [43]. XPS [44] uses a monochromatic Al-Kα-X-ray beam with a photon energy $E_{photon} = h\nu$ = 1486.7 eV (h: *Planck*'s constant, ν: wave frequency) focusing on the sample. The kinetic energies $E_{kinetic}$ of the emitted electrons from the sample surface are measured via an electron energy analyzer, separating the electrons within a given kinetic energy band from the rest of emitted electrons. An electron detector counts these separated electrons. The electron binding energy $E_{binding}$ of each of the emitted electrons is determined by using the photoelectric effect equation [44]

$$E_{binding} = E_{photon} - (E_{kinetic} + \varphi), \tag{3}$$

where φ is a work function-like term for the specific surface of the material and instrument. It is considered as an adjustable instrumental factor for energy calibration. For each element and its chemical states, there are specific discrete energies between the electron levels and hence an individual electron population spectrum of the emitted electrons. Therefore, the intensity CPS (counted electrons per second) at each binding energy band depends on the sample and has specific peaks. Not all the electrons generated by the X-ray irradiated volume escape to the electron analyser without any collision with other atoms. Hence, the peaks in the measured spectra are superimposed on a background of inelastically scattered electrons on the high-binding energy side of each peak [44].

After monitoring the survey spectra, the high-resolution core-level photoemission spectra were collected for iron (Fe $2p_{3/2}$), as the most abundant element on the steel surface, along with oxygen (O 1s) and carbon (C 1s) as the expected constituents, corresponding to iron oxide/hydroxide, iron carbide phases, as well as organic contaminants from sample handling in air, respectively. A small Cr signal is observed especially in the non-damaged region, as well as very weak Cu and Zn signals as contaminations on the top surface, which are neglected here. The XPS data were processed using the *CasaXPS* software package (version 2.3.25) [45], in which the peak position values (Figure 6) were constrained within ±0.1 eV.

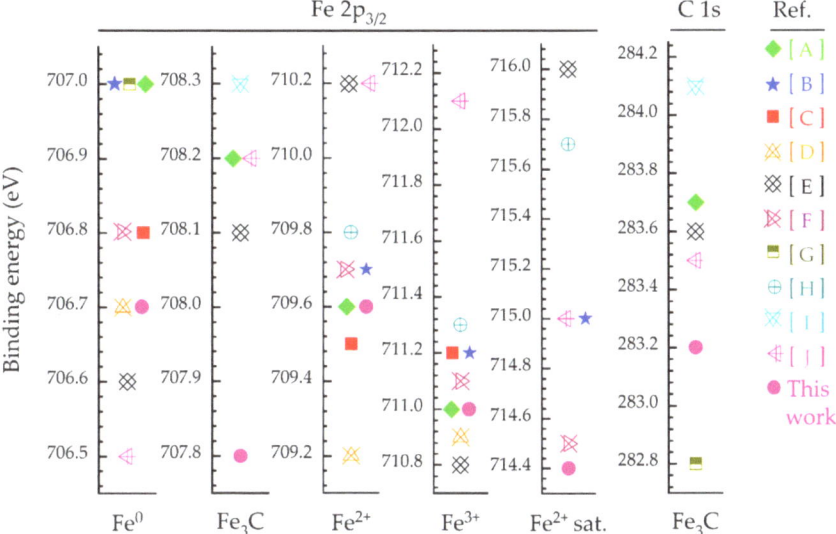

Figure 6. Peak positions of the Fe $2p_{3/2}$ and C 1s spectra used in this work, comparing to the literature [46–55], respectively [A–J].

XPS analysis of iron oxides encounters many difficulties stemming from the complexity of the Fe 2p spectra, e.g., due to the complex multiplet splitting and overlapping of peak positions, similar to other first-row transition metal compounds in the periodic table [56]. Thus, for the sake of simplicity, only broad peaks representing Fe^0, Fe^{2+}, and Fe^{3+} states are shown in Figure 7. The Fe $2p_{3/2}$ spectra of the damaged (D) and non-damaged (ND) regions can be divided into the following two categories:

(i) First in conductive phases, which belong to zero valent Fe^0, and iron carbide Fe_3C, appearing around 706.7 eV and 707.8 eV, respectively.
(ii) Second in low-conductive iron oxide phases at higher binding energies, corresponding to Fe^{2+}, Fe^{3+}, and Fe^{2+} satellite peaks at 709.6 eV, 711 eV, and 714.4 eV, respectively.

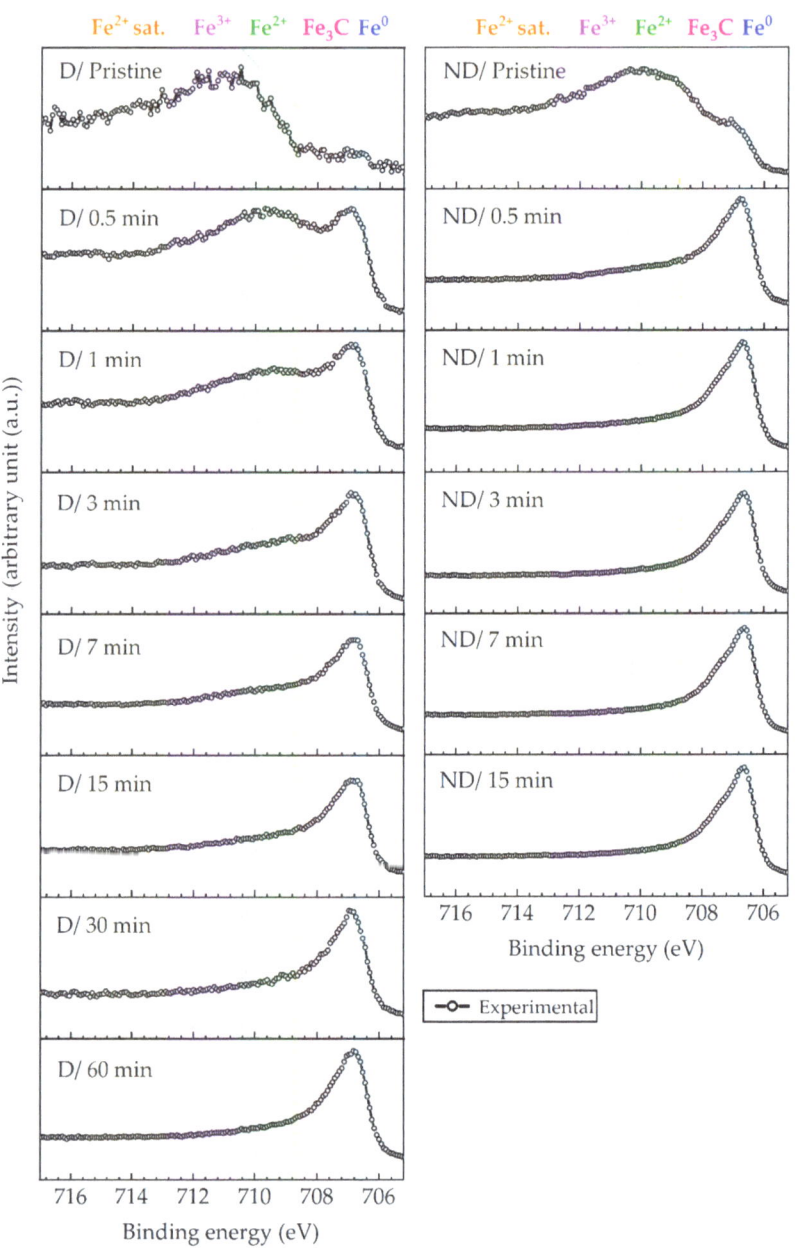

Figure 7. Fe $2p_{3/2}$ high resolution spectra of damaged D (**right panel**) and non-damaged ND (**left panel**) regions, as marked in Figure 4, as a function of sputtering time. The different chemical forms of Fe are specified with colored regions; see Figure 6 for reported peak positions.

Due to the asymmetric line-shape of Fe $2p_{3/2}$, the presence of Fe_3C in the C 1s spectra apparently confirms that iron atoms also exist in carbide form. The main peak positions for C 1s are listed in Figure 6. The components in the high-resolution spectra of Fe $2p_{3/2}$ (region marked) and C 1s (fitted) at the given sputtering times are shown in Figures 7 and 8, respectively.

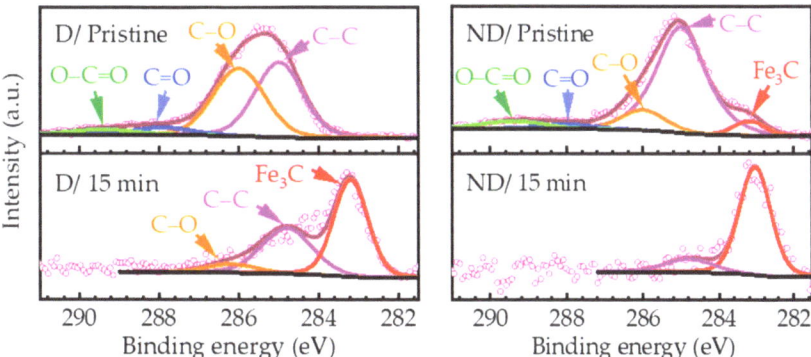

Figure 8. Peak fitting of C 1s high resolution spectra from damaged D (**left panel**) and non-damaged ND regions (**right panel**) in pristine state (0 min) and after 15 min sputtering. See Figure 6 and [46,50] for peak positions. Measured data points are shown with discrete open circles (a. u.: arbitrary unit).

The existence of each element can be recognised on the survey spectra, ranging from 0 eV to 1350 eV, having a step size 0.5 eV, by the received photoelectrons at specific binding energies. In our case, Fe, C, and O were recognized as the main elements in the survey spectra. In the next step, the high-resolution measurements, with a step size 0.075 eV, of the recognized elements at the following binding energy spans are performed: Fe (705.2 eV–716.9 eV) and C (281 eV–290 eV), as shown in Figures 7 and 8, respectively.

The background noise must be subtracted from the spectra before calculating the atomic percentages, obtained from the spectra [44]. Assuming one representative peak for one element, the atomic percentage X of an element g, among m elements in total, can be calculated from the high-resolution spectra [57]:

$$X_g = 100 \times \left(\frac{A_g}{S_g}\right) / \left(\sum_{k=1}^{m} \frac{A_k}{S_k}\right), \qquad (4)$$

where A is the measured peak area in CPS · eV, S is the sensitivity factor of the peak, and k is a counting number. The sensitivity factors S are calculated for each element, considering the ionization cross-sections, the inelastic mean free paths as element specific parameters, and the transmission function and the XPS measurement angle as instrument specific parameters [57]. The used *CasaXPS* program handles most of these adjustments automatically by reading metadata from the XPS data.

The atomic percentages of the total amount of Fe, O, and deconvoluted C peaks are shown in Figure 9. In the Fe $2p_{3/2}$ spectra, besides the peaks associated with metallic iron around 706.7 eV, other resolved peaks are related to iron oxide/carbide phases. For instance, in Figure 7 (left panel) for the damaged region D, before any sputtering (pristine state) there are measured high intensities around 711 eV, which represent iron oxide/oxyhydroxide phases. These high binding energy intensities gradually vanish with longer sputtering times. After 60 min of sputtering, almost no sign of iron oxide phases is seen in Figure 7 at D/60 min. In contrast to the D region, after a short sputtering time of the non-damaged region ND, a significant portion of the surface oxide and of the contaminants (e.g., C-C) are eliminated from the surface, as seen in the right panel of Figure 7. Even after the initial sputtering cycles, for example, after 1 min, a higher metallic iron content exists in the ND region, compared to the D region (Figure 7). Moreover, Figure 7 shows that the increase of metallic content and the decrease of the oxide content reach a saturated state.

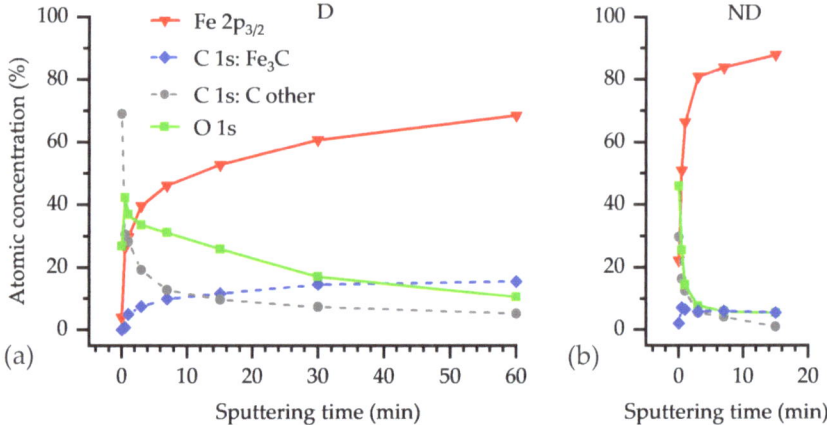

Figure 9. Measured atomic percentage of Fe (total amount), of C (deconvolved in carbide C (Fe$_3$C) and the rest of C), and of O (total amount) for (**a**) the damaged D and (**b**) non-damaged ND region, versus the sputtering time with values 0 min, 0.5 min, 1 min, 3 min, 7 min, 15 min, 30 min (only D), 60 min (only D).

The deconvoluted high-resolution C 1s spectra for D and ND regions at two sputtering times, (a) 0 min (pristine) and (b) 15 min, are displayed in Figure 8. It is revealed that there is a high concentration of iron carbide in the structure after etching the mostly organic carbon contaminations in both regions. Because of using arbitrary units (a. u.) for the peak intensity, the peak areas are only realistically comparable within an overlayered figure or are visible in Figure 9. For instance, the atomic percentage of the Fe$_3$C component of C 1s peak in Figure 8 ND/15 min is significantly smaller than that of Fe$_3$C in Figure 8 D/ 15 min (see Figure 9).

The surface elemental composition as a function of the Ar$^+$-ion sputtering time is plotted in Figure 9. For simplicity, Figure 9a,b shows the atomic percentage of the total amount of Fe, O, and the carbide form of C (Fe$_3$C) and the rest of C content from D and ND regions. A high concentration of organic contaminants (C–C, C–O, and C=O), labeled with "C other", exists at the outermost pristine surfaces for both D and ND regions. The organic contaminants are mostly associated with burnt lubricant in the bearing raceway. The concentration of the contaminants decreases sharply as we approach the bulk from the surface.

We assume that the ND region after 15 min of sputtering (Figure 9b) represents the composition of the bulk steel. Among 100 considered atoms in the ND region at 15 min, there are ca. 88 Fe atoms, from which 18 Fe atoms are in Fe$_3$C form (deduced from three times of carbidic C content). Hence, 18/88 = 20% of the Fe atoms are in Fe$_3$C form. The atomic percentages of Fe and C in the bulk steel 100Cr6 are 93.16% and 4.47%, respectively (see Table 3). Hence, the percentage of Fe atoms in Fe$_3$C form, divided by all Fe atoms in the bulk steel 100Cr6 yields 4.47 × 3/93.16 = 14.4%, which is close to the measured 20% in Figure 9b at 15 min. With further sputtering of the D region (after 60 min), it is expected to reach the same atomic percentages of Figure 9b at 15 min. This is not achieved after 60 min sputtering in Figure 9a on the D region. Among 100 considered atoms in the D region at 60 min, there are 69 Fe atoms, from which 45 Fe atoms are in Fe$_3$C form, giving 48/69 = 70% of the Fe atoms in Fe$_3$C form. It reveals how deep the iron carbide exist in the D region due to the high temperatures caused by the electric current flow. The mechanical sliding of the surface asperities, and probably also particles in the lubricant, ploughs the surface as a plastic abrasion in the *Hertz'*ian area. Hence, the formed oxides and carbides are also relocated in the depth of the raceway surface.

Table 3. Approximate atomic percentage (at%) and mass percentage (m%) of the main elements in the bulk bearing steel 100Cr6, data adapted from [58].

Element	at%	m%	Element	at%	m%
Fe	93.16	96.90	Mn	0.34	0.35
C	4.47	1.00	Cr	1.55	1.50
Si	0.48	0.25			

The bearing steel, having ca. 1% C (m%) in its composition, is already in the hyper eutectoid region of the Iron-Carbon phase diagram [59]. Hence there are hard pro-eutectoid Fe_3C grain boundaries between the pearlite grains. At high temperatures, the ferrite steel can enter the austenite, ledeburite, or even the liquid phase. The "hot" steel can solve more C into its structure from the abundant lubricant available at the ambient of the a-spot. During the rapid cooling down to the ambient temperature, the solved C atoms cannot diffuse out of the grains, resulting in the formation of martensite. As a result, a hard and brittle surface is formed with an increased carbide content [60].

In Figure 9, two substantial differences in the chemical composition of the D and ND regions are observed:

(i) The thickness of the oxide layer, formed on the surface of the D region, is significantly larger than that of the ND region.
(ii) Contrary to the ND region, where the concentration of Fe_3C near the surface closely resembles that of the bulk composition, the Fe_3C concentration in the D region is significantly higher (by 50%), even at depths far away the top surface.

The formation of both iron carbide and iron oxide are expected to be accelerated at elevated temperatures (typically over 1000 °C). The oxidation and carburization processes require O and C atoms, derived from the ambient air and from the lubricant, respectively. The thickness of the layers in the XPS depth profile analysis is not reported here due to the uncertainties in the etching rate of the Ar^+-beam. These uncertainties stem from the chemical composition of the surface and the slightly deviated direction of the Ar^+-beam, due to the magnetic properties of the sample.

Figure 10a shows the typical topography of the D and ND regions before Ar^+-sputtering. The images of the sputtered areas of the D and ND region are shown in Figure 10b,c, respectively. Although the sputtering time for the D region is four times longer than for the ND region, the signs of the electrical damage on the surface are still obvious. This agrees with the XPS measurements, which show that after 60 min sputtering in the D region, the content of Fe_3C is still 3.5-times higher than that of the ND region after 15 min.

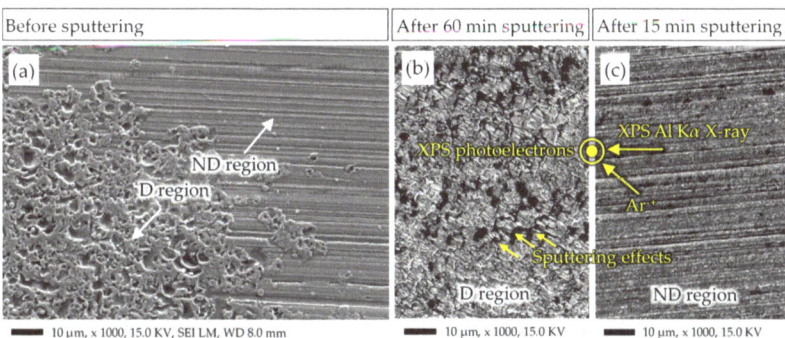

Figure 10. Bearing surface of Figure 4: SEM images of the raceway surface, shown in Figure 4 as "selected areas for XPS". (a) A border between the D and ND region before the Ar^+-sputtering. (b) The D region after 60 min sputtering. (c) The ND region after 15 min sputtering. Directions of the Al-Kα-X-ray beam, of the Ar^+-beam, and of the XPS-photoelectrons, accepted by the electron analyzer, for both (b,c).

The angle between the incident Ar^{+-}-beam and X-ray beam and the surface plane is 45° for each pair of them. The XPS analyzer is located at a normal angle to the sample surface (Figure 10). The effects of material sputtering in Figure 10b, corresponding to the Ar$^+$-beam gun direction, is visible. The 60 min Ar$^+$-sputtering in the D region has drastically changed the morphology of the sample surface. However, 15 min of Ar$^+$-sputtering in (Figure 10c) was not enough to eliminate or alter the surface morphology. The parallel lines in the ND region in Figure 10a are those due to the bearing manufacturing process, with a measured peak-to-valley distance of around several hundred nm.

5. SEM and EDS Results with the FIB Process

In scanning electron microscopy (SEM), a focused beam of electrons with an energy of e.g., E_0 = 15 keV sweeps the sample surface in a raster grid. The scattered electrons with different energies are counted via an electron detector for each point on the sample to determine the contrast of a corresponding pixel in the output image of the SEM. The incident electron from the electron beam either deflects its path due the elastic scattering via the electric field of the atomic nucleus and returns from the surface as "back-scattered electron BSE", or finally strikes an electron of an atom as inelastic scattering, releasing it from its atomic orbit as "secondary electron SE". Leaving the surface, the SE may be detected by the *Everhart–Thornley* electron detector (ETD) and contributes to building up the secondary electron image (SEI) [61]. The SE leaves a vacant position in the orbit of the atom. Other electrons of that atom, in outer orbit positions with higher energies, immediately fill the vacancy, thereby releasing the corresponding energy difference as X-ray photons. The X-ray, measured by an X-ray detector, builds up an X-ray spectrum for each sampling point. Unlike SE and BSE images, the characteristic X-rays give exact information of the composing elements in the energy-dispersive X-ray spectroscopy (EDS).

The information obtained from the sample is not limited to the diameter of the incident beam $d \approx 10$ nm [61]. It is gathered from a volume that is 100 to 1000 times bigger than $33 \cdot d^3$. This drop-shaped "interaction" volume differs in size for SEs, BSEs, and X-rays. The SEs have much lower energy levels < 50 eV with a distribution peak at ca. 2–5 eV and are emanated only from a shallow depth of the surface up to ca. 100 nm [61]. As such, the SEs are more suitable for a topographical contrast image of the surface. The BSEs have much higher energies, from 50 eV to the beam energy E_0, with a distribution peak close to E_0 and a larger penetration depth up to ca. 1 μm. Hence, the BSEs are more for suitable a compositional contrast image. The characteristic X-rays give compositional information of an interaction volume with a depth up to ca. 5 μm [61]. The more electrons are collected by the detector for a pixel, the brighter that pixel is in the SEM images of the SEs and BSEs. A change of contrast of the pixels in an SEM image usually gives useful information on the investigated specimen. Several factors contribute to a good image contrast. The crystal orientation of the topography to the incident beam direction, the tilting of the sample surface with respect to the incident beam direction and to the detector surface, the atomic number Z and density of the composing elements, the electrical conductivity of the surface due to charge accumulation, the magnetic field strength, and other factors that affect the amount and direction of the scattered electrons toward the detectors. Among these influencing factors, SE images are very sensitive to the surface topography, while BSE images are very sensitive to the atomic number of the composing elements. The SE images show generally steep surfaces, edges, and borders of protrusions brighter than flat areas. The BSE images show the areas with higher atomic number Z as brighter than the areas with lower Z.

We inspected the surface of the fluted raceway via SEM (Figure 11). The fluted region D in the SE image of Figure 11a looks brighter due to the higher roughness when compared to the non-damaged region ND. The SE image of Figure 11b, derived from the EDS software (INCA Energy Software 350, version 4.15), shows the border of the damaged and non-damaged region D and ND, respectively. The high contrast between the edges and the flat area is filtered. However, the holes in the damaged area are much darker compared to the asperities due to the lack of incident electrons reaching the holes. The bearing current, passing through the ball-raceway interface, causes locally strong *Joule* heating in the electrically high-resistance *a*-spot contacts due to the very high local current densities, leading to melting and even boiling of the surface at the *a*-spots, hence damaging the surface (Figure 11b). The electrical bearing current, by its magnetic self-field, also slightly magnetized the ferromagnetic ring sample structure. Due to this remanent magnetic field, SEM was challenging. A system with a compensation field was used to cancel the effect of the sample magnetic field on the SEM measurements. The track radius of the raceway and the corresponding curvature form of the track surface makes the SEM challenging as well. To optimally focus the electron beam, many trial-and-errors attempts had to be made.

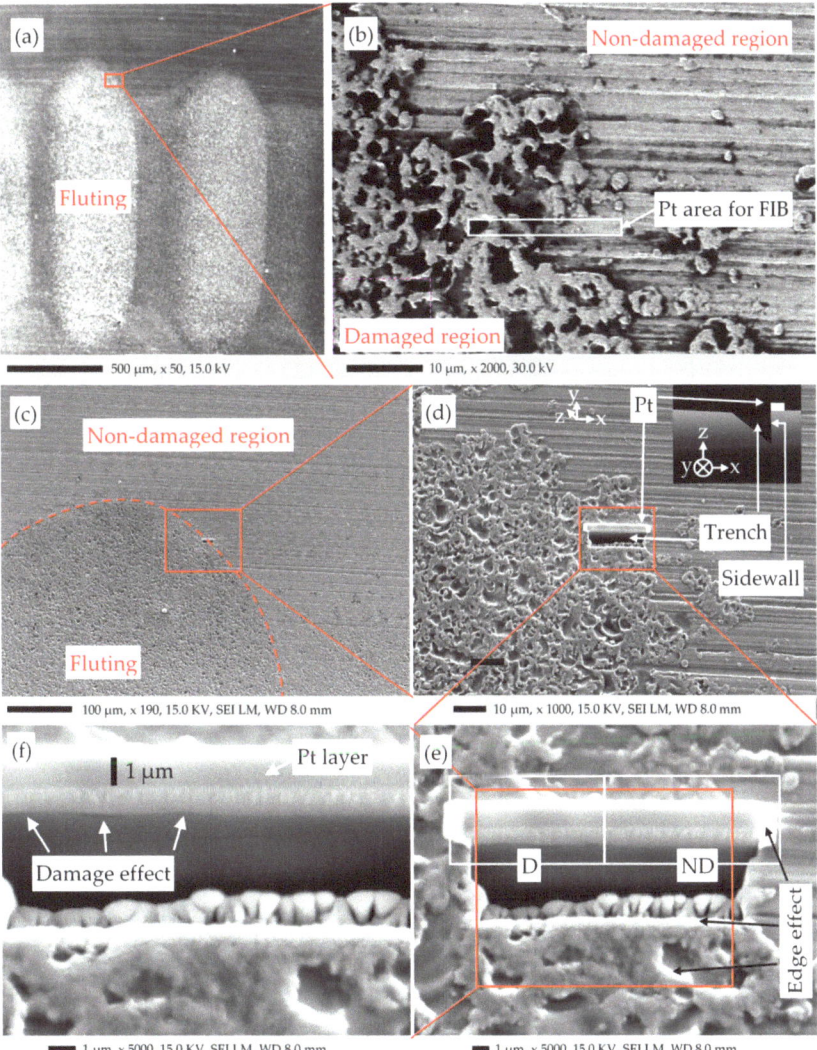

Figure 11. (**a**) SEM image of the sample in Figure 4. (**b**) Magnification of (**a**) with filtering the edge effects. (**c**) SEM image after the FIB process of (**b**). (**d**–**f**) Magnification of (**c**), (**d**) and (**e**), respectively. Measurements with the SEM/FIB instrument 1 in Table 1.

The method of Focused Ion Beam (FIB) excavation was employed (Figure 11b–f) to mill a trench at the border between the damaged and non-damaged regions D and ND. The sidewall of the trench is examined, using SEM to inspect the composing layers on the surface. The FIB-milling proceeds as follows: First, a band of 1 µm thick platinum (Pt) layer is deposited on the surface as a protective layer against the excavating Ga-ion beam. Then the trench is milled via the focused Ga-ion beam up to the Pt-band (Figure 11d). At the beginning of the FIB-process, for a coarse-milling the Ga-ion beam is wide and with high energy in order to speed up the milling process, but this results in a vague SEM image of the sidewall. Hence, with a FIB-fine-milling, the cross-section is polished via a narrow Ga-ion beam with lower energies, giving a clear view of the stacked layers (see Figure 11f). Each FIB process takes about 6 h.

Despite this technique, SEM could not show a distinct oxide layer on the damaged raceway. Between the deposited Pt layers and the bulk steel, only signs of surface damage are visible. Apart from that, the difference between the damaged and non-damaged region is insignificant. A white layer between the Pt layer and bearing surface appeared that could not be explained. Further investigations showed no sign of this layer. The edges in the SE images of Figure 11d–f appeared brighter, as expected, due to the edge effect. The reason that the upper-right edges of the asperities are brighter than the edges on the other directions is due to the position of the electron detector on the upper-right side of the sample.

The EDS is used to determine the chemical composition of the surface in Figures 12–17. The depth of the characteristic X-ray for an element is calculated from [61]:

$$R_x = \frac{0.064}{\rho} \cdot \left(E_0^{1.68} - E_c^{1.68}\right), \tag{5}$$

where R_x is characteristic X-ray penetration depth in µm, ρ is material density in g/cm^3, E_0 is incident electron beam energy, and E_c is the critical excitation energy, both in keV. The "characteristic" X-ray penetration depths for Fe, O, and C are calculated in Table 4. The term "characteristic" is skipped in the following.

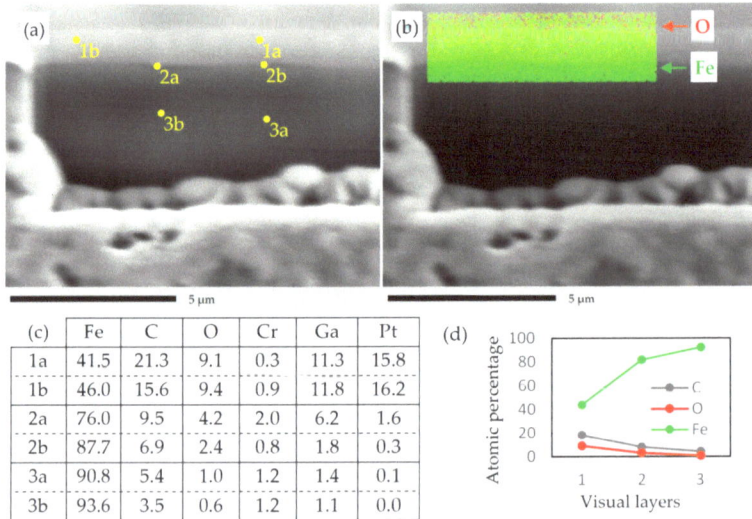

Figure 12. EDS analysis of the damaged part D of the sample in Figure 11e. (a) Selected points for EDS analysis. (b) Elemental mapping of the given area for the elements O (red) and Fe (green). (c) Atomic percentage of the elements for the selected points in (a). (d) Average of C, O, and Fe as atomic percentage from layer 1 to 3 from the data in (c). Measurements with SEM/FIB/EDS instrument 1 in Table 1.

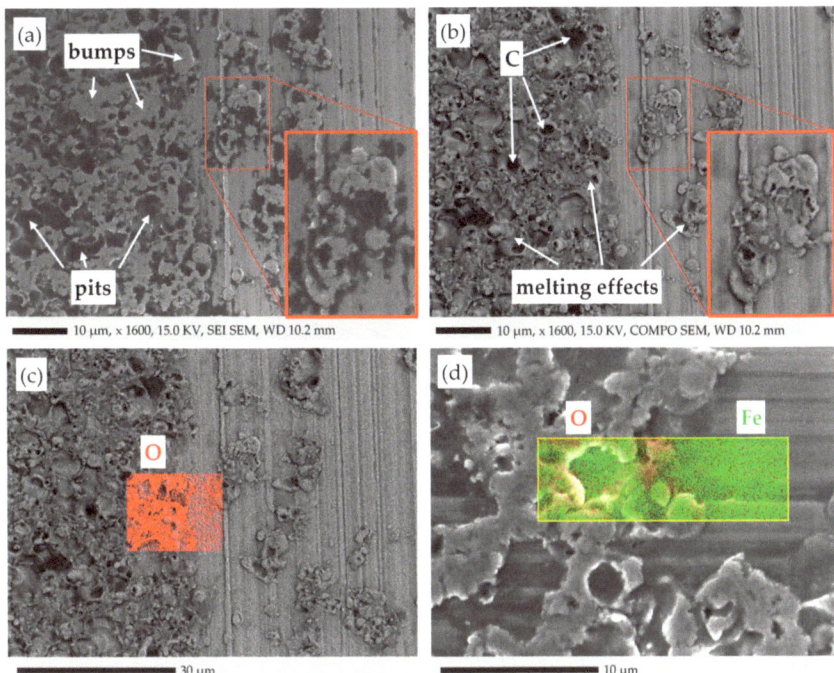

Figure 13. SEM and EDS elemental mapping of the sample surface in Figure 4 on a border between the damaged and non-damaged region D and ND. (**a**) SE image. (**b**) BSE image in COMPO mode. (**c**) EDS elemental mapping of O (red). (**d**) SE image with overlaid EDS elemental mapping of O (red) and Fe (green). Measurements with SEM/EDS instrument 1 in Table 1.

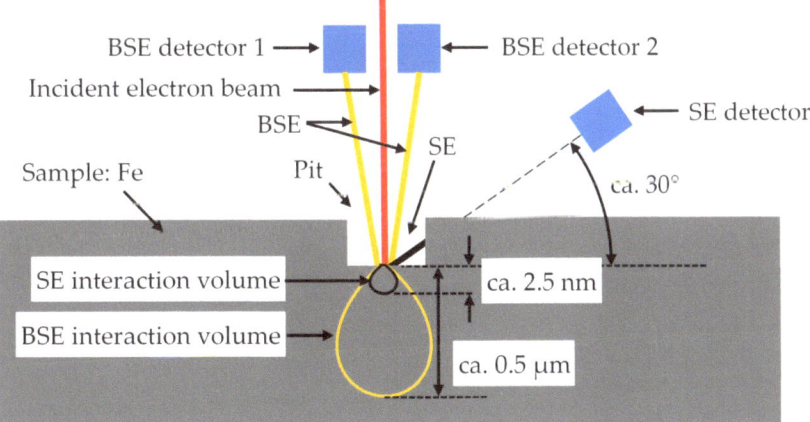

Figure 14. Schematic situation of the incident electron beam, the SE and BSE for a pit (e.g., due to electrical current damage) on a surface, measured with the SEM instrument. The energy and penetration depth of the SEs are much lower than of the BSEs.

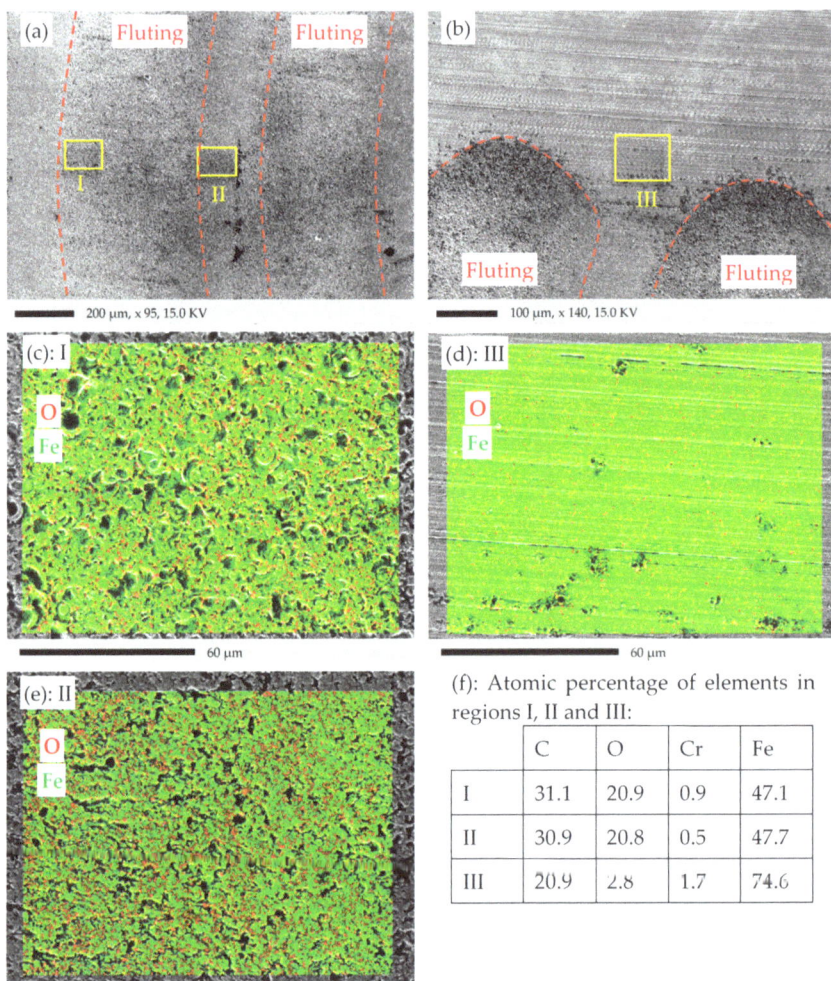

Figure 15. SEM and EDS elemental mapping of the sample surface in Figure 4. (**a**) SEM image of the middle of the raceway in the damaged area D. (**b**) SEM image of the border of damaged and non-damaged area D and ND. (**c**,**d**) Regions I and II in (**a**), respectively. (**e**) Region III in (**b**). (**f**) Atomic percentage of the selected elements in the regions I, II and III. Measurements are done with the SEM/EDS instrument 1 in Table 1.

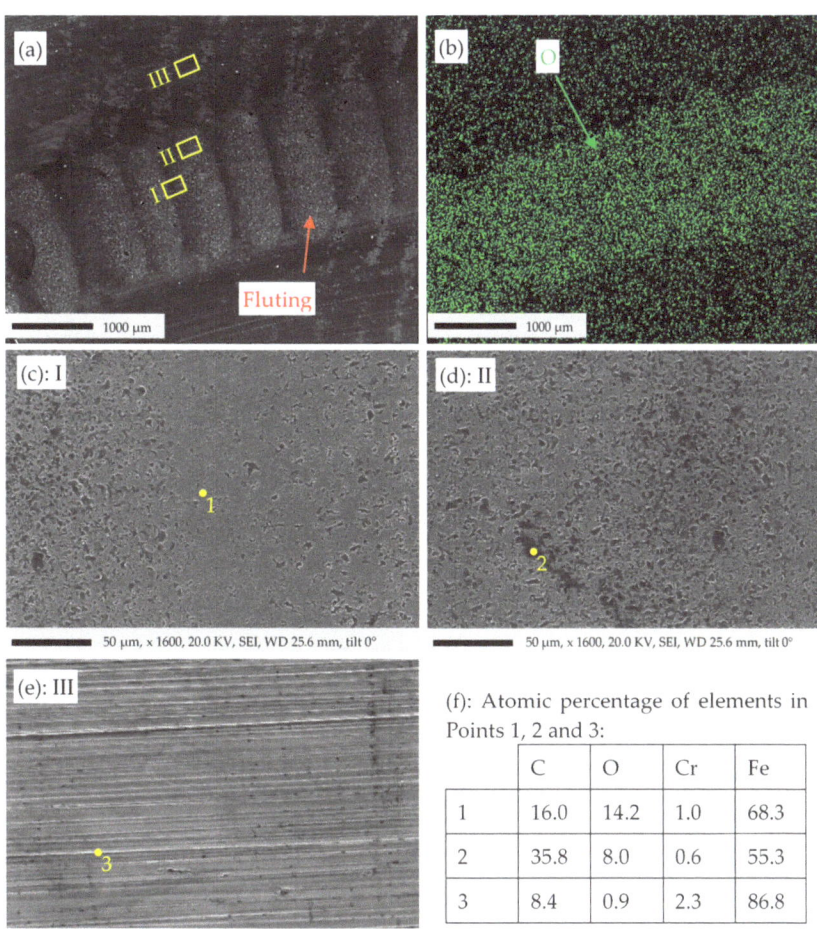

Figure 16. SEM and elemental mapping of the sample surface in Figure 4. (**a**) Filtered SE image of the damaged and non-damaged area D and ND. (**b**) EDS elemental mapping of O. (**c–e**) Region I, II, and III in (**a**), respectively. (**f**) Atomic percentage of the selected elements in Points 1, 2, and 3. Measurements were conducted with the SEM/EDS instrument 2 in Table 1.

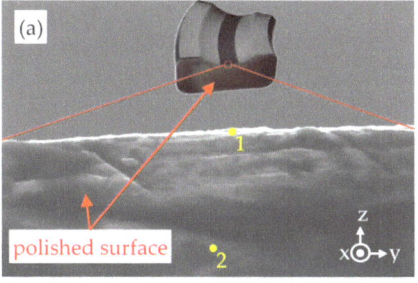

Figure 17. SEM/EDS analysis of the bearing ring cross-section at the damaged raceway D (Figure 4), after polishing with sandpaper (Table 1). (**a**) SEM image of the cut surface (**b**) EDS measurements of Points 1 and 2 from (**a**). Measurements with the SEM/EDS instrument 2 in Table 1.

Table 4. X-ray penetration depth of Fe, O, and C in steel for ρ_{steel} = 7.8 g/cm^3 and E_0 = 15 keV.

Elements	Fe	O	C
Critical excitation energy E_c/keV	7.11	0.532	0.283
X-ray penetration depth R_x/μm	0.55	0.77	0.78

In Figure 12a, six points are selected on the sidewall of the trench in Figure 11e. The visually distinguishable domains are indexed with 1 to 3, from surface-to-bulk, respectively. On the top of layer 1 is the deposited Pt layer, which is beyond the scope of this study. For each visual domain, two points are measured, labeled with a and b, via the EDS. The results are shown in Figure 12c. Note that Pt and Ga are the contaminations of the FIB process. The trend curves of C, O, and Fe with increasing depth are plotted in Figure 12d, indicating that the Fe-concentration increases from-surface-to-bulk, whereas the C-concentration decreases. The changing trend of the concentrations are in good agreement with our XPS measurements.

The EDS elemental mapping of O and Fe in Figure 12b reveals that the O-concentration decreases from-surface-to-bulk. Each EDS data point represents a drop-shaped interaction volume with a depth of ca. 0.7 μm (see Table 4), so the EDS results do not necessarily represent the features on the SE image, but also beneath of it. The SEs have an "escape depth" up to $5\lambda_e \approx 2.5$ nm [61,62], where λ_e is the mean free path of the SEs in Fe.

The topology of the surface was further investigated with the SEM analysis via SEs and BSEs in Figure 13a,b, respectively. Two electron detectors for BSEs were mounted over the sample close to the incident electron beam (Figure 14), each detecting the BSEs of one side of the beam better than the other side [61]. The BSE image has two modes of operation: the Topographical Mode (TOPO) and the Compositional Mode (COMPO). The COMPO mode gives a better atomic number contrast for composition, since the intensities of the BSE detectors are added to intensify the atomic number contrast. The TOPO mode gives a better topographical contrast, since the intensities of the BSE detectors are subtracted from each other to intensify the variation of the atomic number with the moving electron beam. Here only the COMPO mode is used. The pits caused by the electric current in the SE image of Figure 13a are darker than in the BSE image of Figure 13b. As shown in Figure 14, unlike the BSE detector, the SE detector is placed at an angle of ca. 30° with the sample plane [61]. Hence, the SEs, emitted from the pits, may be disturbed by the crater rims of the pits, and do not reach the detector. Due to this "shadow effect", the pits look darker in the SE image. In the BSE image, for the very deep pits the number of emanated BSEs is reduced, contributing to their darkness. Moreover, the disintegrated burnt lubricant yields carbon (Z_C = 6), entrapped in the pits, and appears darker when compared to the metallic iron atoms Fe (Z_{Fe} = 26).

The EDS elemental mapping is employed in Figure 13c,d to detect the surface oxide. It is expected that the oxide covers the whole surface, especially at the damaged areas, where the high temperatures due to the electric current flow occurred. The penetration depth of the EDS technique for O is ca. 0.77 μm (Table 4). The oxide layer exists on the surface in a much thinner depth than 0.77 μm. Hence, the EDS technique cannot inspect the intended oxidized surface exclusively. The EDS results contain information from the deeper regions as well. The O in the elemental mapping is not distributed evenly over the damaged region D. The areas with the higher bumps due to surface damage have a thicker metal layer of previously molten material, resulting in a higher O concentration. In any case, a higher O concentration in the damaged area D is obvious in Figure 13c,d when compared to the non-damaged area ND.

EDS elemental mapping is used for larger areas in Figure 15 to verify the findings of Figure 13c,d. Three areas, one on the fluting valley in D, one on the fluting peak in D, and one outside the raceway in the non-damaged region ND, are selected, as shown in Figure 15a,b. The EDS elemental mappings in Figure 15c–f show that the averaged atomic percentage of O in area I and II is almost equal at around 21%, which is much bigger than that of area III with only 2.8% outside the flutings in the non-damaged region.

To verify the findings of Figure 15, another SEM/EDS investigation was performed in Figure 16. Figure 16b shows a higher concentration of O in the damaged region D. Unlike Figure 15, in Figure 16c–e, EDS spectra are measured at three points (1, 2, and 3) in the regions I, II, and III, respectively. The concentration of C in Point 1 is higher than in Point 2, suggesting that Point 2 is in a deep pit located on the damaged surface D, where a significant amount of C is trapped as a result of the burnt lubricant. A higher O concentration in Point 1 compared to Point 2 is due to the difference in the C concentration and due to the height difference. The bigger heights of the bumps compared to the pits result in their longer exposure time to the high temperatures during the bearing current flow, leading to more oxidation. Point 3 outside the damaged region has a much lower O concentration, a much higher Fe concentration, and a relatively lower C concentration, as listed in Figure 16f.

To investigate the surface characteristics, like in Figures 11 and 12 but without deposited Pt, the whole cross-section of the bearing ring is cut via an EDM wire cut. Thus, the bearing ring is divided into four parts. For one part, the cut surface (Figure 17) is polished with sandpaper (Table 1) to clean the surface of cutting-caused contaminants. To protect the edge of the damaged surface and to get a finer flattening of the cross-section surface, another part of the cut surface (Figure 18) was finely polished with the polishing machine in Table 1 with SiC paper with the following grit numbers: P120, P220, P320, P500, P1200, P500, P4000; each for approx. 2 min.

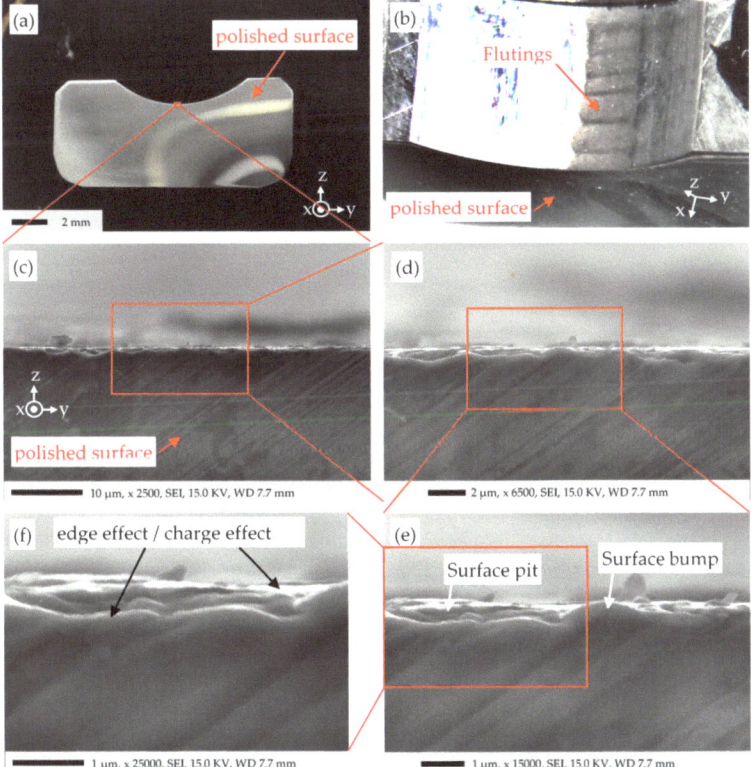

Figure 18. Bearing ring cross-section, cut at the damaged raceway (Figure 4), after polishing with the polishing machine (Table 1). (a) Light microscope image of the cut and polished surface (b–f) SE images of the surface. Measurements were conducted with the SEM instrument 1 in Table 1.

The SEM image of Figure 17a does not show any distinct layer on the surface related to the oxide. The bright color shining on the edge may be due to the "edge effect" of the emitted SEs in SEM technique [61]. The SEs are emitted much more from the edges of a protrusion than from a flat region. This effect arises because the SEs have shorter "escape paths" from the SE interaction volume to the surface. The sample is metallic (conductive), and the sample holder is electrically grounded, so the electrons cannot accumulate in any part, connected electrically to the bulk. Any electrically insulated particle on the surface appears shiny due to the "charging effect" [61]. The insulating oxide may result in a charging effect on the surface, causing the shiny color. Two points are selected for EDS analysis: Point 1 on the edge of the sample targeting the damaged surface and point 2 on the bulk steel. Figure 17b shows, by 5.4 times and 2 times, respectively, higher O and C concentration on the damaged surface (point 1), compared to the bulk steel (point 2).

The SEM image in Figure 18e shows the typical bumps and pits due to the electrical damage. Figure 18f shows no clear margin between the surface layers and the bulk steel. The XPS results in Figure 9 have shown the composing layers on the surface. The iron oxide concentration decreases exponentially, moving inward from surface-to-bulk. A pronounced change and a visible border line between the oxide and bulk steel is therefore not expected in Figure 18f. The edge effect may cause the shiny regions in Figure 18f, and/or the insulating oxide on the surface may accumulate the electrons from the incident beam, resulting in an increased SE emission and the shining color. This charge effect is more probable visible in Figure 19.

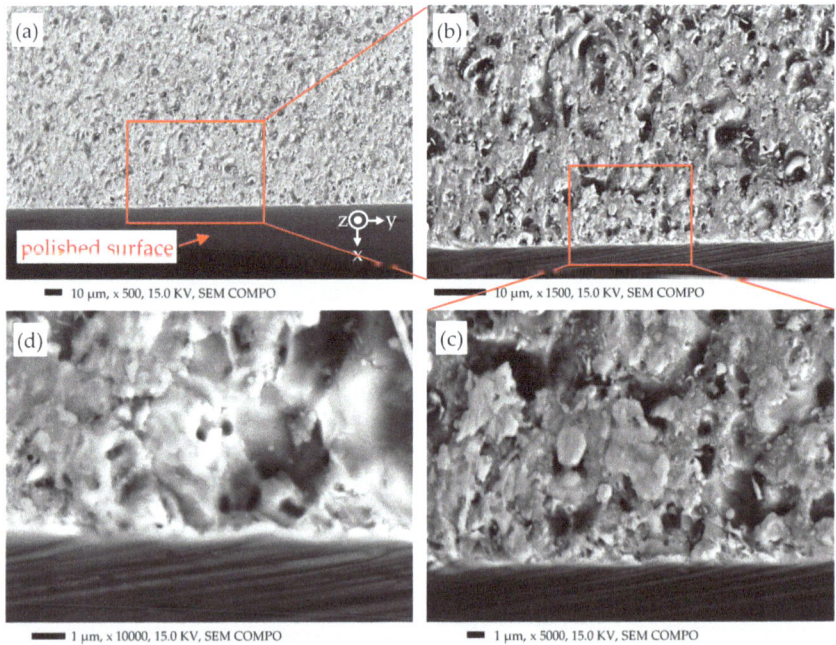

Figure 19. Like Figure 18 but rotated to view the damaged surface and to inspect via BSEs. (**a**–**d**): low to high magnifications, respectively.

Figure 19 is the same sample as in Figure 18 but rotated to take images from the damaged surface and via BSEs. The typical surface damages are seen in Figure 19, similar to those in Figures 11b and 13d. The dark spots on the surface are attributed to the deep holes or the accumulation of C. The generally brighter color of the surface compared to the bulk may be due to the insulating oxide layer (charge effect) and/or due to a reduced amount of incident electrons reaching the bulk steel.

6. Summary

The electrically damaged surface of a fluted raceway of an axial ball bearing type 51208 was studied. An DC apparent bearing current density of 3.85 A/mm^2, which is ten times bigger than the safe limit value 0.3 A/mm^2, at 400 N axial bearing force and a grease lubrication with *Arcanol MULTI3*, had generated severe fluting after 24 h operation at 1500 rpm. The properties of the electrically damaged area of the raceway within the fluting pattern, influenced by the degraded lubricant, in comparison with the non-damaged area was closely investigated by light microscope, WLI, SEM, XPS, and EDS.

WLI showed high roughness peaks on the surface. The motif analysis showed that the surface roughness was irregular and stochastic. These findings support our hypothesis that the very high bearing currents lead to a roughened surface, causing mechanical contact points between the raceway and the ball, even at an elevated speed of e.g., 1500 rpm, where typically full lubrication was expected.

SEM showed the signs of locally molten electrical contact points (*a*-spots). The residual carbons from the degraded lubricant are trapped in the surface pits and can be seen in dark contrasts in the BSE images. These findings support our hypothesis that the locally high current density results in excessive *Joule* heating, which leads to local melting of the steel. This heat obviously burns lubricant molecules, explaining the carbon deposits and the blackening of the lubricant. A distinguishable border between the oxide phases and the bulk of bearing was not seen in the SEM images after the FIB or the cut-and-polish procedures.

The XPS and EDS results proved the existence of surface oxide phases. The XPS also showed the high share of iron carbide on the surface in the damaged area. Skipping the surface contaminations and moving from-surface-to-bulk, the atomic concentration of iron oxide phases decreases, the carbide increases first and then slightly decreases, and the metallic iron increases. This supports our hypothesis that the electrical current flow must punch the insulating oxide layers at distinct changing contact points. The electrical current flow chemically changed both the lubricant and the raceway surface. Figure 20a gives a schematic view of the intact bearing surface without current flow, hence the non-damaged region ND, where the "native oxide" due to exposure to air results in a thin oxide layer. On the contrary, Figure 20b illustrates the compositional surface changes schematically due to current flow in the damaged region D, along with the surface topographies, where the asperity peaks make the mechanical contacts, and the carbon residues are trapped in the surface pits. Not all the mechanical contacts cause electric contacts (*a*-spots) due to the insulating oxide layer, which mostly covers the ball-raceway interfaces. Hence the fluting pattern generation may be influenced by the local abrasion of the oxide layer. Its periodic pattern is so far explained by micro-vibrations of the contact partners with peak-to-valley depths of up to 3 μm in our case.

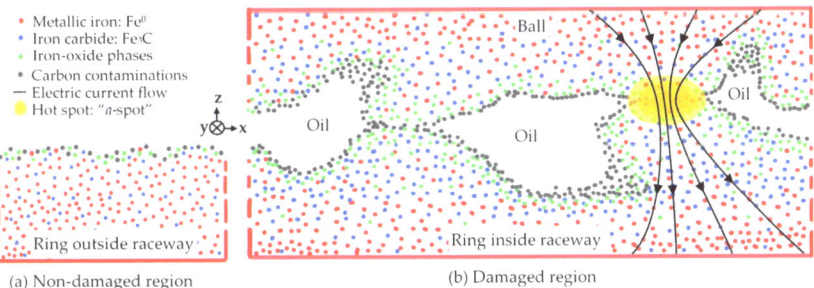

Figure 20. Schematic illustration of the sample surface of Figure 4. (**a**) Non-damaged region ND. (**b**) Damaged region D by current flow.

7. Conclusions

The findings reveal that high bearing current flow of ten times the admissible limit, used as a time accelerator in fluting generation, does not only change the bearing raceway surface topography but also leads to an increased concentration of "electrically insulating" iron oxide layer and "mechanically hard" iron carbide layer on the bearing steel. This electrically insulating oxide layer only allows an electric current flow via small contact spots (*a*-spots), penetrating the insulating layer. The electrical resistance of these fast-changing spots explains the electrical voltage drop, which remains nearly constant close to the melting voltage of steel, even with an increasing current. New contact spots occur in parallel, existing spots are widened due to a local melting. The very high local electrical current density leads to high local temperatures at the electrical contact points, melting and deforming the surface locally and causing an increased surface roughness, which supports a bigger current flow. The high local temperatures accelerate the surface oxidation and, by burning of lubricant molecules, starts the carburization of the iron. This explains the increased iron oxide and iron carbide concentrations in the electrically damaged regions along with the blackened lubricant. The investigated fluting pattern is a periodic peak-valley surface deterioration, may be influenced by this oxide layer. Its periodicity is up to now explained by a perpendicular micro-vibration of the electrical contact partners within the elastic lubrication film. A simulation of this fluting generation process for the starting with the first surface craters to the final fluting pattern is still missing. But our findings provide a closer insight on the fluted surface due to bearing current flow and shall support future simulation models on the fluting generation. Hence, any future simulation to model the process of fluting generation must consider the following components:

(1) a relative movement between the metallic contact partners,
(2) a lubrication film with elastic properties to allow a perpendicular ball vibration,
(3) an electrical current flow via small contact points,
(4) a proper description of the electrically insulating surface layer.

Author Contributions: Conceptualization: O.S., A.F. and A.B.; investigation and experiment: O.S., A.F. and H.S.; software analysis: O.S., A.F. and H.S.; writing—original draft preparation: O.S. and A.F.; writing—review and editing: O.S., A.F., A.B., H.S. and J.P.H. All authors have read and agreed to the published version of the manuscript.

Funding: This research has been supported by the research fund of the Faculty of Electrical Engineering and Information Technology, Technical University of Darmstadt.

Data Availability Statement: All the relevant data were presented in this article. Further inquiries may be directed to the corresponding author.

Acknowledgments: We are grateful to the following institutes and people at TU Darmstadt: (1) The Advanced Electron Microscopy Division for SEM systems, and the measurements of *Ulrike Kunz*. (2) The Institute for Product Development and Machine Elements for the surface profilometer and scientific support of *Steffen Puchtler*.

Conflicts of Interest: The authors declare no conflicts of interest.

Nomenclature

A_{Hz}	Total *Hertz*'ian area per raceway
a.u.	Arbitrary unit
BSE	Back Scattered Electron
COMPO	Compositional mode
CPS	Count per second
D	Damaged region
EDM	Electric discharge machining
EDS	Energy Dispersive Spectroscopy
ETD	*Everhart–Thornley* detector
F_b	Bearing force

FIB	Focused ion beam
HF	High frequency
h	Central lubrication film thickness
I_b	Time-averaged bearing current
i_b	Momentary bearing current
$\hat{i}_b$	Bearing current amplitude
J_b	Apparent bearing current density
LM	Low Magnification
ND	Non-damaged region
n	Shaft rotational speed
P	Electrical power
S	Sensitivity factor
S_q	Surface profile roughness, rms
SEM	Scanning Electron Microscopy
SE	Secondary Electron
SEI	Secondary Electron Image
TOPO	Topographical mode
t	Time
U_b	Time-averaged bearing voltage
u_b	Momentary bearing voltage
u_s	Momentary source voltage
WD	Working Distance
WLI	White Light Interferometry
XPS	X-ray Photoelectron Spectroscopy
X	Atomic percentage
ϑ_b	Average bearing temperature
λ	Lambda ratio of roughness h/σ
λ_e	Mean free path of electron
σ	Composite surface roughness, rms

References

1. Muetze, A. Bearing Currents in Inverter-Fed AC-Motors. Ph.D. Thesis, Technical University of Darmstadt, Darmstadt, Germany, 2004.
2. Furtmann, A.; Poll, G. Electrical Stress and Parasitic Currents in Machine Elements of Drivetrains with Voltage Source Inverters. *Int. VDI Conf. Gears* **2017**, *7*, 109–118. [CrossRef]
3. Puchtler, S.; van der Kuip, J.; Kirchner, E. Analyzing Ball Bearing Capacitance Using Single Steel Ball Bearings. *Tribol. Lett.* **2023**, *71*, 10. [CrossRef]
4. Scheuermann, S.; Hagemann, B.; Brodatzki, M.; Doppelbauer, M. Influence Analysis on the Bearings' Impedance Behavior of Inverter-Fed Motor Drives. In Proceedings of the International Conference Elektrisch-Mechanische Antriebssysteme, Vienna, Austria, 8–9 November 2023; pp. 99–106.
5. Manjunath, M.; Hausner, S.; Heine, A.; De Baets, P. Dieter Fauconnier Electrical Impedance Spectroscopy for Precise Film Thickness Assessment in Line Contacts. *Lubricants* **2024**, *12*, 51. [CrossRef]
6. Martin, G.; Becker, F.M.; Kirchner, E. A Novel Method for Diagnosing Rolling Bearing Surface Damage by Electric Impedance Analysis. *Tribol. Int.* **2022**, *170*, 107503. [CrossRef]
7. Romanenko, A. Study of Inverter-Induced Bearing Damage Monitoring in Variable-Speed-Driven Motor Systems. Ph.D. Thesis, Lappeenranta University, Lappeenranta, Finland, 2017.
8. Muetze, A. Thousands of Hits: On Inverter-Induced Bearing Currents, Related Work, and the Literature. *e&i Elektrotechnik Informationstechnik* **2011**, *128*, 382–388. [CrossRef]
9. Graf, S.; Werner, M.; Koch, O.; Götz, S.; Sauer, B. Breakdown Voltages in Thrust Bearings: Behavior and Measurement. *Tribol. Trans.* **2023**, *66*, 488–496. [CrossRef]
10. Nagata, Y.; Glovnea, R. Dielectric Properties of Grease Lubricants. *Acta Trobologica* **2024**, *10*, 34–41.
11. Bechev, D.; Kiekbusch, T.; Sauer, B. Characterization of Electrical Lubricant Properties for Modelling of Electrical Drive Systems with Rolling Bearings. *Bear. World J.* **2018**, *3*, 93–106.
12. Raadnui, S.; Kleesuwan, S. Electrical Pitting Wear Debris Analysis of Grease-Lubricated Rolling Element Bearings. *Wear* **2011**, *271*, 1707–1718. [CrossRef]
13. Plazenet, T.; Boileau, T.; Caironi, C.; Nahid-Mobarakeh, B. Influencing Parameters on Discharge Bearing Currents in Inverter-Fed Induction Motors. *IEEE Trans. Energy Convers.* **2021**, *36*, 940–949. [CrossRef]
14. Gonda, A.; Capan, R.; Bechev, D.; Sauer, B. The Influence of Lubricant Conductivity on Bearing Currents in the Case of Rolling Bearing Greases. *Lubricants* **2019**, *7*, 108–121. [CrossRef]

15. Tischmacher, H. Systemanalysen Zur Elektrischen Belastung von Wälzlagern Bei Umrichtergespeisten Elektromotoren (System Analysis Regarding Electrical Stress in Rolling Element Bearings of Inverter-Fed Electrical Machines, in German). Ph.D. Thesis, Leibniz University Hannover, Hannover, Germany, 2017.
16. Ortegel, F. Waelzlager in Elektromaschinen (Roller Bearings in Electrical Machines, in German). In *Proceedings of the Conference Waelzlager in Elektromaschinen und in der Buerotechnik, Schweinfurt, Germany*; FAG Publ.-Nr. WL 01 201 DA: Schweinfurt, Germany, 1989.
17. Weicker, M.; Binder, A. Characteristic Parameters for the Electrical Bearing Damage. In Proceedings of the International Symposium on Power Electronics, Electrical Drives, Automation and Motion (SPEEDAM), Sorrento, Italy, 22–24 June 2022; pp. 776–781.
18. Kohaut, A. Riffelbildung in Waelzlagern Infolge Elektrischer Korrosion (Fluting in Roller Bearings due to Electrical Erosion, in German). *Z. Fuer Angew. Phys.* **1948**, *1*, 197–211.
19. Safdarzadeh, O.; Capan, R.; Werner, M.; Binder, A.; Koch, O. Influencing Factors on the Fluting in an Axial Ball Bearing at DC Bearing Currents. *Lubricants* **2023**, *11*, 455. [CrossRef]
20. Andreason, S. Stromdurchgang Durch Waelzlager (Passage of Electric Current through Rolling Bearings, in German). *Die Kugellagerzeitschrift* **1968**, *153*, 6–12.
21. Pittroff, H. Waelzlager Im Elektrischen Stromkreis: Riffelbildung Infolge von Stromdurchgang (Roller Bearings in Electrical Circuits: Fluting due to Current Flow, in German). *Elektr. Bahnen* **1968**, *39*, 54–61.
22. Popov, V.L. *Contact Mechanics and Friction: Physical Principles and Applications*, 2nd ed.; Springer: Berlin/Heidelberg, Germany, 2010; ISBN 978-3-662-53080-1.
23. Hultqvist, T. Transient Elastohydrodynamic Lubrication: Effects of Geometry, Surface Roughness, Temperature, and Plastic Deformation. Ph.D. Thesis, Lulea University of Technology, Luleå, Sweden, 2020.
24. Krupka, I.; Hartl, M.; Matsuda, K.; Nishikawa, H.; Wang, J.; Guo, F.; Yang, P.; Kaneta, M. Deformation of Rough Surfaces in Point EHL Contacts. *Tribol. Lett.* **2019**, *67*, 33. [CrossRef]
25. Safdarzadeh, O.; Weicker, M.; Binder, A. Transient Thermal Analysis of the Contact in Bearings Exposed to Electrical Currents. *Bear. World J.* **2022**, *7*, 51–60.
26. Hansen, J.; Bjoerling, M.; Larsson, R. A New Film Parameter for Rough Surface EHL Contacts with Anisotropic and Isotropic Structures. *Tribol. Lett.* **2021**, *69*, 37. [CrossRef]
27. Safdarzadeh, O.; Binder, A.; Weicker, M. Measuring Electric Contact in an Axial Ball Bearing at DC Current Flow. *IEEE Trans. Ind. Appl.* **2023**, *59*, 3341–3352. [CrossRef]
28. Needelman, W.M.; Madhavan, P.V. Review of Lubricant Contamination and Diesel Engine Wear. In *SAE Technical Paper Series*; SAE International: Warrendale, PA, USA, 1988; 14p.
29. Zhou, Y.; Cai, Z.B.; Peng, J.F.; Cao, B.B.; Jin, X.S.; Zhu, M.H. Tribo-Chemical Behavior of Eutectoid Steel during Rolling Contact Friction. *Appl. Surf. Sci.* **2016**, *388*, 40–48. [CrossRef]
30. Zhou, Y.; Peng, J.; Wang, W.J.; Jin, X.; Zhu, M. Slippage Effect on Rolling Contact Wear and Damage Behavior of Pearlitic Steels. *Wear* **2016**, *362–363*, 78–86. [CrossRef]
31. Holm, R. *Electric Contacts*, 4th ed.; Springer: Berlin/Heidelberg, Germany, 2013; ISBN 9783642057083.
32. Capan, R.; Safdarzadeh, O.; Graf, S.; Weicker, M.; Sauer, B.; Binder, A.; Koch, O. *Schädlicher Stromdurchgang (Harmful Bearing Currents, Research Report in German), Part III*; Forschungsvereinigung Antriebstechnik e.V., (FVA 650 III): Frankfurt am Main, Germany, 2023.
33. Harris, T.A. *Rolling Bearing Analysis*, 4th ed.; Wiley & Sons: New York, NY, USA, 1991; ISBN 9780471513490.
34. Safdarzadeh, O.; Weicker, M.; Binder, A. Measuring of Electrical Currents, Voltage and Resistance of an Axial Bearing. In Proceedings of the International Conference OPTIM-ACEMP, Brasov, Romania, 2–3 September 2021; IEEE: New York, NY, USA, 2021; pp. 1–7.
35. Prashad, H. *Tribology in Electrical Environments*; Elsevier: Hyderabad, India, 2005; ISBN 9780080521589.
36. *ISO 25178-2*; Geometrical Product Specifications (GPS)—Surface Texture: Areal—Part 2: Terms, Definitions and Surface Texture Parameters. ISO: Geneva, Switzerland, 2021.
37. *ISO 16610-61*; Geometrical Product Specification (GPS)—Filtration—Part 61: Linear Areal Filters—Gaussian Filters. ISO: Geneva, Switzerland, 2015.
38. Abbott, E.J.; Firestone, F.A. Specifying Surface Quality—A Method on Accurate Measurement and Comparison. *Mech. Eng. ASME* **1933**, *55*, 569–572.
39. Hamrock, B.J.; Dowson, D. *Ball Bearing Lubrication, the Elastohydrodynamics of Elliptical Contacts*; John Wiley & Sons: New York, NY, USA, 1981; ISBN 9780471035534.
40. Weicker, M.; Gemeinder, Y. *Lagerimpedanz-Berechnungsprogramm V3-0, AxRiKuLa (Bearing Impedance Calculation Program)*; In-House Calculator, Institute for Electrical Energy Conversion, TU Darmstadt: Darmstadt, Germany, 2018.
41. *ISO 12085*; Geometrical Product Specifications (GPS)—Surface Texture: Profile Method: Motif Parameters. ISO: Geneva, Switzerland, 1996.
42. Blateyron, F. The Areal Feature Parameters. In *Characterisation of Areal Surface Texture*; Springer: Berlin/Heidelberg, Germany, 2013; ISBN 9783642364587.
43. Tiwari, S.; Prakash, R.; Choudhary, R.J.; Phase, D.M. Oriented Growth of Fe_3O_4 Thin Film on Crystalline and Amorphous Substrates by Pulsed Laser Deposition. *J. Phys. D Appl. Phys.* **2007**, *40*, 4943–4947. [CrossRef]

44. Moulder, J.F.; Chastain, J.; King, R.C. *Handbook of X-ray Photoelectron Spectroscopy: A Reference Book of Standard Spectra for Identification and Interpretation of XPS Data*; Physical Electronics, Uitgeverij: Eden Prairie, MN, USA, 1995; ISBN 9780964812413.
45. Fairley, N.; Fernandez, V.; Richard-Plouet, M.; Guillot-Deudon, C.; Walton, J.; Smith, E.; Flahaut, D.; Greiner, M.; Biesinger, M.; Tougaard, S.; et al. Systematic and Collaborative Approach to Problem Solving Using X-Ray Photoelectron Spectroscopy. *Appl. Surf. Sci. Adv.* **2021**, *5*, 100112. [CrossRef]
46. Ghods, P.; Isgor, O.B.; Brown, J.R.; Bensebaa, F.; Kingston, D. XPS Depth Profiling Study on the Passive Oxide Film of Carbon Steel in Saturated Calcium Hydroxide Solution and the Effect of Chloride on the Film Properties. *Appl. Surf. Sci.* **2011**, *257*, 4669–4677. [CrossRef]
47. Brundle, C.R.; Chuang, T.J.; Wandelt, K. Core and Valence Level Photoemission Studies of Iron Oxide Surfaces and the Oxidation of Iron. *Surf. Sci.* **1977**, *68*, 459–468. [CrossRef]
48. Jung, R.-H.; Tsuchiya, H.; Fujimoto, S. XPS Characterization of Passive Films Formed on Type 304 Stainless Steel in Humid Atmosphere. *Corros. Sci.* **2012**, *58*, 62–68. [CrossRef]
49. Stambouli, V.; Palacio, C.; Mathieu, H.J.; Landolt, D. Comparison of In-Situ Low-Pressure Oxidation of Pure Iron at Room Temperature in O_2 and in O_2/H_2O Mixtures Using XPS. *Appl. Surf. Sci.* **1993**, *70–71*, 240–244. [CrossRef]
50. Wilson, D.; Langell, M.A. XPS Analysis of Oleylamine/Oleic Acid Capped Fe_3O_4 Nanoparticles as a Function of Temperature. *Appl. Surf. Sci.* **2014**, *303*, 6–13. [CrossRef]
51. Lin, T.C.; Seshadri, G.; Kelber, J.A. A Consistent Method for Quantitative XPS Peak Analysis of Thin Oxide Films on Clean Polycrystalline Iron Surfaces. *Appl. Surf. Sci.* **1997**, *119*, 83–92. [CrossRef]
52. Lopez, D.D.; Schreiner, W.H.; Sanchez, S.R.; Simison, S.N. The Influence of Carbon Steel Microstructure on Corrosion Layers. *Appl. Surf. Sci.* **2003**, *207*, 69–85. [CrossRef]
53. Barbieri, A.; Weiss, W.; Van Hove, M.A.; Somorjai, G.A. Magnetite Fe_3O_4(111): Surface Structure by LEED Crystallography and Energetics. *Surf. Sci.* **1994**, *302*, 259–279. [CrossRef]
54. Shabanova, I.N.; Trapeznikov, V.A. A Study of the Electronic Structure of Fe_3C, Fe_3Al and Fe_3Si by X-Ray Photoelectron Spectroscopy. *J. Electron Spectrosc. Relat. Phenom.* **1975**, *6*, 297–307. [CrossRef]
55. Guo, C.; He, J.; Wu, X.; Huang, Q.; Wang, Q.; Zhao, X.; Wang, Q. Facile Fabrication of Honeycomb-like Carbon Network-Encapsulated $Fe/Fe_3C/Fe_3O_4$ with Enhanced Li-Storage Performance. *ACS Appl. Mater. Interfaces* **2018**, *10*, 35994–36001. [CrossRef] [PubMed]
56. Biesinger, M.C.; Payne, B.P.; Grosvenor, A.P.; Lau, L.W.M.; Gerson, A.R.; Smart, R.S.C. Resolving Surface Chemical States in XPS Analysis of First Row Transition Metals, Oxides and Hydroxides: Cr, Mn, Fe, Co and Ni. *Appl. Surf. Sci.* **2011**, *257*, 2717–2730. [CrossRef]
57. Cocco, F.; Elsener, B.; Fantauzzi, M.; Atzei, D.; Rossi, A. Nanosized Surface Films on Brass Alloys by XPS and XAES. *RSC Adv.* **2016**, *6*, 31277–31289. [CrossRef]
58. AUSA—SPECIAL STEELS—Material Data Sheet of Steel Grade 100Cr6. Available online: https://www.ausasteel.com/fichas/Bearing-Steel-100Cr6-AUSA.pdf (accessed on 11 January 2024).
59. Chipman, J. Thermodynamics and Phase Diagram of the Fe-C System. *Metall. Trans.* **1972**, *3*, 55–64. [CrossRef]
60. Kawulok, R.; Schindler, I.; Sojka, J.; Kawulok, P.; Opela, P.; Pindor, L.; Grycz, E.; Rusz, S.; Sevcak, V. Effect of Strain on Transformation Diagrams of 100Cr6 Steel. *Crystals* **2020**, *10*, 326. [CrossRef]
61. Ul-Hamid, A. *Beginners' Guide to Scanning Electron Microscopy*; Springer Nature: Cham, Switzerland, 2019; ISBN 9783030074982.
62. Sastry, M. On the Correlation between the Inelastic Mean Free Paths of Secondary Electrons and the Secondary-Electron Yield Parameter in Some Metal Films. *J. Electron Spectrosc. Relat. Phenom.* **2000**, *106*, 93–99. [CrossRef]

Disclaimer/Publisher's Note: The statements, opinions and data contained in all publications are solely those of the individual author(s) and contributor(s) and not of MDPI and/or the editor(s). MDPI and/or the editor(s) disclaim responsibility for any injury to people or property resulting from any ideas, methods, instructions or products referred to in the content.

Article

Research on Design and Optimization of Micro-Hole Aerostatic Bearing in Vacuum Environment

Guozhen Fan [1,2], Youhua Li [1], Yuehua Li [2,3], Libin Zang [2,4], Ming Zhao [4], Zhanxin Li [2], Hechun Yu [1], Jialiang Xu [2], Hongfei Liang [2], Guoqing Zhang [1] and Weijie Hou [2,3,*]

[1] School of Mechatronics Engineering, Zhongyuan University of Technology, Zhengzhou 451191, China; 2022004078@zut.edu.cn (G.F.); yhli1994@126.com (Y.L.); 6222@zut.edu.cn (H.Y.); cims@msn.cn (G.Z.)

[2] Tianjin Institute of Aerospace Mechanical and Electrical Equipment, Tianjin 300301, China; yuehua_li88@163.com (Y.L.); zanglibin906@163.com (L.Z.); menglili99@163.com (Z.L.); jialiangxu1027@163.com (J.X.); 13820332637@139.com (H.L.)

[3] Tianjin Key Laboratory of Microgravity and Hypogravity Environment Simulation Technology, Tianjin 300301, China

[4] School of Mechanical Engineering, Tianjin University, Tianjin 300350, China; ming.zhao@tju.edu.cn

* Correspondence: bithouwj@163.com

Abstract: Micro-hole aerostatic bearings are important components in micro-low-gravity simulation of aerospace equipment, and the accuracy of micro-low-gravity simulation tests is affected by them. In order to eliminate the influence of air resistance on the attitude control accuracy of remote sensing satellites and achieve high fidelity of micro-low-gravity simulation tests, in this study, a design and parameter optimization method was proposed for micro-hole aerostatic bearings for a vacuum environment. Firstly, the theoretical analysis was conducted to investigate the impact of various bearing parameters and external conditions on the bearing load capacity and mass flow. Subsequently, a function model describing the variation in bearing load capacity and mass flow with bearing parameters was obtained utilizing a BP neural network. The parameters of aerostatic bearings in a vacuum environment were optimized using the non-dominated sorting genetic algorithm (NSGA-II) with the objectives of maximizing the load capacity and minimizing the mass flow. Subsequently, experimental tests were conducted on the optimized bearings in both atmospheric and vacuum conditions to evaluate their load capacity and mass flow. The results show that in a vacuum environment, the load capacity and mass flow of aerostatic bearings are increased compared to those in standard atmospheric conditions. Furthermore, it has been determined that the optimal solution for the bearing's load capacity and mass flow occurs when the bearing has an orifice aperture of 0.1 mm, 36 holes, and an orifice distribution diameter of 38.83 mm. The corresponding load capacity and mass flow are 460.644 N and 11.816 L/min, respectively. The experimental and simulated errors are within 10%; thus, the accuracy of the simulation is verified.

Keywords: vacuum environment; NSGA-II algorithm; load capacity; mass flow; experimental study

1. Introduction

To ensure the operational precision of aerospace equipment, a micro-low-gravity environment needs to be constructed on the ground to validate the performance of the aerospace equipment, thereby comprehensively assessing various performance indicators of the aerospace equipment. In ambient pressure environments, it is challenging to achieve agile maneuvering and high-precision attitude control of remote sensing satellites due to factors such as air resistance [1]. In contrast, a vacuum environment can reduce the impact of the environment on micro-low-gravity simulation tests. Therefore, remote sensing satellites need to undergo testing in a vacuum to eliminate the effects of air resistance and other environmental factors on ground tests. As a crucial component in ground micro-low-gravity simulation for remote sensing satellites, the aerostatic bearings are essential to

Citation: Fan, G.; Li, Y.; Li, Y.; Zang, L.; Zhao, M.; Li, Z.; Yu, H.; Xu, J.; Liang, H.; Zhang, G.; et al. Research on Design and Optimization of Micro-Hole Aerostatic Bearing in Vacuum Environment. *Lubricants* **2024**, *12*, 224. https://doi.org/10.3390/lubricants12060224

Received: 30 April 2024
Revised: 30 May 2024
Accepted: 11 June 2024
Published: 17 June 2024

Copyright: © 2024 by the authors. Licensee MDPI, Basel, Switzerland. This article is an open access article distributed under the terms and conditions of the Creative Commons Attribution (CC BY) license (https://creativecommons.org/licenses/by/4.0/).

ensure the high fidelity of the experiments. Excessive air discharge from the bearings in a vacuum environment can adversely affect the level of the vacuum, thereby compromising the accuracy of ground-based micro-low-gravity simulation tests. However, the parameters of the bearings directly impact their performance. In order to design a micro-hole aerostatic bearing with a high-load capacity and low-mass flow in a vacuum, optimization of the bearing parameters is necessary.

Miyatake et al. [2] studied aerostatic thrust bearings with a diameter smaller than 0.05 mm, and the static and dynamic characteristics of the bearings were investigated using computational fluid dynamics (CFD) in comparison with those of composite throttling aerostatic thrust bearings. Gao et al. [3] examined the pressure distribution, load capacity, and stiffness of aerostatic bearings under various speeds and eccentric conditions, discussing the coupling of aerostatic and aerodynamic effects within the bearings. Chakraborty et al. [4] analyzed the performance of porous alumina film aerostatic bearings by combining theory and experiment under atmospheric pressure, and the results showed that the bearing bore diameter, air supply pressure, and radial load had a great influence on its performance. Khim et al. [5] analyzed the phenomenon of pressure increase in aerostatic bearings moving under vacuum conditions through theoretical analysis and experimental validation, attributing the pressure rise to additional leakage caused by stage velocity and air molecules adsorption and desorption from the guide rail surface. Fukui et al. [6] investigated the characteristics of externally pressurized bearings under high Knudsen number conditions at an environmental pressure of 0.1 kPa. Experimental results showed similarities with lubrication equation simulations, confirming the applicability of the lubrication equation. Trost et al. [7] studied small orifice throttling aerostatic bearings under vacuum conditions, determining the critical pressure and critical radius at which viscous air flows within the bearing transition to molecular flow and establishing the relationship. Schenk et al. [8] researched the characteristics of aerostatic bearings in a vacuum environment, observing a decrease in stiffness and an increase in load capacity compared to atmospheric pressure conditions, and proposed a flow field calculation model at any supply air pressure. Gan et al. [9] investigated the slip effect of air lubrication at arbitrary Knudsen numbers using the Boltzmann equation. These studies indicated that the performance of bearings in a vacuum environment differs from that in atmospheric pressure conditions, yet the lubrication equation remains applicable in a vacuum. Chan et al. [10] optimized the structure of bearings using a particle swarm optimization algorithm to maximize stiffness and minimize friction. Wang et al. [11] employed a genetic algorithm to optimize the structure of porous bearings to maximize load capacity and stiffness, enhancing computational efficiency. Nenzi et al. [12] optimized aerostatic bearings using the cubic partitioning method (HDM), demonstrating that the HDM method could obtain multiple optimal solutions with less computational effort. Yifei et al. [13] established mathematical models for stiffness and dynamic stability, conducted optimization calculations under different load conditions, and verified the optimization results. Hsin et al. [14] utilized the two-stage grid-interpolated fitting (GIF) method for bearing optimization design, obtaining a larger Pareto front distribution with reduced computational effort. Zhang et al. [15] studied the application of machine learning with neural networks and genetic algorithms in multi-objective optimization of heat exchangers, developing prediction models for heat transfer coefficients and pressure drops. By designing models with different parameters and training datasets, they utilized neural networks for prediction and employed multi-objective genetic algorithms for global optimization to obtain the Pareto boundary solution set for optimal combinations of key parameters. Few researchers have focused on optimizing bearing parameters based on mass flow, and optimization of aerostatic bearings parameters has mainly been conducted under atmospheric pressure conditions, thus lacking optimal parameters for bearings in a vacuum environment. In conclusion, there is a need to propose a design and optimization method for aerostatic bearings in a vacuum environment.

According to the aforementioned research, most scholars have focused on the study of aerostatic bearings in ambient pressure conditions, with limited research on bearing performance in vacuum environments. Furthermore, studies conducted in vacuum environments have only explored bearings under specific parameters, without providing principles for selecting and optimizing parameters for bearings in vacuum conditions. In order to mitigate the impact of air resistance on the precision of attitude control for remote sensing satellites, micro-low-gravity experiments and the use of bearings must be carried out in a vacuum. In a vacuum environment, the Reynolds equation is solved using the five-point difference method in this study. The effects of bearing parameters (air film thickness, number of orifices, aperture of orifice, and distribution diameter of orifice) and external environment (air supply pressure and vacuum level) on the load capacity and mass flow are analyzed. With the objectives of maximizing load capacity and minimizing mass flow, the target function is nonlinearly fitted using a BP neural network. The NSGA-II algorithm is then employed to determine the Pareto front optimal solution of the bearing in a vacuum environment. The optimal parameters of the bearing in a vacuum environment are obtained by applying the maximum–minimum value method to multiple sets of solution strategies. The accuracy of the optimization method is demonstrated through a comparison of numerical analysis and experimental verification. A comparison between numerical analysis and experimental verification demonstrates the accuracy of the optimization method, providing a theoretical and experimental approach for the application of bearings in vacuum settings.

2. Theory and Methods
2.1. Structure of Air Bearing

The structure of aerostatic thrust bearing is shown in Figure 1. The aerostatic thrust bearing mainly consists of the thrust shell and the thrust bearing. There is an annular groove on the thrust shell, and this structure allows for the air to enter from the supply hole to be uniformly distributed to the orifices of the thrust bearing. The outer diameter of the aerostatic thrust bearing is R; the distribution diameter of the orifice is r; the aperture of the orifice is d_0, and the air film thickness is h. When the bearing is in operation, high-pressure external air flows into the annular groove of the thrust shell from the supply hole and then into the uniformly distributed orifices. Due to the pressure inside the air film thickness being higher than the external ambient pressure, the pressure difference between them generates the load capacity, causing the bearing to float.

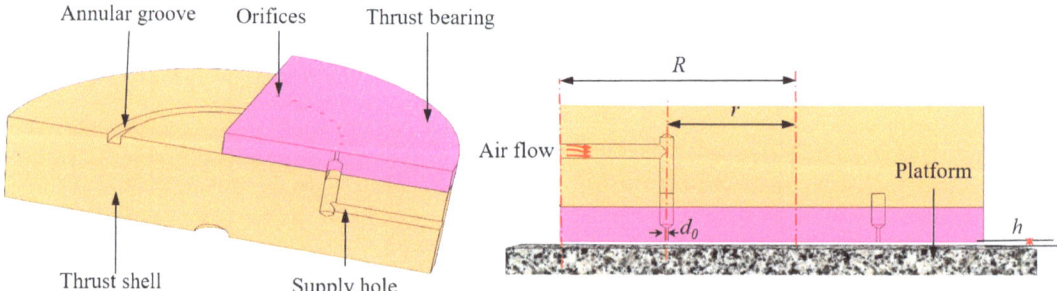

Figure 1. The structure of aerostatic thrust bearing.

2.2. Theory Model

In the vacuum environment, the flow of air in the flow field satisfies the Navier–Stokes equations. Simplifying the flow field, the Reynolds equation for the stable flow of air in a thrust-bearing flow field is expressed as [16]:

$$\frac{1}{r}\frac{\partial}{\partial r}(rh^3\frac{\partial p^2}{\partial r}) + \frac{\partial}{r^2\partial \theta}(h^3\frac{\partial p^2}{\partial \theta}) + Q\delta_i = 0 \tag{1}$$

Let $\bar{r} = \frac{r}{r_0}, \bar{p} = \frac{p}{p_a}, \bar{h} = \frac{h}{h_0}, \bar{\zeta} = \ln \bar{r}, \overline{Q} = \frac{24\eta r_0^2}{h_0^3 p_a \rho_a} p\tilde{v}$, and the dimensionless Reynolds equation can be written as

$$\frac{\partial}{\partial \bar{\zeta}}(\bar{h}^3 \frac{\partial \bar{p}^2}{\partial \bar{\zeta}}) + \frac{\partial}{\partial \theta}(\bar{h}^3 \frac{\partial \bar{p}^2}{\partial \theta}) + \bar{r}^2 \overline{Q} \delta_i = 0 \tag{2}$$

where r is the radial length of the bearing; θ is the circumferential angle of the bearing; h is the air film thickness; p is the internal pressure of the air; η is the dynamic viscosity of the air; r_0 is the reference radius; p_a is the ambient pressure, and h_0 is the reference film thickness. At the orifice $\delta_i = 1$, and at the non-orifice $\delta_i = 0$.

The mass flow of air outflow at the orifice is expressed as [16]:

$$\dot{m}_{out} = \varphi P_s A \cdot \sqrt{\frac{2\rho_a}{P_a}} \cdot \psi \tag{3}$$

where φ is the flow coefficient; P_S is the air supply pressure; A is the area of the orifice; p_a and ρ_a are the environmental pressure and atmospheric density respectively, and the air density in vacuum is

$$\rho_a = \frac{\rho_0 T_0}{P_A T_a} p_a \tag{4}$$

where ρ_0 is the density of air in the standard state; T_0 is the temperature of the air in the standard state; p_A is the standard atmospheric pressure, and T_a is the ambient temperature.

The flow function ψ is represented as follows [16]:

$$\psi = \begin{cases} \frac{k}{2}\left(\frac{2}{k+1}\right)^{\frac{k+1}{k-1}} & \frac{p_d}{p_s} \leq \beta_k \\ \left[\frac{k}{k-1}\left(\left(\frac{p_d}{p_s}\right)^{\frac{2}{k}} - \left(\frac{p_d}{p_s}\right)^{\frac{k+1}{k}}\right)\right]^{1/2} & \frac{p_d}{p_s} > \beta_k \end{cases} \tag{5}$$

where k is the adiabatic coefficient; p_d is the orifice outlet pressure, and β_k is the critical pressure ratio. The mass flow of the air into the orifice should be equal to the mass flow out, as shown in Equation (6).

$$\dot{m}_{in} = \rho \tilde{v} s = \dot{m}_{out} \tag{6}$$

By solving the partial differential equation of Equation (2), the pressure distribution of the bearing can be obtained; the bearing load can be obtained by integrating the pressure against the bearing surface, and the mass flow of the bearing can be obtained by Equation (3).

2.3. Solution Algorithm

The finite difference method has the advantages of high precision and fast convergence, etc. Therefore, the finite difference method is used to solve the partial differential equation of Equation (2), and the solving process is shown in Figure 2. $m + 1$ nodes are set in direction θ, and $n + 1$ nodes are set in direction $\bar{\zeta}$. The $\theta = 1$ and $\theta = m + 1$ boundaries are periodic boundaries, and the $\bar{\zeta} = 1$ and $\bar{\zeta} = n + 1$ boundaries are atmospheric boundaries.

Using the difference Equation and the pressure of neighboring points, $\bar{p}(i,j)$ is obtained as follows:

$$\bar{p}_{i,j} = \sqrt{\frac{E_{i,j}}{A_{i,j} + B_{i,j} + C_{i,j} + D_{i,j}}} \tag{7}$$

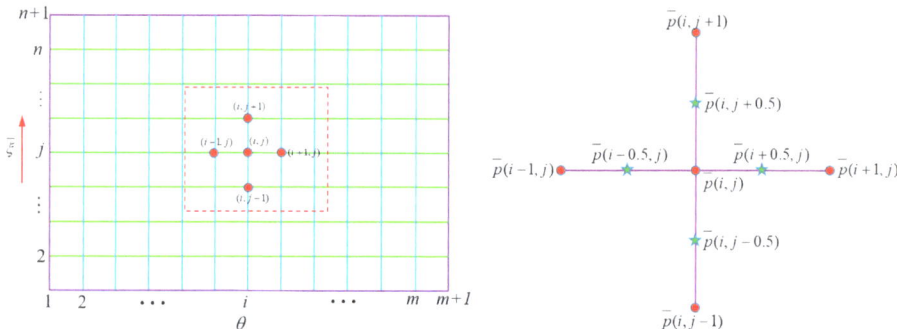

Figure 2. Finite difference method.

The coefficients in the Equation are expressed as follows:

$$A_{i,j} = \frac{(\bar{h}_{i,j}+\bar{h}_{i,j+1})^3}{8\Delta\bar{\xi}^2}$$
$$B_{i,j} = \frac{(\bar{h}_{i,j}+\bar{h}_{i,j-1})^3}{8\Delta\bar{\xi}^2}$$
$$C_{i,j} = \frac{(\bar{h}_{i+1,j}+\bar{h}_{i,j})^3}{8\Delta\theta^2} \quad (8)$$
$$D_{i,j} = \frac{(\bar{h}_{i-1,j}+\bar{h}_{i,j})^3}{8\Delta\theta^2}$$
$$E_{i,j} = A_{i,j}\bar{p}_{i,j+1}^2 + B_{i,j}\bar{p}_{i,j-1}^2 + C_{i,j}\bar{p}_{i+1,j}^2 + D_{i,j}\bar{p}_{i-1,j}^2 + \bar{r}^2\overline{Q}\delta_i$$

The pressure is adjusted by Equation (9):

$$\bar{p}_{i,j} = k\bar{p}_{i,j} + (1-k)\overline{pk}_{i,j} \quad (9)$$

where k is the adjustment coefficient and its value is 0 to 1; $\bar{p}_{i,j}$ is the pressure of the previous iteration, and $\overline{pk}_{i,j}$ is the pressure of the next iteration. The iteration stops when the difference between the two is sufficiently small. Based on the above methods, calculate the pressure distribution of the bearing and obtain the bearing capacity and mass flow as follows:

$$W = \iint_D (\bar{p}-1)p_a ds$$
$$Q = \sum \varphi P_s A \cdot \sqrt{\frac{2\rho_a}{P_a}} \cdot \psi \quad (10)$$

2.4. Optimization Method

According to Equation (10), it can be obtained that the bearing capacity and mass flow vary with the position, quantity, and diameter of the bearing orifices. In order to achieve high bearing capacity, a greater number of orifices and a larger mass flow are required, resulting in coupling between variables. Therefore, in order to obtain a bearing structure with high bearing capacity and low mass flow, it is necessary to establish the functional relationship between bearing capacity and mass flow concerning the number, diameter, and position of the orifices. By identifying its relative minimum value, the optimal solution can be obtained.

As a nonlinear fitting algorithm, the BP neural network can process and fit complex functional relationships with strong applicability. Therefore, the BP neural network is chosen to fit the model [17,18]. The constructed neural network is shown in Figure 3. The input layer consists of three variables, namely, the distribution circle radius r, the number of orifices n, and the aperture of orifice d_0. Two hidden layers are selected for nonlinear mapping, and the output layers are bearing capacity and mass flow.

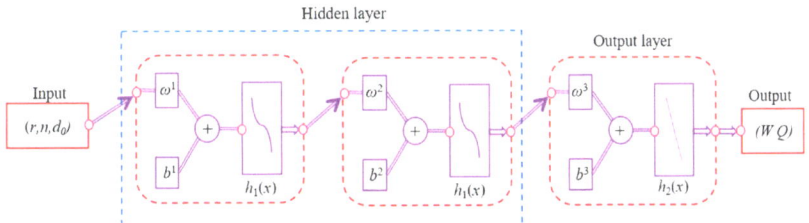

Figure 3. BP neural network structure.

The weights and biases obtained from the training of the BP neural network are passed through the transfer functions of the hidden and output layers, which then undergo nonlinear mapping to ultimately derive the output results. These output results are compared with the actual target values to calculate the loss function. Finally, adjustments are made to the parameters through the backpropagation of the neural network. The training of the neural network stops when the sum of the overall error squares is minimized.

Given the input variable $X = (r, n, d_0)$, the weights and biases of the hidden layer are denoted as ω^{1i}, ω^{2i}, b^{1i}, b^{2i}, with the activation function of the hidden layer being $h_1(x) = \text{tansig}(x)$. The weights and biases of the output layer are represented by ω^{3i} and b^{3i}, with the activation function of the output layer being $h_2(x) = \text{purelin}(x)$. According to the principle of the BP neural network, the output of the output layer can be expressed as

$$\begin{cases} W = h_2(\omega^{31}h_1(\omega^{21}h_1(\omega^{11}X + b^{11}) + b^{21}) + b^{31}) \\ Q = h_2(\omega^{32}h_1(\omega^{22}h_1(\omega^{12}X + b^{12}) + b^{22}) + b^{32}) \end{cases} \quad (11)$$

The determination coefficient is usually used to judge the fit degree, as shown in Equation (12). In this equation, F_{ij}^{predict} is the predicted value of the objective function; F_{ij}^{actual} is the true value of the objective function; $\overline{F}_{ij}^{\text{actual}}$ is the average value of the objective function, and R is the determination coefficient. The closer the determination coefficient is to 1, the better the fitting effect is proved.

$$R^2 = 1 - \frac{\sum_{i=1}^{m} \left(F_{ij}^{\text{predict}} - F_{ij}^{\text{actual}}\right)^2}{\sum_{i=1}^{m} \left(F_{ij}^{\text{actual}} - \overline{F}_{ij}^{\text{actual}}\right)^2} \quad (12)$$

According to Equation (11), the relationship between bearing capacity and mass flow as functions of the position, quantity, and diameter of the bearing orifices can be obtained. This represents the optimized model, as shown in Equation (13). $r_{\min}$, $n_{\min}$, and $d_{0\min}$ represent the minimum values of design variables, while $r_{\max}$, $n_{\max}$, and $d_{0\max}$ represent the maximum values of design variables, with $F(X)$ being the objective function.

$$\begin{cases} \min \quad F(X) = (-W(X), Q(X)) \\ s.t. \quad r_{\min} \leq r \leq r_{\max} \\ \quad n_{\min} \leq n \leq n_{\max} \\ \quad d_{0\min} \leq d_0 \leq d_{0\max} \end{cases} \quad (13)$$

The non-dominated sorting genetic algorithm (NSGA-II) is a multi-objective optimization algorithm based on domination [19,20], which can obtain better optimization results when solving complex multi-objective optimization problems. In this paper, the NSGA-II algorithm is used to optimize the structure of bearings to obtain multiple sets of Pareto front optimal solutions [21]. Subsequently, a decision is made on the multiple sets of Pareto front optimal solutions based on Equation (14) [22], leading to the ultimate determination of the optimal bearing parameters.

$$\begin{cases} L_i^W = \frac{|W_i - W_{min}|}{W_{max} - W_{min}} \\ L_i^Q = \frac{|Q_i - Q_{min}|}{Q_{max} - Q_{min}} \\ G = \min(\max(L_i^W, L_i^Q)) \end{cases} \quad (14)$$

where L_i^W and L_i^Q are the deviation of bearing capacity and mass flow from the optimal solution; W_i is the size of bearing capacity in the Pareto frontier optimal solution, and L_i^Q is the size of each mass flow in the Pareto frontier optimal solution. W_{max}, W_{min}, Q_{max}, Q_{min} are the maximum and minimum values of bearing capacity and mass flow in the Pareto frontier optimal solution. By calculating G from Equation (14), the optimal bearing capacity and mass flow of the bearing under vacuum can be obtained, and the corresponding solution is the optimal design parameter of the bearing. The calculation process is shown in Figure 4.

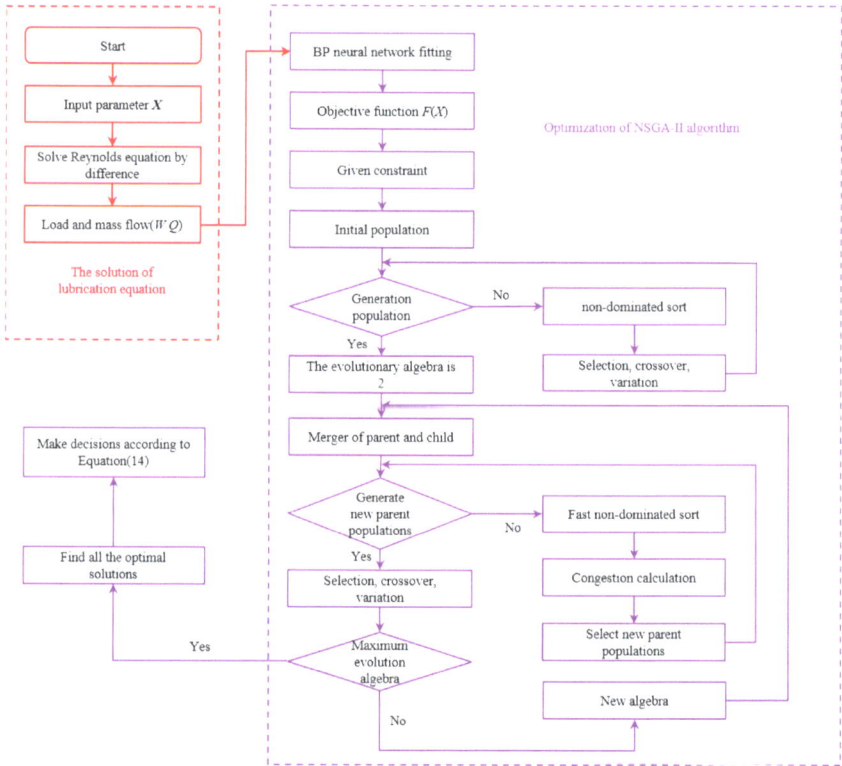

Figure 4. Calculation flow chart.

The solution process is as follows:

1. Input bearing parameters (distribution diameter of orifice, number of orifices, aperture, etc.) and solve Reynolds equation according to Equations (1)–(9) to obtain pressure distribution;
2. According to the pressure distribution, the bearing capacity W and mass flow Q are solved according to Equation (10);
3. The objective function $F(X)$ is obtained by using the BP neural network and Equation (11);
4. Constraints of the independent variables are given, and the population is initialized;

5. Select, cross, and mutate the population, then perform non-sequential sorting to obtain the next generation of the population;
6. Merge the offspring and parent populations to form a new population, then conduct a fast, non-dominated sorting, calculate crowding distance, and rank the population based on non-inferior levels;
7. Repeat the process of selecting, crossing, and mutating the population, then evaluate if the evolutionary generation criteria are met; if so, output the results; if not, continue the process until the criteria are met;
8. Make decisions on multiple solutions to obtain the optimal bearing parameters.

3. Experiment

3.1. Experimental Apparatus

According to the simulation results, the optimal structural parameters of the bearing are obtained, and the bearing is manufactured according to the simulation parameters. It is necessary to test the static characteristics of bearings in atmospheric pressure and a vacuum environment to verify the accuracy of simulation results. The experimental apparatus for the static characteristics of the bearing in an atmospheric pressure environment is shown in Figure 5, consisting mainly of an air pump (1), LION PRECISION CPL592 non-contact capacitive displacement sensor is sourced from Lion, Dayton, OH, USA, (2), table stand (3), NS-WL1 force sensor is sourced from Tianmu, Shanghai, China (4), bracket (5), linear guide rod (6), weights (7), computer (8), data acquisition processing system (9), granite platform (10), SMC PFM711 flow meter is sourced from Gantuo, Shanghai, China (11), test bearing (12), pressure gauge (13), power supply (14), valve (15), etc. The entire test apparatus is leveled and placed on a vibration isolation platform.

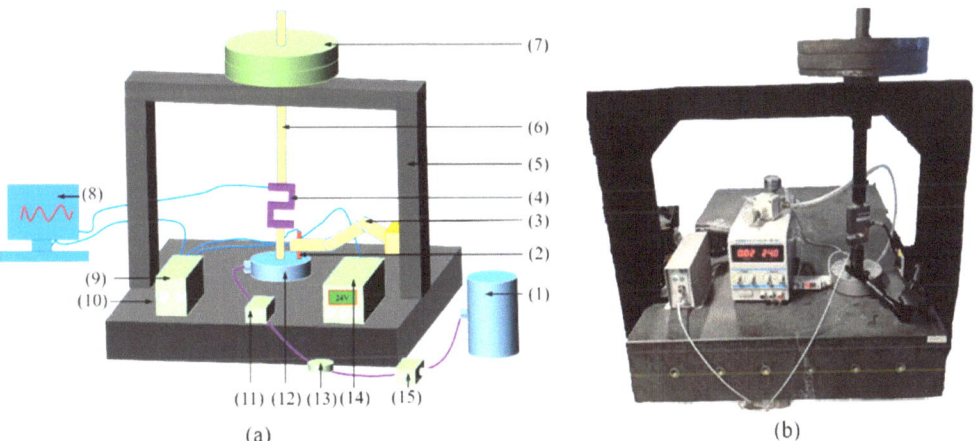

Figure 5. Static characteristic test of atmospheric pressure: (**a**) schematic diagram; (**b**) test equipment.

The air generated by the air pump is filtered and then connected to the supply orifice of the bearing to be tested through the pressure regulating valve, pressure gauge, flow meter, and micro-orifice restrictor. The bearing is lifted by the micro-orifice restrictor, forming an air film between the bearing and the platform to provide the bearing with load capacity. During the experiment, the bearing load can be changed by adjusting the mass of the weights, and the magnitude of the load is measured by a force sensor. The table frame is fixed on the column and clamps a non-contact capacitive displacement sensor. By measuring the displacement changes between the sensor and the bearing, the variation in the air film gap can be obtained. The measurement data are then transmitted to the computer to obtain the curve of the load capacity with the change in the air film thickness. The mass flow of the bearing is directly measured by the flow meter.

To test the static characteristics of aerostatic bearings under different air supply pressures and loads in a vacuum environment, a test platform was constructed, as shown in Figure 6. The test platform mainly consists of a measuring platform (1), a cushion block (2), a bearing to be tested (3), a weight (4), a gauge frame (5), a dial gauge (6), a vacuum tank (7), a ZJ51-T resistance gauge is sourced from Rui Bao Technology, Chengdu, China (8), a data acquisition and processing system (9), a vacuum pump (10), a computer (11), an air pump (12), a valve (13), a pressure gauge (14), and a flow meter (15).

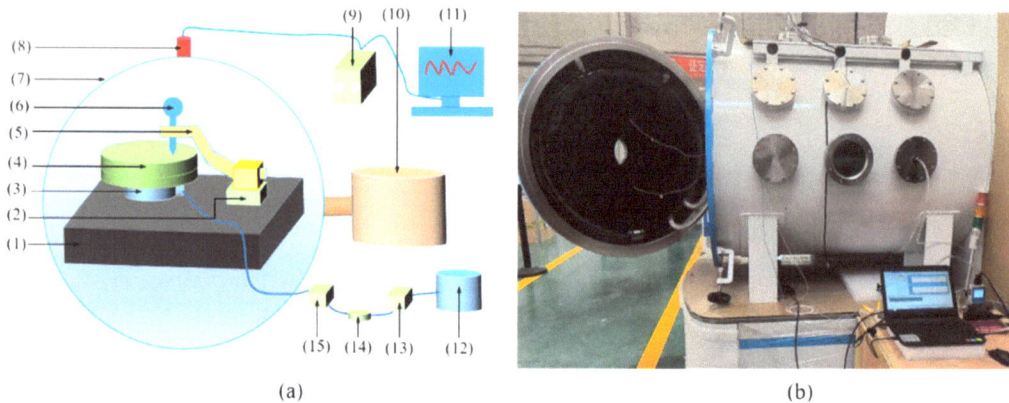

Figure 6. Static characteristics test of different environmental pressures: (**a**) schematic diagram; (**b**) test equipment.

The bearings are placed into a vacuum chamber, where the internal pressure is adjusted by extracting the air with a vacuum pump. The pressure inside the chamber is monitored by a resistance gauge, which feeds the values back to the computer. The test bearings are placed on a measuring platform, and the load capacity is adjusted by varying the mass of the weights. A dial gauge is used to measure the changes in the air film thickness of the bearings, while the mass flow is measured in real time using a flow meter. During the experiment, the working environment of the bearings is controlled by adjusting the air supply pressure and the vacuum level of the vacuum chamber.

3.2. Experimental Methods

3.2.1. Experimental Conditions

Different working conditions are achieved by changing the external environmental pressure and air supply pressure. The setting of working conditions is shown in Table 1. In order to study the influence of vacuum degree on bearing performance, according to the experimental conditions and the suction rate of the vacuum tank, environmental pressures are set at 40,000 Pa, 37,000 Pa, 3500 Pa, 2800 Pa, 2500 Pa, 2300 Pa, 1600 Pa, 1100 Pa, 890 Pa. To ensure the accuracy of experimental data, multiple operating conditions are set to explore the performance of bearings. Air supply pressures are set at 0.1 MPa, 0.2 MPa, 0.3 MPa, 0.4 MPa, 0.5 MPa, and 0.6 MPa according to the pressure provided by the pump source.

Table 1. Test conditions.

Environmental pressure (Pa)	40,000	37,000	3500	2800	2500	2300	1600	1100	890
Air supply pressure (MPa)	0.1	0.2	0.3	0.4	0.5	0.6			

3.2.2. Variables of Experimental Measurement

Bearing capacity and mass flow are the variables measured in atmospheric and vacuum environments. In atmospheric conditions, the weight of the weight is changed; the change in bearing displacement is measured by a capacitance sensor; the change in air film thickness is obtained, and the mass flow is measured by the flow meter. Under different environmental pressures, the weight of the weight is changed; the change in bearing displacement is measured by the dial meter, and the mass flow is also measured by the flowmeter.

Considering the experimental error, three measuring points were taken for each working condition to measure the change in the air film thickness of the bearing, and the loading process and unloading were respectively measured and repeated three times. The average value of the obtained results is taken as the final experimental result, and finally, the experiment to obtain the influence of various parameters on bearing performance is conducted.

4. Results and Discussion

In addition to design variables, parameters set by simulation are shown in Table 2.

Table 2. Setting of simulation parameters.

Parameters	Value
Bearing diameter D (mm)	64
Distribution diameter of orifice d (mm)	32
Number of orifices n	36
Aperture of orifice d_0 (mm)	0.1
Air film thickness h (μm)	15
Air supply pressure P_s (MPa)	0.5
Ambient pressure P_e (MPa)	0.1
Viscosity of air η (N·s·m^{-2})	1.82×10^{-5}
Ambient temperature T_a (k)	293.15
Adiabatic coefficient k	1.4
Critical pressure ratio β_k	0.528
Throttle coefficient Ψ	0.8

4.1. Influence of Geometric Parameters on Bearing Performance

4.1.1. Influence of Number of Orifices on Bearing Performance

Figure 7 shows the variation in bearing capacity and mass flow with air film thickness and a number of orifices under normal pressure and 4000 Pa. In a vacuum environment, it can be observed from Figure 7c,e that the bearing load capacity decreases gradually with increasing air film thickness and increases with an increase in the number of orifices. When the number of orifices is 22, the bearing capacity is the smallest, and when the number of orifices is 50, the bearing capacity is the highest. According to Figure 7d, it can be seen that the mass flow of the air bearing increases with the increase in the air film thickness. When the air film thickness is 25 μm, the mass flow changes very slowly with the air film thickness. Therefore, when the air film thickness is large, the number of orifices has a more significant effect on the bearing mass flow, but when the air film thickness is between 10 μm and 20 μm, the number of orifices has a more significant effect on the bearing capacity. According to Figure 7f, it can be seen that the mass flow of the air bearing gradually increases with an increase in the number of orifices.

In atmospheric pressure conditions, it can be deduced from Figure 7a,b,e,f that while the load capacity and mass flow of the bearing follow a similar trend as the thickness of the air film and the number of orifices in the vacuum environment, the bearing load capacity decreases by about 150 N, and the mass flow decreases by about 0.2 L/min compared with the vacuum environment. The reason for this is as follows: at an ambient pressure of 4000 Pa, the outlet pressure of the bearing is lower, resulting in a greater difference between the internal and external pressures of the bearing and a stronger load capacity. As the outlet pressure of the bearing decreases, the outlet flow rate is higher than atmospheric pressure, so the mass flow is also higher than atmospheric pressure.

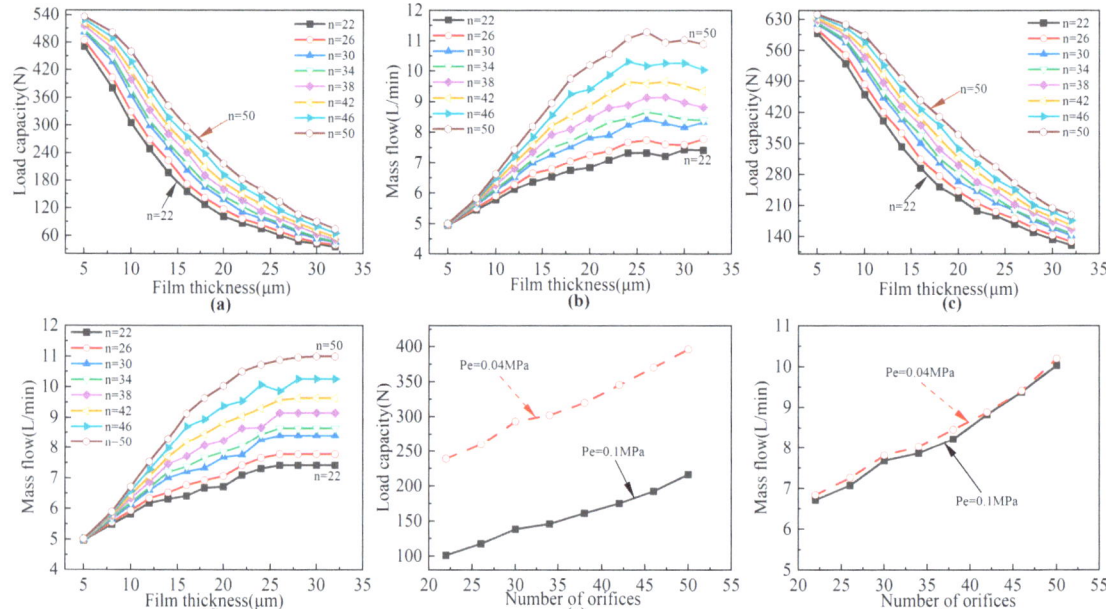

Figure 7. Influence of air film thickness and number of orifices on bearing performance under different environmental pressures: (**a**) influence of air film thickness and orifice number on bearing capacity under atmospheric pressure; (**b**) influence of air film thickness and orifice number on mass flow under atmospheric pressure; (**c**) influence of 4000 Pa environmental air film thickness and the number of orifices on the bearing capacity; (**d**) influence of air film thickness and orifice number in 4000 Pa environment on mass flow; (**e**) influence of the number of orifices under different environmental pressure on the bearing capacity when the air film thickness is 20 μm; (**f**) influence of the number of orifices under different ambient pressure on the mass flow when the air film thickness is 20 μm.

4.1.2. Influence of Distribution Diameter of Orifice on Bearing Performance

Figure 8 shows the variations in bearing load capacity and mass flow of the bearing with changes in air film thickness and orifice distribution diameter at atmospheric pressure and 4000 Pa.

In a vacuum environment, it can be observed from Figure 8c,e that with the increase in the orifice distribution diameter, the load capacity of the bearing first increases and then decreases. When the orifice distribution diameter is 38 mm, the bearing has the maximum load capacity, while at 55 mm, it has the minimum load capacity. The reason behind this phenomenon lies in the formation of a stable high-pressure zone as the air flows from the orifice toward the center of the bearing. When the orifice is closer to the center position, the proportion of the intermediate high-pressure zone is relatively small, resulting in a lower bearing load capacity. As the orifice distribution diameter gradually increases, the proportion of the intermediate high-pressure zone also increases, leading to an enhancement in the bearing load capacity. However, as the orifice distribution diameter continues to increase, the orifice approaches the bearing outlet, causing the air to rapidly overflow from the outlet, thereby reducing the bearing load capacity. According to Figure 8d, it is evident that when the air film thickness is less than 25 μm, the mass flow of the bearing increases slowly with the increase in orifice position. When the air film thickness exceeds 25 μm, the mass flow of the bearing is 9.15 L/min and remains constant as the orifice distribution diameter no longer changes.

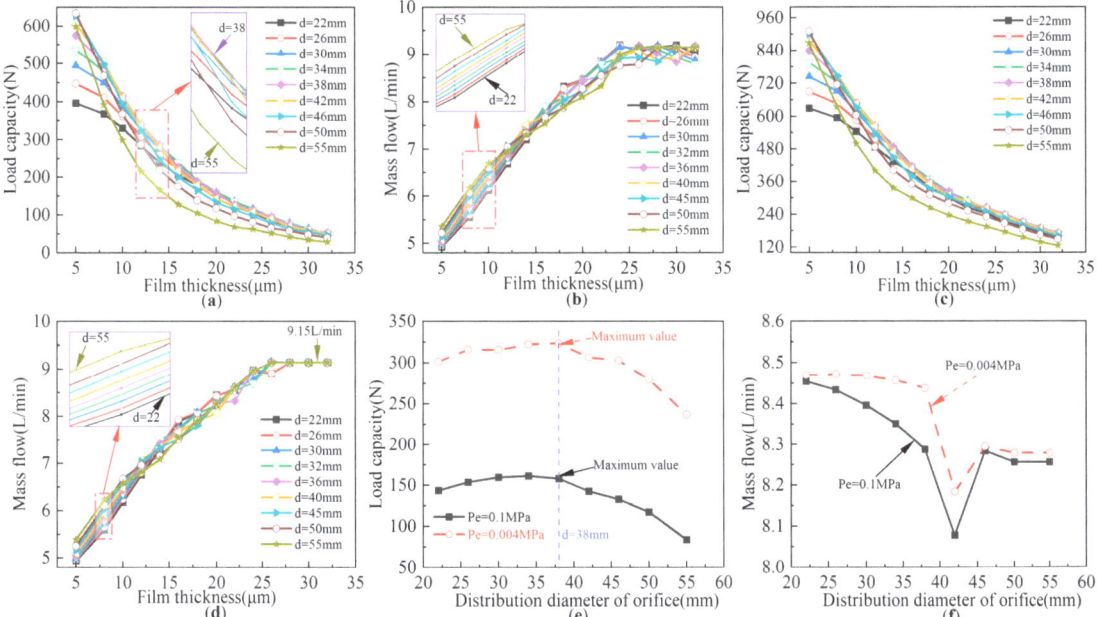

Figure 8. Influence of air film thickness and orifice distribution diameter on bearing performance under different environmental pressures: (**a**) influence of air film thickness and orifice distribution diameter on bearing capacity under atmospheric pressure; (**b**) influence of air film thickness and orifice distribution diameter on mass flow under atmospheric pressure; (**c**) influence of air film thickness and orifice distribution diameter on bearing capacity in 4000 Pa environment; (**d**) influence of air film thickness and orifice distribution diameter on mass flow in 4000 Pa environment; (**e**) influence of the distribution diameter of orifice under different ambient pressure on the bearing capacity when the air film thickness is 20 μm; (**f**) influence of different ambient pressure distribution diameters on mass flow when the air film thickness is 20 μm.

In the atmospheric environment, it can be inferred from Figure 8a,b,e,f that the overall trends of load capacity and mass flow follow a similar pattern to that of the vacuum environment. However, the load capacity value decreases by approximately 160 N compared to the vacuum environment, while the maximum mass flow decreases by around 0.2 L/min. Based on the analysis above, when the orifice distribution diameter is 38 mm, the bearing exhibits a higher load capacity.

4.1.3. Influence of Orifice Aperture on Bearing Performance

Figure 9 shows the variation in bearing capacity and mass flow with air film thickness and orifice aperture under atmospheric pressure and 4000 Pa.

According to Figure 9, it can be seen that at atmospheric and environmental pressures of 4000 Pa, as the orifice aperture increases, the bearing capacity and mass flow gradually increase. Moreover, the bearing capacity in a vacuum is 200 N higher than atmospheric pressure, and the mass flow is 0.1–0.2 L/min higher than atmospheric pressure. When the air film thickness is between 10 μm and 20 μm, the bearing's load capacity uniformly changes with the orifice aperture, while the mass flow of the bearing increases slowly. When the air film thickness exceeds 20 μm, the gradient of load capacity change remains consistent with before, but the mass flow increases rapidly. The reason is that the larger the orifice aperture, the larger the throttling area, and when the air film thickness is large, the outlet pressure of the bearing is very small, resulting in a higher flow rate, ultimately leading to a rapid increase in the mass flow of the bearing. When the load is constant, the air film thickness of the bearing in a vacuum is higher than that under atmospheric pressure. Therefore, the air film thickness of the bearing should not be too high. Based on the influence of the number and diameter of the orifices on the bearing performance, the air film thickness should be selected as 20 μm.

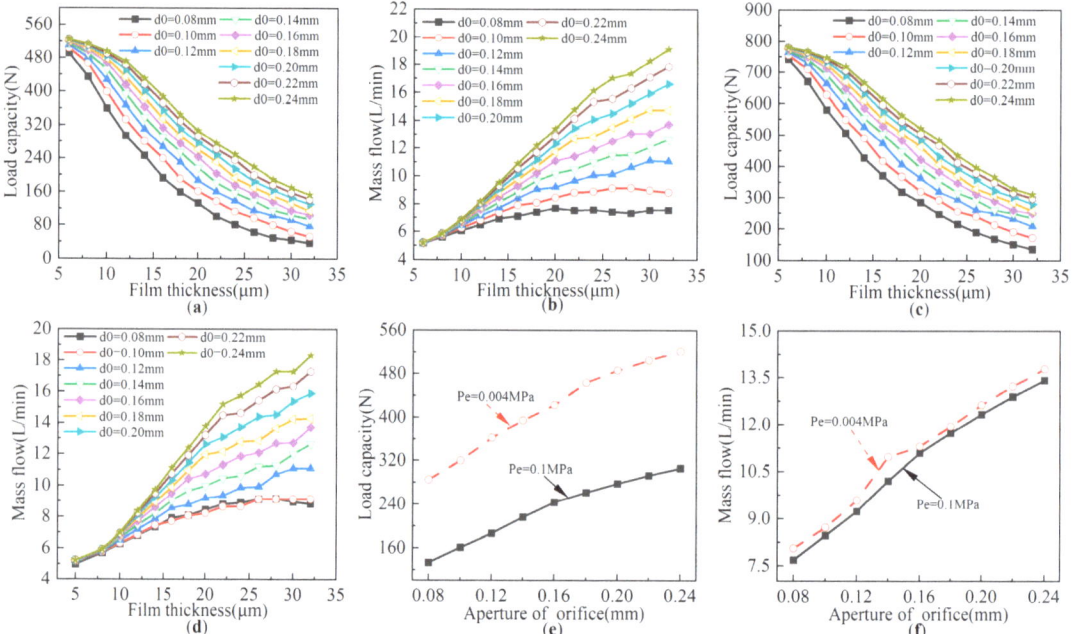

Figure 9. Influence of air film thickness and orifice aperture on bearing performance under different environmental pressures: (**a**) influence of air film thickness and orifice aperture on load capacity in atmospheric pressure conditions; (**b**) influence of air film thickness and orifice aperture on mass flow in atmospheric pressure conditions; (**c**) influence of air film thickness and orifice aperture on load capacity under 4000 Pa environmental pressure; (**d**) influence of air film thickness and orifice aperture on mass flow under 4000 Pa environmental pressure; (**e**) influence of orifice aperture on load capacity under varying environmental pressures with an air film thickness of 20 μm; (**f**) influence of orifice aperture on mass flow under varying environmental pressures with an air film thickness of 20 μm.

4.2. Influence of Environmental Conditions on Bearing Performance

4.2.1. Influence of Air Supply Pressure on Bearing Performance

Figure 10 shows the influence of air supply pressure on the performance of bearings under atmospheric pressure and an ambient pressure of 4000 Pa.

According to Figure 10, it can be seen that at atmospheric and environmental pressures of 4000 Pa, the bearing capacity and mass flow of the bearing increase with the increase in air supply pressure. When the air supply pressure is 0.2 MPa, the bearing capacity increases slowly with the decrease in air film thickness, and the mass flow also increases slowly. When the air supply pressure is increased to 0.6 MPa, the bearing capacity increases rapidly with the decrease in film thickness, and the mass flow increases rapidly. When the environmental pressure is 4000 Pa, the bearing capacity is increased by about 20% compared with atmospheric pressure, and the mass flow is about 1.05 times atmospheric. In the vacuum environment, the floating amount increases compared with the normal pressure. To mitigate the self-excited vibration of the bearings and enhance their stability, the air supply pressure of 0.4 MPa is recommended based on Figure 10.

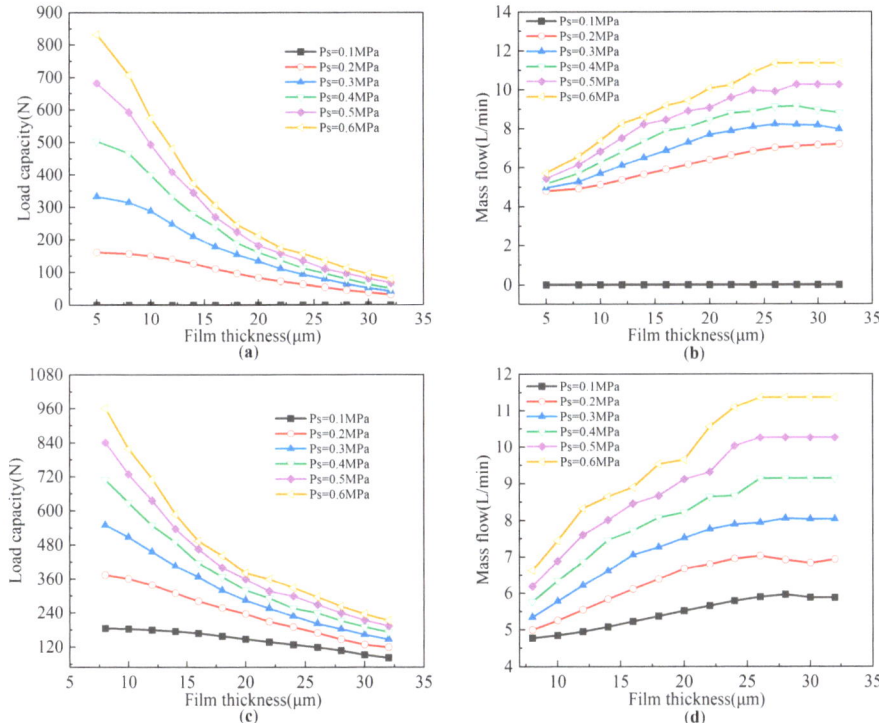

Figure 10. Influence of air supply pressure on bearing performance under different environmental pressures: (**a**) influence of atmospheric air supply pressure on the bearing capacity; (**b**) influence of atmospheric air supply pressure on mass flow; (**c**) influence of 4000 Pa ambient air supply pressure on the bearing capacity; (**d**) influence of 4000 Pa ambient air supply pressure on mass flow.

4.2.2. Influence of Environmental Pressure on Bearing Performance

Figure 11 shows the variation in bearing capacity and mass flow with air film thickness under different environmental pressures. According to Figure 11a, bearing capacity gradually increases with the reduction in environmental pressure. When the environmental pressure is 37 kPa to 40 kPa, the bearing capacity increases by about 30% compared with atmospheric pressure; when the environmental pressure is reduced below 3.5 kPa, the bearing capacity is increased by about 50% compared to the atmospheric pressure environment. The bearing capacity increases very slowly by continuing to reduce the environmental pressure. According to Figure 11b, it can be seen that the mass flow of the bearing in vacuum increases by about 5% compared with the atmospheric pressure environment, but the mass flow increases very slowly with the increase in vacuum degree. Therefore, when the vacuum degree is about 3.5 kPa, the bearing capacity and mass flow reach the best parameters.

4.3. Pareto Frontier Optimal Solution

Transfer functions such as trainlm, trainbr, trainscg, and traingdx are used for fitting $F(X)$. The determination coefficients obtained according to Equation (12) are 0.9621, 0.9997, 0.9977, and 0.9650, respectively. Therefore, trainbr is selected to train the data to obtain $F(X)$.

When the ambient pressure is 3500 Pa, the air supply pressure is 0.4 MPa, and the air film thickness is 20 μm; the optimal solution of the Pareto front of the objective function is shown in Figure 12. According to Figure 12, a total of 29 groups of optimal solutions were obtained by using the optimization algorithm, and the G value was the 28th solution. In this case, the distribution diameter of the orifice corresponding to the 28th solution is

38.83 mm; the number of orifices is 36, and the aperture of the orifice is 0.1 mm. At this time, the corresponding bearing capacity in the vacuum environment is 460.644 N, and the mass flow is 11.816 L/min. According to this parameter, the static test of the bearing is carried out under different air supply pressure and environmental pressure.

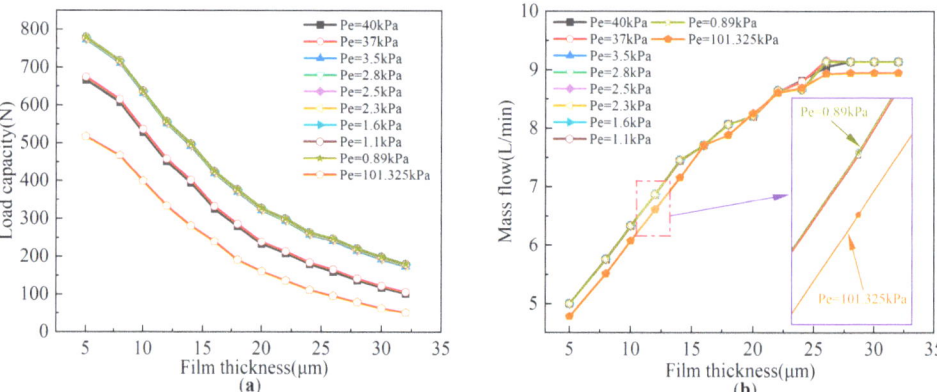

Figure 11. Influence of different environmental pressures on bearing performance: (**a**) influence of different environmental pressures on the bearing capacity; (**b**) influence of different environmental pressures on mass flow.

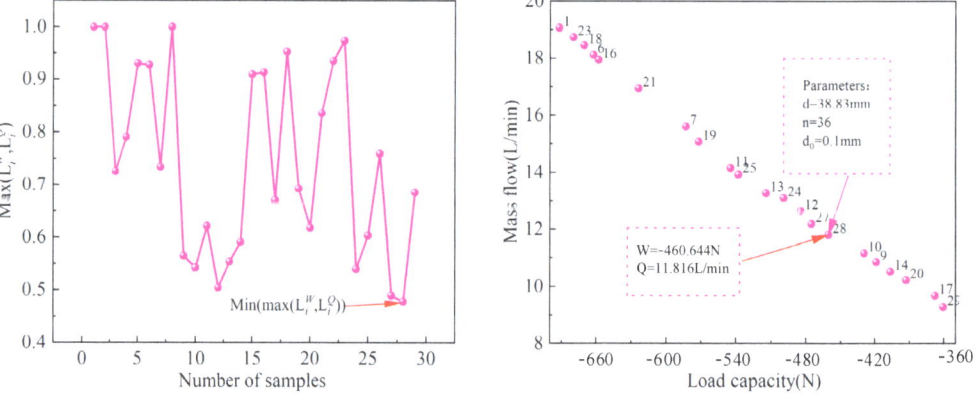

Figure 12. Pareto frontier optimal solution.

In order to verify the accuracy of the NSGA-II multi-objective optimization algorithm, particle swarm optimization algorithm, simulated annealing algorithm, and Tabu search algorithm were used to obtain bearing capacity and mass flow. The results are shown in Table 3. It can be seen from Table 3 that compared with other algorithms, the ratio of carrying capacity to mass flow obtained by the NSGA-II algorithm is the largest, and the solution is optimal under this algorithm.

Table 3. Algorithm comparison.

Algorithms	NSGA-II Algorithm	Particle Swarm Optimization	Simulated Annealing Algorithm	Tabu Search Algorithm
Bearing capacity W (N)	460.644	480.598	500.987	470.325
Mass flow Q (L/min)	11.816	12.658	14.658	12.126
Ratio	38.98	37.97	34.18	38.79

4.4. Comparison between Simulation and Experiment

The bearing material is S136 stainless steel made in Tianjin, China. In the case of high load and small mass flow, the optimized bearing parameters are shown in Table 4. The physical picture of the bearing is shown in Figure 13.

Table 4. Bearing parameters.

Parameters	Value
Bearing diameter D (mm)	64
Distribution diameter of orifice d (mm)	38.8
Number of orifices n	36
Aperture of orifice d_0 (mm)	0.1

Figure 13. Bearing picture.

When the ambient pressure is 0.1 MPa, and the air supply pressure is 0.4 MPa, 0.5 MPa, and 0.6 MPa, respectively, the bearing capacity and mass flow obtained by simulation and test are shown in Figure 14. According to Figure 14, under different air supply pressures, bearing capacity decreases with the increase in air film thickness, and mass flow increases with the increase in air film thickness. The load capacity and mass flow obtained by simulation are in good agreement with the experimental values. To further verify the correctness of the simulation, the influence of different environmental pressure and supply pressures on the bearing air film thickness when the bearing load is 280 N was tested, and the results are shown in Figure 15.

According to Figure 15a, the air film thickness and mass flow gradually increase with the decrease in environmental pressure. The reason is that the bearing outlet flow rate gradually increases with the reduction in pressure, so when the load is constant, the air film thickness is higher, and the flow rate is larger. According to Figure 15b, when the load is constant, the air film thickness and mass flow increase with the increase in air supply pressure. The maximum error of the whole test is about 10%. The main reasons for the difference between the theoretical value and the test value are as follows: First, the flatness error of the bearing surface and the inclination of the bearing during measurement are not considered in the simulation. Second, the external environment, such as noise and equipment vibration, will cause fluctuations in the thickness of the air film. In addition, according to the previous analysis, the orifice aperture has a great impact on the bearing performance. Because the orifice aperture is small and difficult to process, it will produce errors with the simulated theoretical value. Considering the influence of processing and assembly errors, we have taken a safety factor of 1.5 times under rated load conditions, so it still meets the requirements for use under this error.

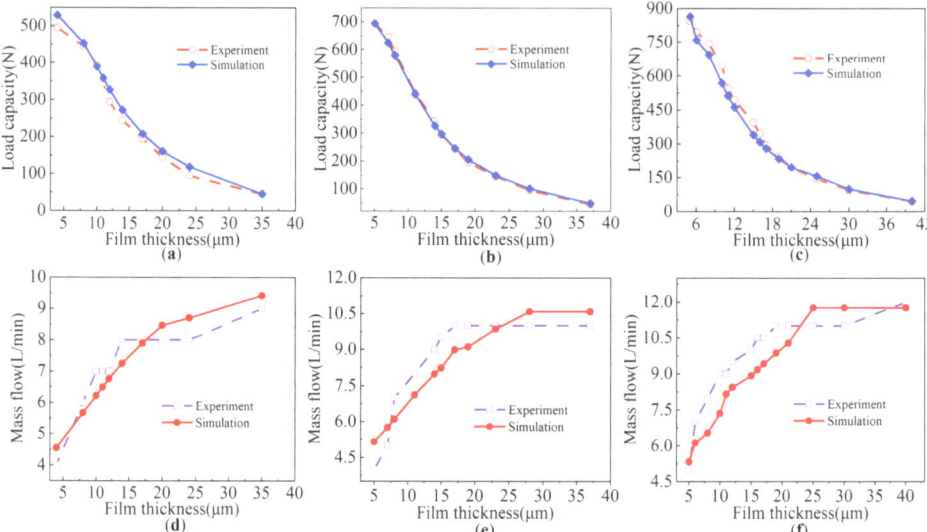

Figure 14. Comparison of simulation and experiment in atmospheric pressure environment: (**a**) comparison of bearing capacity when $Ps = 0.4$ MPa; (**b**) comparison of bearing capacity when $Ps = 0.5$ MPa; (**c**) comparison of bearing capacity when $Ps = 0.6$ MPa; (**d**) mass flow comparison when $Ps = 0.4$ MPa; (**e**) mass flow comparison when $Ps = 0.5$ MPa; (**f**) mass flow comparison when $Ps = 0.6$ MPa.

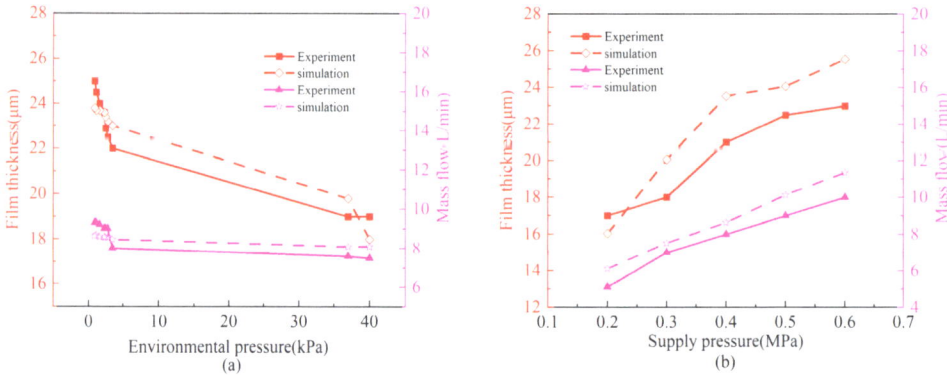

Figure 15. Comparison of simulation and experiment under different environmental pressures and different air supply pressures: (**a**) comparison of simulation and experiment of different environmental pressures; (**b**) comparison of simulation and experiment of different air supply pressures.

5. Conclusions

In this paper, the influence of different parameters and external environments on the bearing capacity and mass flow of air bearings applied in a vacuum environment were analyzed by combining numerical simulation and experimental verification, and the bearing structure was optimized. The following conclusions were reached:

1. In the vacuum environment, because the outlet pressure of the bearing is lower than that of the atmospheric pressure, the pressure difference between the inside and outside of the bearing increases, and the outlet flow rate increases, resulting in the bearing capacity and flow rate being increased compared with the atmospheric pressure environment and gradually increasing with the decrease in vacuum degree;

2. With the optimization goals of maximum load capacity and minimum mass flow of bearings in the vacuum, the objective function was fitted using a BP neural network and combined with the NSGA-II multi-objective optimization algorithm to optimize the bearing parameters. When the bearing's aperture of the orifice is 0.1 mm, the number of orifices is 36, and the distribution diameter of the orifice is 38.83 mm, the bearing has a maximum load capacity of 460.644 N and a minimum mass flow rate of 11.816 L/min;
3. Experiments are carried out on the optimized bearing parameters in atmospheric pressure and vacuum environments. The maximum error between the experimental and theoretical values is 10%, which verified the correctness and feasibility of the simulation and provided a theoretical and experimental method for the design of the aerostatic thrust bearing in a vacuum;
4. When designing bearings for application in a vacuum environment, the optimization and experimental methods proposed in this paper can quickly design ideal parameters, greatly improving the efficiency of design. This is of great significance for the design, manufacturing, and maintenance of bearings;
5. Although the optimization method of bearing parameters presented in this paper is carried out under specific fields, materials, and experimental conditions, the analysis and calculation methods presented in this paper are still of popularization significance for other fields, materials, and experimental conditions. In future studies, we will try to obtain the output values for how to use numerical methods in other experimental conditions.

Author Contributions: Conceptualization, G.F. and Y.L. (Youhua Li); methodology, G.F. and H.Y.; software, J.X. and G.F.; validation, Z.L. and M.Z.; formal analysis, Y.L. (Yuehua Li) and H.L.; investigation, H.L. and W.H.; data curation, G.F. and W.H.; writing—original draft preparation, G.F. and H.Y.; writing—review and editing, Z.L. and L.Z.; visualization, H.Y. and G.Z.; funding acquisition, H.Y. and M.Z. All authors have read and agreed to the published version of the manuscript.

Funding: This research was funded by the National Nature Science Foundation of China, grant number 51875586; the Tianjin Natural Science Foundation grant number 22JCZDJC00910.

Data Availability Statement: The original contributions presented in the study are included in the article material, further inquiries can be directed to the authors.

Conflicts of Interest: The authors declare no conflicts of interest.

References

1. Wapman, J.D.; Sternberg, D.C.; Lo, K.; Wang, M.; Jones-Wilson, L.; Mohan, S. Jet Propulsion Laboratory Small Satellite Dynamics Testbed Planar Air-Bearing Propulsion System Characterization. *J. Spacecr. Rockets* **2021**, *58*, 954–971. [CrossRef]
2. Miyatake, M.; Yoshimoto, S. Numerical investigation of static and dynamic characteristics of aerostatic thrust bearings with small feed holes. *Tribol. Int.* **2010**, *43*, 1353–1359. [CrossRef]
3. Gao, S.; Cheng, K.; Chen, S.; Ding, H.; Fu, H. Computational design and analysis of aerostatic journal bearings with application to ultra-high speed spindles. *Proc. Inst. Mech. Eng. Part C* **2017**, *231*, 1205–1220. [CrossRef]
4. Chakraborty, B.; Bhattacharjee, B.; Chakraborti, P.; Biswas, N. Evaluation of the performance characteristics of aerostatic bearing with porous alumina (Al_2O_3) membrane using theoretical and experimental methods. *Proc. Inst. Mech. Eng. Part J* **2024**, 13506501241245760. [CrossRef]
5. Khim, G.; Park, C.H.; Lee, H.; Kim, S.W. Analysis of additional leakage resulting from the feeding motion of a vacuum-compatible air bearing stage. *Vacuum* **2006**, *81*, 466–474. [CrossRef]
6. Fukui, S.; Kaneko, R. Experimental Investigation of Externally Pressurized Bearings Under High Knudsen Number Conditions. *J. Tribol.* **1988**, *110*, 144–147. [CrossRef]
7. Trost, D. Design and analysis of hydrostatic gas bearings for vacuum applications. In Proceedings of the ASPE Spring Topical Meeting on Challenges at the Intersection of Precision Engineering and Vacuum Technology, Orlando, FL, USA, 2 April 2006.
8. Schenk, C.; Buschmann, S.; Risse, S.; Eberhardt, R.; Tünnermann, A. Comparison between flat aerostatic gas-bearing pads with orifice and porous feedings at high-vacuum conditions. *Precis. Eng.* **2008**, *32*, 319–328. [CrossRef]
9. Gans, R. Lubrication theory at arbitrary Knudsen number. *J. Tribol.* **1985**, *107*, 431–433. [CrossRef]
10. Chan, C.-W. Modified Particle Swarm Optimization Algorithm for Multi-Objective Optimization Design of Hybrid Journal Bearings. *J. Tribol.* **2015**, *137*, 021101. [CrossRef]

11. Wang, N.; Chang, Y.Z. Application of the Genetic Algorithm to the Multi-Objective Optimization of Air Bearings. *Tribol. Lett.* **2004**, *17*, 119–128. [CrossRef]
12. Wang, N.; Cha, K.-C. Multi-objective optimization of air bearings using hypercube-dividing method. *Tribol. Int.* **2010**, *43*, 1631–1638. [CrossRef]
13. Yifei, L.; Yihui, Y.; Hong, Y.; Xinen, L.; Jun, M.; Hailong, C. Modeling for optimization of circular flat pad aerostatic bearing with a single central orifice-type restrictor based on CFD simulation. *Tribol. Int.* **2017**, *109*, 206–216. [CrossRef]
14. Wang, N.; Chen, H.-Y. A two-stage multiobjective optimization algorithm for porous air bearing design. *Tribol. Int.* **2016**, *93*, 355–363. [CrossRef]
15. Zhang, T.; Chen, L.; Wang, J. Multi-objective optimization of elliptical tube fin heat exchangers based on neural networks and genetic algorithm. *Energy* **2023**, *269*, 126729. [CrossRef]
16. Wang, G.; Li, W.; Liu, G.; Feng, K. A novel optimization design method for obtaining high-performance micro-hole aerostatic bearings with experimental validation. *Tribol. Int.* **2023**, *185*, 108542. [CrossRef]
17. Chang, L.; Tenghui, J.; Xing, H.; Xiansheng, J. Study on parameter optimization of laser cladding Fe60 based on GA-BP neural network. *J. Adhes. Sci. Technol.* **2023**, *37*, 2556–2586.
18. Gao, F.; Zheng, Y.; Li, Y.; Li, W. A back propagation neural network-based adaptive sampling strategy for uncertainty surfaces. *Trans. Inst. Meas. Control.* **2024**, *46*, 1012–1023. [CrossRef]
19. Ding, C.; Ding, Y.; Yuan, Z.; Li, J. Structural Optimization Design of Electromagnetic Repulsion Mechanism Based on BP Neural Network and NSGA–II. IEEJ Trans. *Electr. Electron. Eng.* **2023**, *18*, 1914–1922.
20. Guofang, L.; Xing, L.; Meng, L.; Tong, N.; Shaopei, W.; Wangcai, D. Multi-objective optimisation of high-speed rail profile with small radius curve based on NSGA-II Algorithm. *Veh. Syst. Dyn.* **2023**, *61*, 3111–3135.
21. Kim, I.Y.; de Weck, O.L. Adaptive weighted-sum method for bi-objective optimization: Pareto front generation. *Struct. Multidiscip. Optim.* **2005**, *29*, 149–158. [CrossRef]
22. Xu, X.F.; Xu, S.; Zhou, T. An Improved Spectral Clustering Algorithm Using Minimum Maximum Principle. *Appl. Mech. Mater.* **2012**, *1810*, 1881–1884. [CrossRef]

Disclaimer/Publisher's Note: The statements, opinions and data contained in all publications are solely those of the individual author(s) and contributor(s) and not of MDPI and/or the editor(s). MDPI and/or the editor(s) disclaim responsibility for any injury to people or property resulting from any ideas, methods, instructions or products referred to in the content.

Article

Study on the Friction Characteristics and Fatigue Life of Carbonitriding-Treated Needle Bearings

Yong Chen [1,*], Xiangrun Pu [2], Lijie Hao [2], Guangxin Li [2] and Li Luo [1]

1. State Key Laboratory of Featured Metal Materials and Life-Cycle Safety for Composite Structures, Guangxi University, Nanning 530004, China; luoli178050512@163.com
2. Tianjin Key Laboratory of Power Transmission and Safety Technology for New Energy Vehicles, School of Mechanical Engineering, Hebei University of Technology, Tianjin 300130, China; a17854190711@163.com (X.P.); jiejiehao2021@163.com (L.H.); ligx1229@163.com (G.L.)
* Correspondence: chenyong1585811@163.com

Abstract: Being a key component of the transmission system, the needle bearing's performance and service life affects the overall service life of mechanical equipment. This study takes needle bearings composed of AISI 52100 steel as the research object and studies the effect of carbonitriding surface strengthening treatment on the bearing friction, wear, and fatigue life. The carbon and nitrogen co-infiltration surface-strengthening method was employed to prepare cylindrical and disc samples. The surface hardness, residual austenite content, microscopic morphology and organization composition, coefficient of friction, and wear scar were studied to analyze the effect on the wear performance of the material. The bearing fatigue wear comparison test was conducted on a test bench to compare the actual fatigue life and surface damage of the needle bearing through conventional martensitic quenching heat treatment and carbonitriding treatment. The results demonstrate that the carbonitriding strengthening method enhances the toughness of the material while improving its surface hardness. It also improves the wear resistance of the needle roller bearings, and the fatigue life of the bearings is significantly improved. In conclusion, carbon and nitrogen co-infiltration treatment is a strengthening method that effectively extends the service life of needle roller bearings, indicating its high practical value.

Keywords: carbonitriding treatment; needle bearing; friction characteristics; fatigue life; experimentation; simulation

1. Introduction

As an important type of mechanical component, needle bearings have the function of speed and torque transmission. In the working process, the needle rollers repeatedly make contact with the shaft and the outer bore. The contact surface is subjected to high alternating stress, which generates a significant amount of heat due to mutual friction. This leads to a reduction in hardness and plastic deformation, thereby reducing the service life of the material [1]. The chosen rolling bearings must not only display the required strength, but must also have excellent fatigue resistance [2].

Contact fatigue wear is one of the most prevalent forms of friction failure in roller bearings. A considerable number of scholars have conducted research into the contact fatigue and friction characteristics of bearings [3–5]. Wang et al. [6] established a fatigue damage accumulation model to analyze the subsurface crack evolution and fatigue life of rolling contact fatigue. Morales-Espejel et al. [7] investigated the effect of roughness on contact bearings. Smolnicki et al. [8] adjusted the bearing finite element model's mesh size to study and analyze load distribution. Lostado et al. [9] carried out a stress analysis of a bearing using a numerical simulation model and compared the theoretical model with the results of contact sensor measurements. Kania et al. [10] studied the variation rules of the contact stresses and strains in various bearing parts. Bakolas et al. [11] proposed a method for the estimation of the frictional energy loss of a specific bearing type on a global scale. Ilie [12] conducted a study to analyze the diffusion and mass transfer mechanisms in a friction couple comprising steel and bronze. Previous studies have pointed out that the failure due to contact fatigue mainly occurs on the surface or in the subsurface. A mere improvement in the manufacturing precision or enhancement in the material composition cannot adequately increase the surface properties of the material. These measurements need to be combined with surface strengthening treatment technology.

Regarding the AISI 52100 bearing's carbonitriding process, the research analysis is mainly based on the test results, focusing on the influence of the furnace temperature, quenching temperature, carbon potential, and other factors on the organizational characteristics. Huan et al. [13] explored the improvement of the properties of AISI 52100 steel through two strengthening treatments, namely pre-carburizing + quenching and tempering versus, carbonitriding + quenching and temperature. The results showed that the latter could effectively reduce the friction coefficient and wear rate of AISI 52100 steel. Wan et al. [14] investigated the effect of laser impact shot peening and vacuum carburizing on the strength and plasticity of a 20CrMn2Mo alloy. The results showed that the yield and tensile strength of the treated material were improved. Rajan et al. [15] concluded that an AISI 52100 ball bearing's reliability was effectively improved after carbonizing. Liu et al. [16] conducted a detailed analysis of the black structure produced by carbonitriding the inner and outer rings of AISI 52100 bearings. Zhao et al. [17] investigated the bearing inner and outer rings of the quenching structure and the surface hardness under carbonitriding pretreatment. Karamis et al.'s [18] analysis showed that carbon–nitrogen strengthening increased the AISI 1020 mild steel's surface hardness by two times; its inherent toughness was not reduced, and the tribological properties were improved significantly. Kanchanomai et al. [19] analyzed the carbonitriding alloy steel's fatigue performance via residual compressive stress, concluding that the fatigue life of the carbonitriding steel was approximately five times that of a sample without carbonitriding treatment. Jiang et al. [20] used gas carbonitriding technology to strengthen the surface of martensitic bearing steel, significantly improving the residual compressive stress and austenite content of the sample. Co-infiltration heat treatment with carbon and nitrogen can effectively improve the material properties. Further research is required to know the effect of this process on the anti-wear performance and fatigue life of needle bearings.

In this study, research was conducted on carbonitride-strengthened needle bearings. The effect of the carbonitriding heat treatment on their surface friction and wear performance and fatigue life is discussed. Conventional martensitic quenching and tempering treatment and carbonitriding surface strengthening treatment were carried out on the bearing material AISI 52100 to produce test specimens and test bearings, respectively. The surface hardness, residual austenite content, microscopic morphology and organization composition, coefficient of friction, and wear scar were used to investigate the effectiveness of the conventional heat treatment. The fatigue life test was carried out on a fatigue test rig. A comparison of the fatigue life results and surface failure status of conventionally heat–treated needle bearings with those of carbonitriding-treated needle bearings is presented.

2. Materials

2.1. Preparation of Test Disc Samples

Before carrying out the friction wear and fatigue life tests, AISI 52100 steel was selected as the basic material to prepare the required specimens and needle bearings. The chemical composition of AISI 52100 steel is shown in Table 1.

Table 1. Chemical composition of AISI 52100 bearing steel materials.

Element	C	Si	Mn	P	S	Cr	Ti	H	O
Mass fraction (%)	0.98	0.22	0.33	0.010	0.003	1.48	0.0025	0.0001	0.0006

AISI 52100 bearing steel is an American standard high-strength alloy structural steel. The reference standard is ASTM A 29/A 29M-04 [21]. The corresponding ISO grade of this steel is 100Cr6, and the corresponding Chinese standard steel type is GCr15. The reference standard is ISO 683-17: 2014 [22]. These are designated as AISI 52100.

Since the contact form in needle bearings is primarily linear, tests were conducted with a disc that had a diameter of 24 mm and a thickness of 7.9 mm, as well as a cylinder with a diameter of 6 mm and a length of 8 mm. The test specimens are shown in Figure 1.

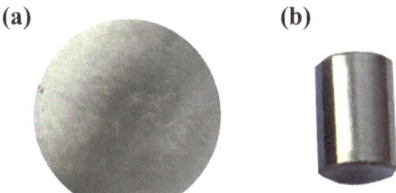

Figure 1. Disc and cylindrical specimens for testing: (**a**) the surface of the test disc (view); (**b**) the cylindrical specimen.

2.2. Carbonitriding Process

A controlled-atmosphere multi-purpose gas chamber furnace heat treatment production line was utilized for the corresponding carbon and nitrogen co-infiltration process. The furnace atmosphere consisted of carbon potential (CP) = 0.9~1.2%, CO_2 = 0.2~0.4%, NH_3 = 0.1~0.5%. After the product's quenching out of the furnace, it was placed in a vacuum cleaning machine before tempering.

The following chemical reactions occur during carbonitriding:

$$CH_4 + NH_3 \rightarrow HCN + 3H_2 \tag{1}$$

$$CO + NH_3 \rightarrow HCN + H_2O \tag{2}$$

$$2HCN \rightarrow H_2 + 2[C] + 2[N] \tag{3}$$

The depth of the carbonitride layer on the AISI 52100 bearing steel after carbonitriding treatment was 0.4–0.7 mm. The carbon concentration of the carbonitriding bearing surfaces was approximately 1.2%, and the carbon concentration of ordinary bearing steel is approximately 0.9–0.95%. After the carbonitriding treatment of the bearing steel, the infiltration layer's organization consisted of a small quantity of nitrogen and carbon compounds, nitrogenous martensite, residual austenite, and a small quantity of carbides.

2.3. Lubrication Conditions

To ensure the accuracy and consistency of the test results, the same lubricant was used for the wear tests and life tests. Table 2 shows the parameters of the utilized lubricants. The same amount of lubricant was added to each group before testing.

Table 2. Lubricant parameters.

Parameters	L-HM 32
Density at 20 °C (kg·m^{-3})	854.0
Dynamic viscosity at 40 °C (mm^2·s^{-1})	32.11
Dynamic viscosity at 100 °C (mm^2·s^{-1})	5.580
Viscosity index	112

2.4. Needle Bearing Models

In this study, the needle bearing K28 × 33 × 19 composed of AISI 52100 bearing steel, was selected as the main research object. The physical object is shown in Figure 2a,b, the geometrical parameters are shown in Table 3, and its three-dimensional model is shown in Figure 2c.

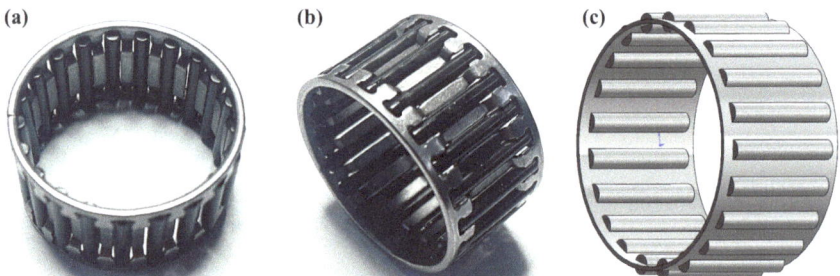

Figure 2. Needle bearing K28 × 33 × 19 (**a**,**b**) needle bearing object; (**c**) needle bearing 3D model.

Table 3. The parameters of needle bearing K28 × 33 × 19.

Parameter	Value
Dimension (mm)	28 × 33 × 19
Needle diameter (mm)	2.5
Needle roller length (mm)	15.8
Number of rolling bodies (Pcs)	21
Needle roller tolerance grade	G2
Pitch circle diameter (mm)	30.5
Ultimate RPM (rpm)	16,000
Rated dynamic load (kN)	19.6
Rated static load (kN)	32.4

3. Methods

3.1. Line Contact Friction Test

The Schwing-Reib-Verschleiss-5 (SRV-5) tester, manufactured by the Optimol Instruments Prueftechnik GmbH (München, Germany), is a physical property testing instrument used for friction and wear testing in the physical field. The SRV-5 test setup is shown in Figure 3a and the motion shape of the specimen under test is shown in Figure 3b. During the test, the lower specimen (disc) was held in place, while the upper specimen (cylinder) was subjected to reciprocating frictional movement. Prior to the test, the lubricant was injected through a burette into the contact area between the upper and lower specimens. The load was applied vertically to the disc through a transducer.

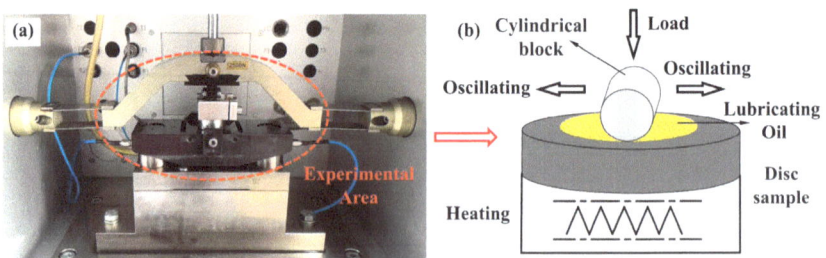

Figure 3. SRV-5 friction and wear test rig: (**a**) SRV-5 test equipment; (**b**) test motion principle.

Taking the automatic transmission bearings as the research object, the maximum input torque was usually between 250 Nm and 350 Nm, and the contact stress was between

1000 and 2500 MPa. Taking the working conditions of the bearing test as the reference, in combination with the D7421-11 [23] of the American Society for Testing and Materials (ASTM), the test loads were set at 100 N, 200 N, and 300 N. The specific test parameters are shown in Table 4 [24].

Table 4. SRV friction and wear test parameters.

Test Parameter	Line Contact Experiments
Load (N)	100/200/300
Experimental period (s)	3600
Temperature (°C)	80
Frequency (Hz)	50
Sliding velocity (m/s)	0.1
Oil volume (mL)	0.3

3.2. Surface Inspection of Test Samples

The subject of the carbonitriding needle bearing surface inspection project, which was the subject of this study, involved the following equipment:

(1) The HMV-G series micro-Vickers hardness tester produced by QATM GmbH (Mammelzen, Germany), was used for the surface hardness measurement of the samples;

(2) An AreX-L diffraction analyzer (X-ray diffractometer, XRD) from GND S.R.L. (Agrate Conturbia, Italy) was employed to investigate the variations in the residual austenite content. During the measurements, Mo radiation and a 1 mm collimator were applied to detect the magnitude of the diffraction angle and the intensity of the diffraction peaks at (200) α, (211) α, (200) γ, (220) γ, and (311) γ, respectively;

(3) The surface was analyzed using a scanning electron microscope (SEM) manufactured by JEOL K.K. (Tokyo, Japan). An energy spectrometer (Energy-Dispersive Spectrometer, EDS) from Oxford Instruments PLC (Oxfordshire, England) were used to observe the morphology and composition of the surface of each sample;

(4) A SURFCOM NEX 001SD-12 surface roughness measurement instrument manufactured by Tokyo Seimitsu CO., LTD (Tokyo, Japan) was employed to assess the surface roughness of the disc samples. The three-dimensional morphology and roughness were analyzed using white light interferometry, provided by Bruker Nano Surfaces (Beijing, China).

3.3. Fatigue Life Simulation

The needle and rotating shaft contact of needle bearings can be considered as line contact and calculated using Hertz contact theory. The two-cylinder contact was simplified as an equivalent cylinder and infinite-length plane contact, and the empirical formula proposed by Palmgren was used to calculate the elastic deformation convergence δ:

$$\delta = 3.81 \left(\frac{1-v_1^2}{\pi E_1} + \frac{1-v_2^2}{\pi E_2} \right)^{0.9} \frac{Q^{0.9}}{l^{0.8}} \quad (4)$$

where Q is the rolling load, and l is the rolling body length. For steel bearings, $v_1 = v_2 = 0.3$, $E_1 = E_2 = 2.06 \times 10^5$ N/mm². For rolling bearings, the rated life can be calculated using Equation (5):

$$L_{10} = \left(\frac{C}{P} \right)^\varepsilon \quad (5)$$

where L_{10} is the rated life in 10^6 revolutions, P is the equivalent dynamic load, and C is the dynamic load rating of the bearing. When the outer ring is stationary and the inner ring is rotating, the rated life of the bearing is calculated in terms of the number of revolutions and the continuous load in one direction that it can withstand. C can be obtained from Equation (6).

$$C = b_m f_c (i l \cos \alpha)^{\frac{7}{9}} Z^{\frac{3}{4}} D_w^{\frac{29}{27}} \tag{6}$$

In the formula, b_m is the material correction factor; for needle bearings, it usually takes a value of 1. f_c is the rated basic life calculation coefficient of the bearing, which is usually closely related to the bearing's geometric parameters. i refers to the rolling bearing with i rows of rolling elements. a refers to the nominal contact angle of the bearing. Z refers to the number of rolling elements. D_w is the diameter of the needle rollers.

The equivalent dynamic load P of a bearing describes the value of the load that the bearing can support while operating. The equivalent dynamic load is defined as a preset load of constant magnitude and direction, under which the fatigue service life of a bearing is equivalent to its service life under an actual load. In the case of needle bearings subjected to radial loads, the equivalent dynamic load can be determined using Equation (7).

$$P = x F_r + y F_a \tag{7}$$

F_r represents the radial load, while F_a is the axial load. It can be demonstrated that a needle bearing is capable of resisting only radial forces and is, therefore, unable to withstand axial forces. Consequently, the values of x and y are both equal to 1. Moreover, x is assigned a numeric value of 1, and y is assigned a value of 0. During the calculation process, the fatigue life of needle bearings is commonly expressed in terms of the total number of hours for which the bearing has been in operation. If the rotational speed of the bearing is known, the correction factor in Equation (5) is adjusted to read as follows.

$$L_h = \frac{10^6}{60n} \left(\frac{C}{P}\right)^{\varepsilon} \tag{8}$$

where L_h is the bearing life calculated via the number of working hours, and n is the rotational speed of the bearing, where the unit is r/min.

To ascertain the impact of the carbonitriding treatment process on the lifespan of the needle bearings, the Romax DT 2023.1 bearing professional analysis software was used. This enabled the set-up for construction of a needle bearing test bench model, which was then used to compare and analyze the lifespans of the needle bearings before and after the carbonitriding treatment. The three-dimensional model of the axle system of the needle bearing test bench and the physical figure are presented in Figure 4.

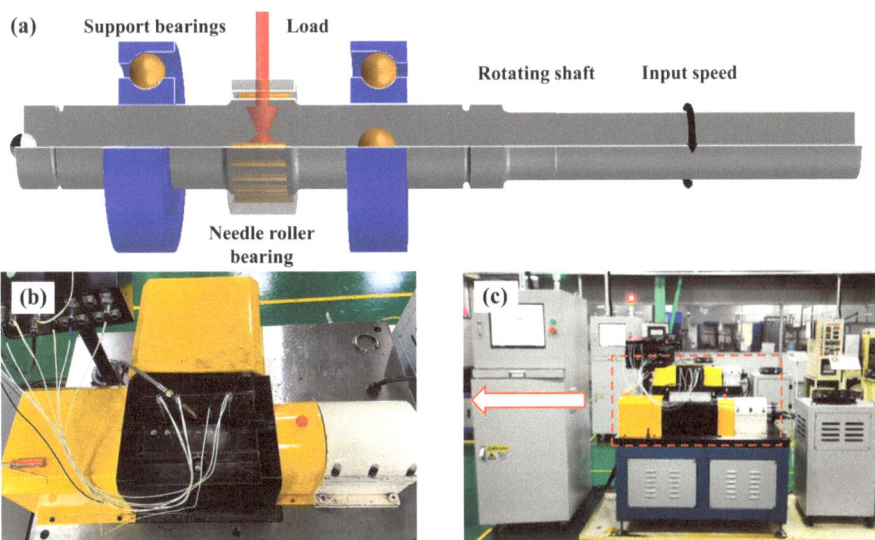

Figure 4. Needle bearing life simulation and experiment: (**a**) needle bearing test rig model; (**b**,**c**) schematic diagram of needle bearing test rig.

3.4. Fatigue Life Test

The needle bearing life test adopted a radial load of 2~4 tons in a type II testing machine. The equipment was programmed to operate automatically in accordance with the pre-set test load, speed, and running time, as well as other main parameters. It employed a vibration and temperature over-run control mode, and utilized the hydraulic step-by-step loading method to subject the bearing to the test load in stages.

A number of sets of needle bearings were prepared for the conventional heat treatment and carbonitriding treatment. These tests utilize two test conditions, normal load and high load, with each condition being prepared with two sets of conventional heat treatment and two sets of carbonitriding treatment of the needle bearings. The preparation of the test samples and test conditions is presented in Table 5 below.

Table 5. Carbonitriding needle bearing test conditions.

Test Conditions	Rotation Speed (rpm)	Radial Load (kN)	Number of Bearings for Test (Pieces)	
			Conventional Heat Treatment	Carbonitriding Treatment
Normal load	4000	4.9	2	2
High load	4000	8.6	2	2

The most prevalent forms of wear in rolling bearings are surface fatigue wear, abrasive wear, and adhesive wear. Adhesive wear and fatigue wear occur when the conditions are inadequate to form the lubricant film required to separate the bearing's contacting surfaces.

The C/P values were 4.0 and 2.28, respectively, for the normal and high load conditions, with a viscosity ratio of k = 0.8. The experimental bearings exhibited high resistance to adhesive wear due to their elevated hardness, which was evidenced by the results of the conducted experiments. The low viscosity of the utilized lubricant may have resulted in the formation of slight adhesive wear between the contacting surfaces. It is possible that initial fatigue at the surface may result from surface defects caused by adhesive wear [25].

The standards used in this life test were the GB/T 6391-2010 [26] Rolling Bearings-Rated Dynamic Load and Rated Life and the GB/T 24607-2009 [27] Rolling Bearings-Life and Reliability Test and Evaluation.

4. Results

4.1. Effect of Carbonitriding Strengthening Treatment on Material Properties

4.1.1. Effect on Surface Hardness

The surface hardness of the conventionally heat-treated disc samples and carbonitriding-treated disc samples was evaluated using the HMV-G series micro-Vickers hardness tester. Five distinct positions on the surface of each test sample were selected for analysis, and the results are presented in Table 6.

Table 6. Comparison of samples' hardness.

	Treatment	HV1	HV2	HV3	HV4	HV5	Average
Hardness (HV)	General	757	788	757	757	804	772.6
	Carbonitriding	804	8740	780	884	865	841.4

The test results demonstrate that the average surface hardness of the carbonitriding treatment samples was significantly higher than that of the samples subjected to the conventional heat treatment, where the hardness value increased by 8.9%. It has been demonstrated that carbon and nitrogen co-infiltration heat treatment improves the nitrogen content of the surface material of AISI 52100 steel. The subsequent quenching and heat treatment of the samples resulted in the penetration of nitrogen atoms into the lattice of the material. This caused the lattice to expand and distort, and increased the strength and hardness of the steel along with the increase in the nitrogen content [28]. Furthermore, the yield strength of the material was improved.

Equation (9) defines the linear relationship between the tensile strength and the hardness of a material. The Vickers hardness (HV) is converted into the Brinell hardness (HB) and then incorporated into Equation (9), calculating tensile strength σ_b.

$$\sigma_b = 237.9 + 1.394 HB \tag{9}$$

The yield strength of the conventionally heat-treated sample was 2030 MPa, while the ultimate tensile strength was 2240 MPa in the life simulation. Table 6 demonstrates that the hardness of the surface of the needle bearings after the carbonitriding treatment was increased by 8.9%. Consequently, the yield strength of the carbonitriding-treated sample was 2210 Mpa, and the ultimate tensile strength was 2439 Mpa.

4.1.2. Effect on Residual Austenite

The diffraction peaks at (200) α, (200) γ, (211) α, (220) γ, and (311) γ were selected according to the XRD patterns. The residual austenite volume fraction was calculated using Equation (10):

$$V_\gamma = \frac{1}{n}\sum_{j=1}^{n}\frac{I_\gamma^j}{R_\gamma^j} \Big/ \left(\frac{1}{n}\sum_{j=1}^{n}\frac{I_\gamma^j}{R_\gamma^j} + \frac{1}{n}\sum_{j=1}^{n}\frac{I_\alpha^j}{R_\alpha^j}\right) \tag{10}$$

$$R = \frac{1}{V^2}F^2\rho\frac{1+\cos^2 2\theta}{\sin\theta \times \sin 2\theta}\exp(-2M) \tag{11}$$

where V_γ is the volume fraction (%) of the austenite phase; n is the number of diffraction peaks; R is the material scattering factor, which is calculated as shown in Equation (11); I is the integrated intensity of the diffraction peaks; θ is the diffraction angle; V is the volume of the single cell; F is the structure factor; ρ is the multiplicity factor; and exp(-2M) is the temperature factor.

Figure 5 illustrates the results of surface residual austenite detection in the conventionally heat-treated discs and carbonitriding-treated discs. The diffraction peaks of ferrite, indicated by (200) α and (211) α, and the diffraction peaks of austenite, indicated by (200) γ, (220) γ, and (311) γ, are shown. A comparison of Figure 5a,b reveals that the diffraction peak value of ferrite in the carbonitriding samples is significantly lower than that in the conventional samples, while the diffraction peak value of residual austenite is higher. This indicates that the residual austenite content of the sample is increased following the carbonitriding treatment. The residual austenite volume fractions of the conventionally heat-treated samples and the carbonitriding-treated samples calculated using Equation (10), were 7.37 ± 0.04% and 25.91 ± 0.12%, respectively. The residual austenite content increased by 251.58 ± 3.54%.

The greater the hardness of a material, the greater its resistance to wear indentation and the less likely it is to exhibit an indentation on the contact surface when contact is created. However, as the hardness of the material increases, its toughness is reduced, and its resistance to cracking is also reduced. Furthermore, an increase in the residual austenite content enhances the toughness of the material itself, which enhances the resistance to crack extension of the material. The carbonitriding treatment of materials resulted in the formation of nitrogen-containing martensite, which improved the hardness of the material's surface. This resulted in a comprehensive improvement in the material's wear resistance and resistance to damage. Consequently, the carbonitriding treatment of needle bearings effectively inhibited the generation and development of initial micro-cracks and improved the contact fatigue life of needle bearings.

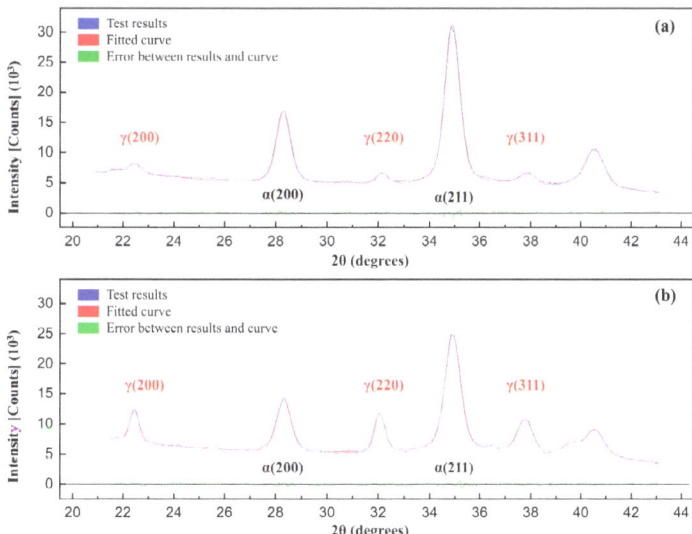

Figure 5. XRD diffraction intensity patterns of the samples: (**a**) conventional heat treatment sample; (**b**) carbonitriding treatment sample.

4.2. Effect of Carbon and Nitrogen Co-Infiltration Strengthening Methods on Microstructure

4.2.1. Before Friction Wear Test

Figure 6 shows the surface morphology and energy spectrum analysis of the conventionally heat-treated samples and carbonitriding-treated samples prior to the friction wear test. The EDS spectrum of randomly selected regions A and B are illustrated in Figure 6b,d. As depicted in Figure 6c, following the carbonitriding treatment, machining traces can still be observed on the surface of the test specimen. The surface morphology and the types and content of the constituent elements were nearly identical to those of the conventionally heat-treated specimen.

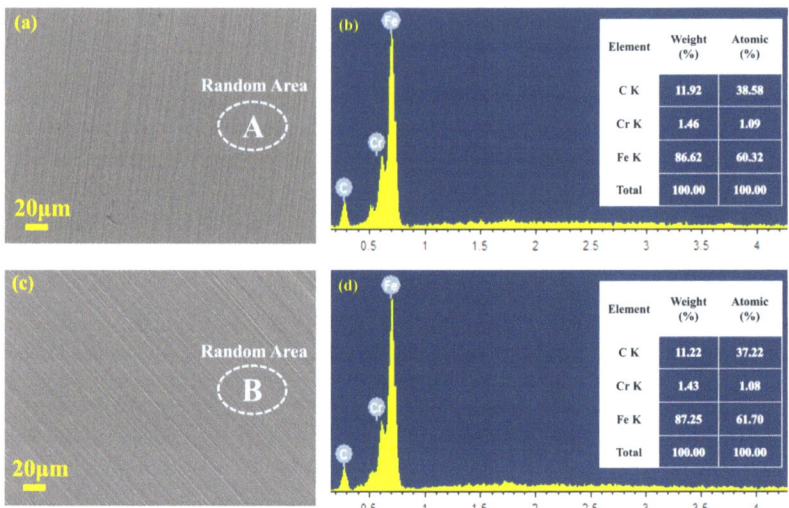

Figure 6. Surface morphology and energy spectrum analysis before the friction wear test: (**a**) SEM morphology of the conventionally heat-treated sample; (**b**) EDS spectrum of region A; (**c**) SEM morphology of the carbonitriding-treated sample; (**d**) EDS spectrum of region B.

4.2.2. After Friction Wear Test

Following the friction wear tests at 100 N and 200 N, the initial machining scratches were still visible on the surfaces of the conventionally heat-treated specimens, with only slight traces of the test wear being evident. After the friction and wear test at 300 N, the initial machining scratches disappeared, and were replaced by test scratches. Figure 7 illustrates the surface morphology and energy spectrum analysis results for the sample following the friction and wear tests at 100 N, 200 N, and 300 N loads, respectively. The EDS spectrum of randomly selected amplification regions A, B, and C are illustrated in Figure 7b,d,f.

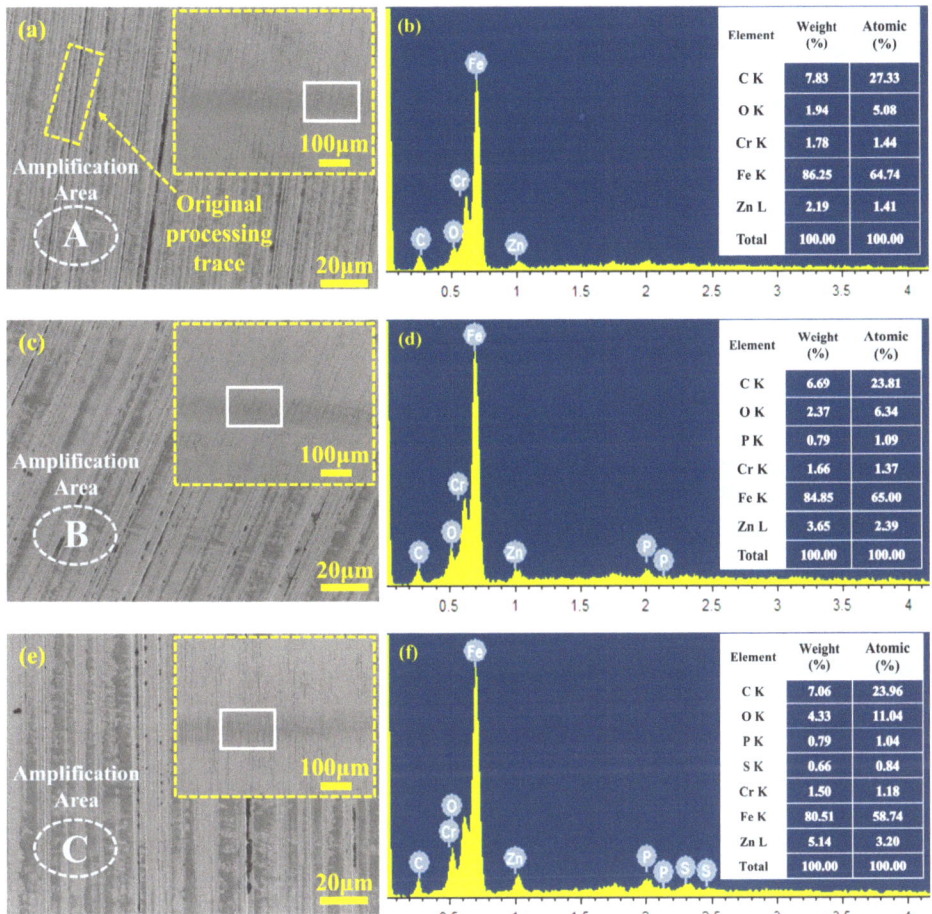

Figure 7. Surface morphology and energy spectrum analysis of the carbonitriding-treated samples after the friction wear test: (**a**) SEM morphology at 100 N; (**b**) EDS spectrum of region A; (**c**) SEM morphology at 200 N; (**d**) EDS spectrum of region B; (**e**) SEM morphology at 300 N; (**f**) EDS spectrum of region C.

Figure 7a demonstrates that after the 100 N friction and wear test, the surface of the sample exhibited only slight marks of movement, with the original scratches remaining intact. From Figure 7c,e, it can be observed that the surfaces of the carbonitriding-treated samples which were subjected to friction and wear tests at 200 N and 300 N loads did not exhibit significant friction and wear marks. Instead, the predominant scratches were those present prior to the tests. From Figure 7b,d,f, it can be observed that the surface element composition and mass fraction of the carbonitriding-treated samples were comparable to those of the conventionally heat-treated samples.

The sample was subjected to a carbonitriding and quenching heat treatment, during which the nitrogen atoms formed a solid solution with the martensite matrix, thereby forming a nitrogen-containing martensite. The incorporation of nitrogen into the martensite matrix resulted in an effect similar to that of applying solid solution strengthening on the material. Concurrently, the dispersed carbon/nitrogen compounds on the martensitic matrix reinforced the bearing steel through the dispersion strengthening effect, and the comprehensive performance of the carbonitriding treatment enhanced the wear resistance of the material [13]. The above phenomenon demonstrates that the carbon and nitrogen co-infiltration treatment of the sample confers a certain degree of wear resistance, which effectively reduces the frictional wear of the bearings.

4.3. Effect of Carbonitriding Strengthening Method on Friction and Wear Performance

4.3.1. Effect on Coefficient of Friction

Figure 8a shows the coefficient of friction (COF) of the conventionally heat-treated samples under different loads with time. The first 300 s of the friction wear experiment constituted the break-in period, during which the COF varied greatly. After a sharp increase, the COF started to decrease and then entered a stabilization period, during which it slowly tended to stabilize. Under a load of 300 N, it was finally stabilized at 0.1499 ± 0.0004. Figure 8b depicts the time-dependent change curve of the COFs of the carbonitriding-treated samples under different loads. The COF was finally stabilized at 0.1454 ± 0.0002 under a load of 300 N. In comparison to the conventional samples, the COF of the carbonitriding-treated samples exhibited a notable decline under loads of 100 N and 200 N, with a more pronounced reduction observed under the load of 300 N. In comparison to the conventionally heat-treated samples, the reduction in the COF was more pronounced for the 100 N and 200 N test loads. Conversely, the reduction in the COF of the carbonitriding-treated samples was less pronounced at the 300 N load.

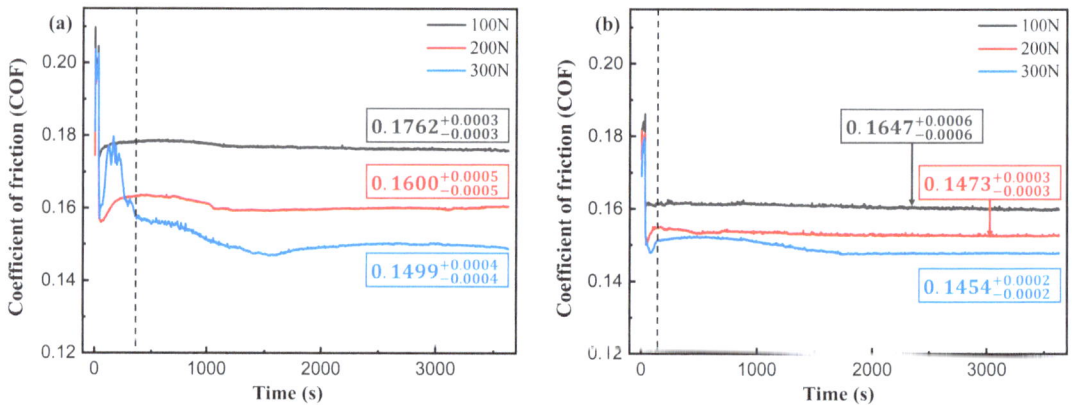

Figure 8. Coefficients of friction of samples: (**a**) conventionally heat-treated sample; (**b**) carbonitriding-treated sample.

4.3.2. Effect on Surface Wear Scar

Figure 9 compares the wear scar depths of the conventionally heat-treated samples. The depth of the conventionally heat-treated test samples after friction wear testing at a 100 N load was 0.176 μm. The wear scar depth in the friction wear test at a 300 N load was 0.623 μm, representing an increase of 0.447 μm, with a growth rate of 253.97%. A comparison of the depths of carbonitrided samples is presented in Figure 10. The wear scar depth of the test samples after the friction wear test at a 100 N load was 0.196 μm, and the depth of the test samples after friction wear testing at a 300 N load was 0.253 μm, representing an increase of 0.057 μm or 29.08%.

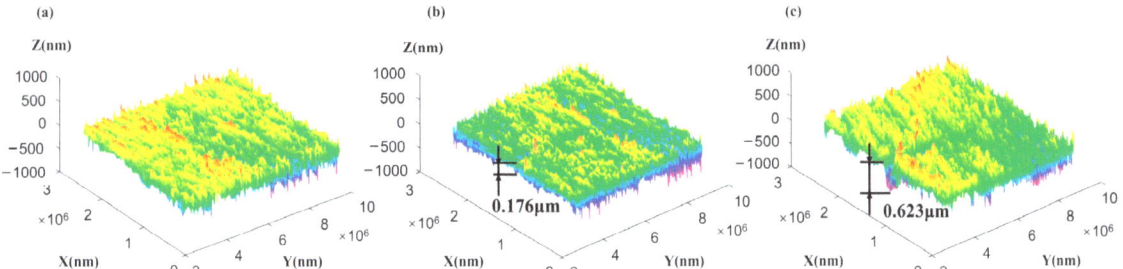

Figure 9. Comparison of the wear scar depths of conventionally heat-treated samples: (**a**) before the experiment; (**b**) under a load of 100 N; (**c**) under a load of 300 N.

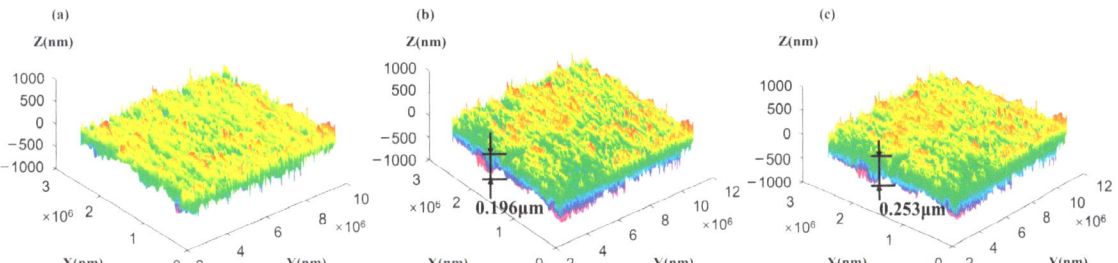

Figure 10. Comparison of the wear scar depths of carbonitriding-treated samples: (**a**) before the experiment; (**b**) under a load of 100N; (**c**) under a load of 300N.

4.3.3. Friction and Wear Performance Analysis

The presence of residual tensile stresses in the surface layers of needle bearing components has been demonstrated to reduce the performance and life of bearings [29]. The carbon and nitrogen co-infiltration treatment of the AISI 52100 bearing steel, whereby C and N atoms penetrate into the surface layer of the work piece, has been shown to effectively reduce the temperature of the onset of martensitic transformation. Concomitant with enhanced hardness in the infiltration layer, significant residual compressive stress is generated in the surface layer. Consequently, the strength and contact fatigue life of the bearing steel are enhanced.

Following the application of conventional heat treatment to a material, it becomes evident that the surface exhibits a notable degree of wear after undergoing a series of friction experiments. As the experiments progress, the original peaks of the surface become progressively worn down, either as wear chips or pressed into the surface. This results in a discernible alteration in the surface morphology and affects the values of the surface roughness and COF. As a consequence of the repeated application of alternating stresses between the contact surfaces, fatigue cracks gradually emerge at the surface or subsurface locations of the material and gradually expand into fatigue craters [30]. Due to the low surface hardness and toughness of conventionally heat-treated samples, fatigue cracks are readily produced and expand rapidly, resulting in localized furrow-like wear marks. The undulation of this surface varies significantly, as evidenced by the high growth rate of the wear marks and the high COF of the contact surface.

Following the carbonitriding treatment, the samples were subjected to a friction wear test, which revealed that the overall change to the surface was relatively minor. The initial roughness peak was flattened, the surface became flat, and the growth of the wear marks was low. The surface pits exhibited no significant expansion and the furrow-like distribution on the surface was uniform and narrow in width. The above analysis demonstrates that carbonitriding can effectively enhance the wear resistance of the bearing steel and material surface.

4.4. Fatigue Life Analysis

4.4.1. Simulation Results

The requisite parameters were entered into Romax and the results of the damage simulation of the needle bearings were obtained, as shown in Table 7. In comparison to the fatigue damage of the conventionally heat-treated needle bearings, the fatigue damage rate of the needle bearings simulated under the ISO 281 [31] decreased by 3.3%. Furthermore, the fatigue damage rate calculated under the ISO/TS 16281 [32] decreased by 3.5%. The fatigue life simulated under the ISO 281 standard increased by 3.44%, while the fatigue life simulated under the ISO/TS 16281 standard increased by 3.79%.

Table 7. Needle bearing life simulation results.

Computational Item	Conventional	Carbonitriding
ISO 281 Damage (%)	33.4	32.3
ISO/TS 16281 Damage (%)	74.6	72.0
ISO 281 Lifetime (h)	1269.73	1313.43
ISO/TS 16281 Lifetime (h)	568.41	589.98

4.4.2. Test Results

A total of eight sets of bearings were prepared for the life comparison test of carbonitrided needle bearings. Four sets each of conventionally heat-treated and carbonitriding-treated needle bearings were subjected to normal-load and high-load tests. The fatigue life comparison results of the bearings subjected to conventional heat treatment and carbonitriding in normal load conditions are shown in Figure 11. The fatigue life of the bearings subjected to the different heat treatment processes was averaged, and it was determined that the fatigue life of carbonitrided needle bearings was increased by 384% and 463%, respectively, under normal-load conditions and high-load conditions. Additionally, the fatigue limit was increased by 31.9% under normal-load conditions and high-load conditions.

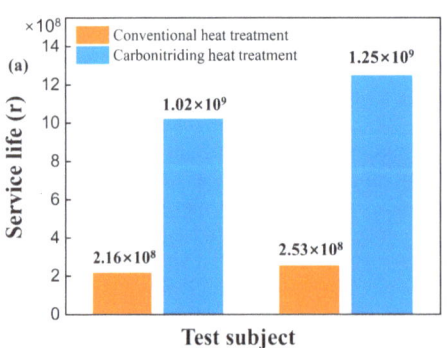 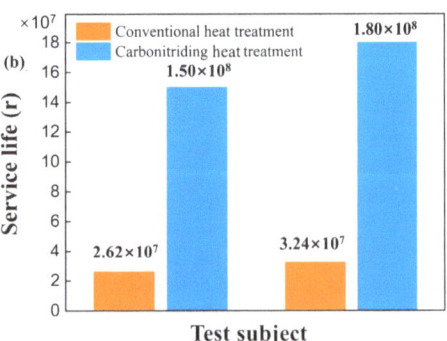

Figure 11. Service life of needle bearings: (**a**) normal-load conditions; (**b**) high-load conditions.

A conventionally heat-treated needle bearing and a carbonitriding-treated needle bearing were selected. The failure regions were observed using a magnifier, as shown in Figure 12.

When the needle bearing obtained by conventional heat treatment failed, it was observed that obvious contact fatigue pits appeared on the contact surface. The areas of the pits were smaller and the depths were larger, indicating that these positions were the weak points of the material. This may be due to the residual compressive stress distribution being non-uniform or the hardness being lower. In the case of needle bearings subjected to carbonitriding heat treatment, the area of the failure region was larger, the number of pits was greater, and the depths were shallower when contact fatigue failure occurred. This phenomenon serves to illustrate that the surface residual stress distribution of needle bearings after carbonitriding treatment is more uniform, the comprehensive fatigue strength is improved, and the ability of each position to resist fatigue spalling is effectively enhanced.

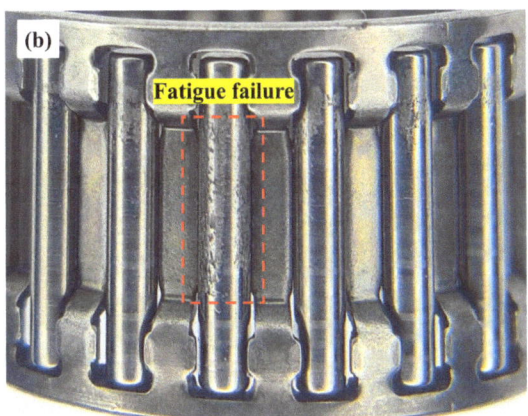

Figure 12. Needle bearing failure locations: (**a**) conventionally heat-treated needle roller; (**b**) carbonitrided needle roller.

5. Conclusions

In order to investigate the impact of carbonitriding surface strengthening treatment on the friction and wear performance and fatigue life of needle roller bearing surfaces, this study employed K28 × 33 × 19-type needle roller bearings for research. The principal findings are as follows:

1. After the carbonitriding treatment, the hardness of the AISI 52100 bearing samples increased by 8.9%, while the residual austenite content increased by 251.58 ± 3.54%. The enhancement in the fatigue characteristics of the bearing is a consequence of the combination of high surface hardness and residual austenite;
2. After performing the SRV friction and wear test, the surfaces of the conventionally heat-treated samples exhibited clear evidence of test wear, whereas the surfaces of the carbonitriding-treated samples did not. The surface scratches observed on the latter were primarily the initial scratches present prior to the test. The change in the surface COF of the carbonitriding samples was minimal, and the depth of the wear scar was significantly smaller than that of the conventionally heat-treated samples. This observation provides a clear indication of the impact of the carbonitriding treatment on the anti-wear performance;
3. The results of the test bench simulation indicate that the degree of damage due to fatigue to the bearings treated with carbonitriding was approximately 3.3% lower than that of a conventionally heat-treated bearing. Upon comparison with conventionally heat-treated bearings, the fatigue life of needle bearings that undergo carbonitriding was enhanced by 384% in the case of normal loads and by 463% in the context of high loads. Following the implementation of a uniform residual stress distribution on the carbonitrided bearing's surface, there was a notable enhancement in comprehensive fatigue strength. The capacity of each position to resist fatigue wear can also be observed.

Author Contributions: Conceptualization, Y.C.; methodology, Y.C.; software, X.P.; validation, X.P., L.H. and G.L.; formal analysis, L.L.; investigation, X.P., L.H., G.L. and L.L.; data curation, Y.C. and X.P.; writing—original draft preparation, X.P.; writing—review and editing, Y.C. and L.H.; visualization,

G.L. and L.L.; funding acquisition, Y.C. All authors have read and agreed to the published version of the manuscript.

Funding: This research was funded by the Guangxi Key Research and Development Plan Special Project through grants No.2023AB07038.

Data Availability Statement: The data presented in this study are available upon request from the corresponding author.

Acknowledgments: Guangxi University and Hebei University of Technology are gratefully acknowledged for their support in producing the samples and providing technical services, respectively.

Conflicts of Interest: The authors declare no conflicts of interest.

References

1. Hao, L.J.; Chen, Y.; Li, G.X.; Zhang, M.; Wu, Y.M.; Liu, R.; Chen, G. Study on the Friction Characteristics and Fatigue Life of Manganese Phosphate Coating Bearings. *Lubricants* **2023**, *11*, 99. [CrossRef]
2. Schönbauer, B.M.; Mayer, H. Effect of small defects on the fatigue strength of martensitic stainless steels. *Int. J. Fatigue* **2019**, *127*, 362–375. [CrossRef]
3. Jouini, N.; Revel, P.; Thoquenne, G. Influence of surface integrity on fatigue life of bearing rings finished by precision hard turning and grinding. *J. Manuf. Processes* **2020**, *57*, 444–451. [CrossRef]
4. Azianou, A.E.; Debray, K.; Bolaers, F.; Chiozzi, P.; Palleschi, F. Fatigue behaviour FEM modeling of deep groove ball bearing mounted in automotive alternator submitted to variable loading. In Proceedings of the Conference on Fatigue Design and Material Defects, France, Paris, 11–13 June 2014.
5. Yessine, T.M.; Fabrice, B.; Fabien, B.; Sébastien, M. Study of ball bearing fatigue damage using vibration analysis: Application to thrust ball bearings. *Struct. Eng. Mech.* **2015**, *53*, 325–336. [CrossRef]
6. Wang, L.W.; Sheng, X.Y.; Luo, J.B. A peridynamic damage-cumulative model for rolling contact fatigue. *Theor. Appl. Fract. Mec.* **2022**, *121*, 103489. [CrossRef]
7. Morales-Espejel, G.E. Surface roughness effects in elastohydrodynamic lubrication: A review with contributions. *P. I. Mech. Eng. J-J. Eng.* **2014**, *228*, 1217–1242. [CrossRef]
8. Smolnicki, T.; Rusiński, E. Superelement-based modeling of load distribution in large-size slewing bearings. *J. Mech. Design* **2007**, *129*, 459–463. [CrossRef]
9. Lostado, R.; Martinez, R.F.; Mac Donald, B.J. Determination of the contact stresses in double-row tapered roller bearings using the finite element method, experimental analysis and analytical models. *J. Mech. Sci. Technol.* **2015**, *29*, 4645–4656. [CrossRef]
10. Kania, L. Modelling of rollers in calculation of slewing bearing with the use of finite elements. *Mech. Mach. Theory* **2006**, *41*, 1359–1376. [CrossRef]
11. Bakolas, V.; Roedel, P.; Koch, O.; Pausch, M. A first approximation of the global energy consumption of ball bearings. *Tribol T* **2021**, *64*, 883–890. [CrossRef]
12. Ilie, F. Diffusion and mass transfer mechanisms during frictional selective transfer. *Int. J. Heat Mass Tran.* **2018**, *116*, 1260–1265. [CrossRef]
13. Huan, Q.T.; Du, S.M.; Wang, M.D.; He, T.T.; Zhang, Y.Z. Effects of different chemical heat treatment on mechanical properties and tribological behavior of GCr15 steel. *Trans. Mater. Heat Treat.* **2021**, *42*, 117–123.
14. Wan, H.Y.; Lu, H.; Ren, Y.P.; Ma, C.; Chen, Y.; Xin, Z.D.; Cheng, L.; He, K.; Tu, X.C.; Han, Q. Strengthening mechanisms and tensile properties of 20Cr2Mn2Mo processed by laser shock peening and vacuum carbonitriding. *Surf. Coat. Tech.* **2022**, *439*, 128462. [CrossRef]
15. Rajan, K.; Joshi, V.; Ghosh, A. Effect of carbonitriding on endurance life of ball bearing produced from SAE 52100 bearing steels. *J. Surf. Eng. Mater. Adv. Technol.* **2013**, *3*, 172–177. [CrossRef]
16. Liu, M.D. Metallographic black microstructure analysis of GCr15 steel bearing rings after carbonitriding. *Heat Treat.* **2019**, *4*, 60–61.
17. Zhao, K.; Zhang, G.W.; Zhu, H.F. Effects of carbonitriding pretreatment on life of ball bearings. *Bearing* **2018**, *10*, 34–37.
18. Karamiş, M.B.; Odabaş, D. A simple approach to calculation of the sliding wear coefficient for medium carbon steels. *Wear* **1991**, *151*, 23–34. [CrossRef]
19. Kanchanomai, C.; Limtrakarn, W. Effect of residual stress on fatigue failure of carbonitrided low-carbon steel. *J. Mater. Eng. Perform.* **2008**, *17*, 879–887. [CrossRef]
20. Jiang, G.H.; Li, S.X.; Pu, J.B.; Wang, H.X.; Chen, Y.J. Rolling contact fatigue failure mechanism of martensitic bearing steel after carbonitriding. *China Surf. Eng.* **2022**, *35*, 12–23.
21. *ASTM A. 29/A 29M-04*; Standard Specification for Steel Bars, Carbon and Alloy, Hot-Wrought, General Requirements. ASTM International: West Conshohocken, PA, USA, 2017.
22. *ISO 683-17*; Heat-Treated Steels, Alloy Steels and Free-Cutting Steels—Part 17: Ball and Roller Bearing Steels. ISO: Geneva, Switzerland, 2023.

23. *ASTM D7421-11*; Standard Test Method for Determining Extreme Pressure Properties of Lubricating Oils Using High-Frequency, Linear-Oscillation (SRV) Test Machine. ASTM International: West Conshohocken, PA, USA, 2017.
24. Balarini, R.; Diniz, G.A.S.; Profito, F.J.; Souza, R.M. Comparison of unidirectional and reciprocating tribometers in tests with MoDTC-containing oils under boundary lubrication. *Tribol. Int.* **2020**, *149*, 105686. [CrossRef]
25. Vencl, A.; Gašić, V.; Stojanović, B. Fault tree analysis of most common rolling bearing tribological failures. *IOP Conf. Ser. Mater. Sci. Eng.* **2017**, *174*, 012048. [CrossRef]
26. *GB/T 6391-2011*; Rolling Bearings—Dynamic Load Ratings and Rating Life. Standardization Administration of China: Beijing, China, 2011.
27. *GB/T 24607-2009*; Rolling Bearings—Test and Assessment for Life and Reliability. Standardization Administration of China: Beijing, China, 2009.
28. Johnson, K.L. Failure atlas for Hertz contact machine elements. *Tribol. Int.* **1993**, *26*, 65–66. [CrossRef]
29. Zhai, Y.Z.; Xia, H.; Zhang, X.; Yang, X.L.; Wang, H.B.; Feng, J.H. Effect of nitrogen content on microstructure and mechanical properties of V-N microalloyed high strength steel bar. *Heat Treat. Met.* **2018**, *43*, 31–34.
30. Javadi, H.; Jomaa, W.; Texier, D.; Brochu, M.; Bocher, P. Surface roughness effects on the fatigue behavior of as-Machined Inconel718. *Solid State Phenom.* **2017**, *258*, 306–309.
31. *ISO 281: 2007*; Rolling Bearings—Dynamic Load Ratings and Rating Life. ASTM International: West Conshohocken, PA, USA, 2007.
32. *ISO/TS 16281*; Rolling Bearings—Methods for Calculating the Modified Reference Rating Life for Universally Loaded Bearings. ASTM International: West Conshohocken, PA, USA, 2008.

Disclaimer/Publisher's Note: The statements, opinions and data contained in all publications are solely those of the individual author(s) and contributor(s) and not of MDPI and/or the editor(s). MDPI and/or the editor(s) disclaim responsibility for any injury to people or property resulting from any ideas, methods, instructions or products referred to in the content.

Article

Dynamic Modeling and Behavior of Cylindrical Roller Bearings Considering Roller Skew and the Influence of Eccentric Load

Yang Yang [1], Jiayu Wang [2], Meiling Wang [3] and Baogang Wen [2,*]

[1] China North Vehicle Research Institute, Beijing 100072, China
[2] School of Mechanical Engineering and Automation, Dalian Polytechnic University, Dalian 116034, China
[3] College of Locomotive and Rolling Stock Engineering, Dalian Jiaotong University, Dalian 116028, China
* Correspondence: wenbg@dlpu.edu.cn; Tel.: +86-411-86328129

Abstract: At high speeds, skew and skid may frequently occur for the rollers in cylindrical roller bearings, especially when under eccentric load, as the uneven load distribution along the generatrix of the roller further aggravates this phenomenon. In this paper, a dynamic model of a cylindrical roller bearing was established, taking into account roller skewing and interactions with the cage. Firstly, the interaction between the roller and the raceway was calculated by slicing the roller along its generatrix. Furthermore, the computation of the interaction between the roller and the cage is based on elastic theory, taking into account pocket clearance. Subsequently, the dynamic equations for both rollers and cage were derived. Based on this foundation, an investigation was conducted to reveal how rotational speed, radial loads, and moment loads affect roller slipping, skewing characteristics, and interactions with the cage under uneven load conditions. The findings indicate a direct proportionality between roller slipping and bearing speed while exhibiting an inverse relationship with load magnitude. Additionally, it was observed that both bearing speed and load have a direct influence on roller skewing angle. Moreover, normal interaction force between the roller and cage demonstrates a direct proportionality to bearing speed while inversely correlating with load magnitude.

Keywords: cylindrical roller bearing; dynamics; uneven loads; sliding velocity; roller skew

Citation: Yang, Y.; Wang, J.; Wang, M.; Wen, B. Dynamic Modeling and Behavior of Cylindrical Roller Bearings Considering Roller Skew and the Influence of Eccentric Load. *Lubricants* **2024**, *12*, 317. https://doi.org/10.3390/lubricants12090317

Received: 28 July 2024
Revised: 4 September 2024
Accepted: 5 September 2024
Published: 14 September 2024

Copyright: © 2024 by the authors. Licensee MDPI, Basel, Switzerland. This article is an open access article distributed under the terms and conditions of the Creative Commons Attribution (CC BY) license (https://creativecommons.org/licenses/by/4.0/).

1. Introduction

Cylindrical roller bearings play a critical role as foundational components in mechanical equipment, characterized by the line contact between rollers and raceways. They primarily endure radial loads and often operate under uneven load conditions caused by misalignment during assembly and overturning moments during operation. The non-uniform distribution of loads along the roller's generatrix resulting from this uneven loading leads to issues such as roller skewing, sliding, and frequent impacts with the cage. Therefore, it is imperative to investigate the mechanical characteristics of cylindrical roller bearings under uneven load conditions while considering roller skewing and its internal interactions.

Walters [1] first proposed a dynamic analysis method for ball bearings with four degrees of freedom, utilizing the Runge–Kutta method to solve differential equations and analyzing the transient characteristics of the cage and steel balls. Gupta [2] developed a dynamic model with full degrees of freedom for bearing components, integrating and solving dynamic equations to obtain the time-varying dynamic performance of bearings. However, this approach faces challenges such as strict initial value selection and lengthy computation times. Therefore, Meeks [3–5] employed an equivalent interaction between the rolling elements and cage using a spring-damping system, reducing computational complexity and simplifying Gupta's model.

In terms of ball bearing dynamics, Xi [6] developed a dynamic analysis model for ball bearings, investigating the effects of radial load and inner ring misalignment on the contact characteristics between balls and raceways in angular contact ball bearings. Ye [7] established a dynamic model for ball bearings, analyzing the impact of axial and radial loads during startup on the stability of the cage. Shan [8] introduced a hysteresis damping coefficient into the Hertz contact model to establish a dynamic analysis model for deep groove ball bearings with clearance, studying the effects of clearance on bearing contact and sliding characteristics. Zhao [9] flexibly treated the cage to establish an improved dynamic model for ball bearings, investigating the interaction between balls and the cage. Ye [10] established a four-degree-of-freedom dynamic model for angular contact ball bearing cages, studying the transient characteristics of the cage under axial loads, radial loads, and speed variations.

In the research field of cylindrical roller bearing dynamics, Wang [11] established a dynamic model for cylindrical roller bearings, analyzing the influence of roller profiling on load distribution, roller posture, and contact characteristics. Han [12] proposed a cylindrical roller bearing dynamic model considering radial clearance and roller profiling, analyzing roller slipping behavior under time-varying loads. Chen [13] conducted experimental studies to determine the effects of bearing acceleration and deceleration on the dynamic characteristics of the cage. Liu [14] developed a dynamic model for cylindrical roller bearings by slicing rollers, investigating the impact of non-uniform loads due to gear meshing, rotor eccentricity, and external excitation on bearing life. Gao [15–17] studied the effects of axial load, combined loads, cage instability, and roller slipping on the dynamic performance of bearings. Cao [18] investigated the influence of structure and lubrication on bearing motion characteristics under acceleration conditions.

Although extensive research has been conducted on the dynamic characteristics of bearings, the majority of studies have primarily focused on aspects such as the contact between rollers and raceways, roller sliding, and cage slipping under varying conditions. However, there remains a relatively limited amount of research addressing issues like roller slipping, skewing, and roller–cage collisions under uneven load conditions.

Therefore, it is imperative to study the dynamic characteristics of cylindrical roller bearings under uneven load conditions and roller skewing. A dynamic model of cylindrical roller bearings was established, taking into account the effects of roller skewing and its interaction with the cage. The dynamic characteristics of cylindrical roller bearings under uneven load conditions were investigated, which is crucial for the design and prevention of failure in cylindrical roller bearings.

2. Dynamic Model for Cylindrical Roller Bearings

The differential equations of motion for rollers and cages, as well as the static equilibrium equation for the inner ring, are established in this section based on an analysis of the bearing's movement, relative position, and mechanical relationship of its components [19–21]. Furthermore, specific expressions for the interaction forces between parts and the calculation process are provided.

2.1. Coordinates and Fundamental Assumptions

To facilitate the analysis of the motion and force distribution of each component of the bearing, the coordinate system is established as follows:

The origin O of the bearing inertial coordinate system $OXYZ$ is located at the center of the outer race plane. The X-axis aligns with the rotation centerline of the outer race, while the Y-axis is radial. The Z-axis follows the right-hand screw rule to determine its direction, as specifically illustrated in Figure 1. The roller rotation coordinate system $oxyz$ is fixed on the roller, with the origin o located at the centroid of the roller. The x-axis is parallel to the X-axis of the inertial coordinate system, while the y-axis extends outward along the revolution direction of the roller. The roller rotation coordinate system undergoes a rotational motion around the X-axis at a speed corresponding to its revolution rate.

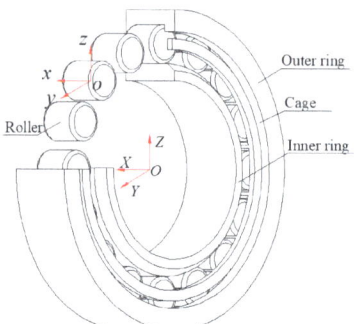

Figure 1. Cylindrical roller bearing coordinate systems.

The following assumptions are made when establishing the dynamic model of cylindrical roller bearings:

(1) The centroids and geometric centers of all components in the bearing coincide.
(2) All components are rigid bodies, and flexible deformation is not considered.
(3) The outer ring is fixed, the inner ring rotates about its own centroid, rollers revolve about the bearing axis and rotate about their own axes with skewing, and the cage rotates about its own centroid.
(4) The cage pocket shape is rectangular, and the centroids of all bearing components coincide with their centers of mass.
(5) Each individual roller has three degrees of freedom: θ_x, θ_z, and θ_X. These correspond to rotation about the X-axis, θ_X rotation about its own axis θ_x, and angular displacement due to skewing about the Z-axis θ_z.
(6) The cage has a rotational angular displacement about the X-axis θ_c.
(7) The cylindrical roller bearing has $3N+1$ degrees of freedom, where N denotes the total number of rollers; its displacement is denoted as $x = [\theta_{x1} \ldots \theta_{xN}, \theta_{z1} \ldots \theta_{zN}, \theta_{X1} \ldots \theta_{XN}, \theta_c]^T$.

The bearing's force state is shown in Figure 2. Assuming external forces $F = \{F_Z, M\}$ on the inner ring, the displacement of the inner ring relative to the inertial coordinate system is denoted as $\{\delta_r, \theta\}$.

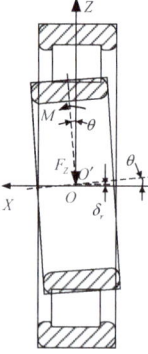

Figure 2. Loads applied on the bearing.

2.2. Differential Equations of Motion

(1) Roller Motion Differential Equations

The forces exerted on a roller in a cylindrical roller bearing under combined radial and moment loads at a specific moment during operation are illustrated in Figure 3. The roller primarily withstands forces exerted by the inner and outer raceways, the cage, and the resistance of the lubricant.

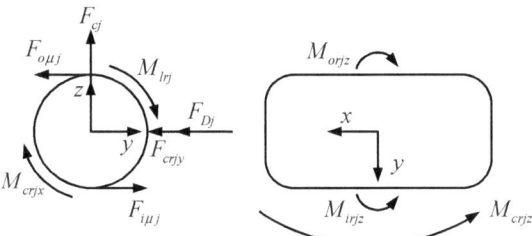

Figure 3. Load applied to the roller.

The differential equations of the roller are as follows:

$$\begin{aligned} I_x \ddot{\theta}_x &= \tfrac{1}{2} D_w (F_{i\mu j} + F_{o\mu j}) - M_{crjx} - M_{lrj} \\ I_z \ddot{\theta}_z &= M_{irjz} - M_{orjz} + M_{crjz} \\ I_X \ddot{\theta}_X &= \tfrac{1}{2} D_i F_{i\mu j} - \tfrac{1}{2} D_o F_{o\mu j} - \tfrac{1}{2} d_m F_{Dj} - \tfrac{1}{2} d_m F_{crjy} \end{aligned} \quad (1)$$

where D_w is the diameter of the inner raceway, $F_{i\mu j}$ and $F_{o\mu j}$ are the traction forces exerted on the roller by the inner and outer raceways, M_{crjx} is the frictional torque exerted on the roller by the cage around the x-axis, M_{lrj} is the rotational resistance torque exerted on the roller by the lubricant [2], M_{irjz} and M_{orjz} are the torques exerted on the roller by the inner and outer raceways around the z-axis direction, M_{crjz} is the torque exerted on the roller by the cage around the z-axis, $F_{i\mu j}$ and $F_{o\mu j}$ are the frictional forces exerted on the roller by the inner and outer raceways, F_{Dj} is the turbulence resistance of the lubricant to the roller revolution [2], F_{crjy} is the torque of the cage on the roller, D_i and D_o are the diameters of the inner and outer raceways, d_m is the diameter of the bearing pitch circle, and I_x, I_z, and I_X are the moment of inertia of the rotation, skew, and revolution of rollers, respectively.

The cage is mainly subjected to the force of rollers and lubricants, and the force is shown in Figure 4. In Figure 4, F_{crj} is the force of the roller on the cage, M_{lc} is the resistance of the lubricant to the cage, and ω_c is the revolution speed of the cage.

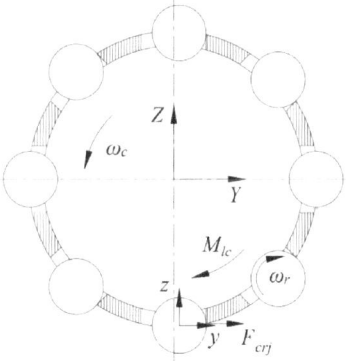

Figure 4. Load applied to the cage.

The differential equation of cage motion is as follows:

$$I_c \ddot{\theta}_c = \frac{1}{2} d_m \sum_{j=1}^{Z} F_{crj} - M_{lc} \quad (2)$$

where I_c is the moment of inertia of the cage about the X-axis and M_{lc} is the resistance of the lubricant to the cage.

The analysis process for solving the differential equation of bearing motion begins with the initial determination of the interaction force among bearing internal parts, which is crucial in establishing the initial value for the differential equation.

2.3. Internal Forces

(1) Forces between the roller and the raceway

As shown in Figure 5, the static equilibrium equation of the roller's radial direction is as follows:

$$Q_{ij} - Q_{oj} + F_{cj} = 0 \tag{3}$$

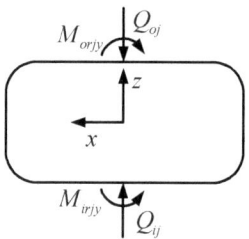

Figure 5. Static equilibrium of roller.

The balance equation of the roller's moment around the y-axis is

$$M_{iry} - M_{ory} = 0 \tag{4}$$

where Q_{ij} and Q_{oj} are the contact forces between the inner and outer raceways with the roller, F_{cj} is the centrifugal force on the roller, and M_{iriy} and M_{orjy} are the deflection torque of the inner and outer raceway to the direction of the roller around the y-axis.

Q_{ij}, Q_{oj}, M_{iriy}, and M_{orjy} are generated by the contact between the roller and the inner and outer raceways, and can be obtained by the slicing method [22].

When the radial load and moment act on the cylindrical roller bearing, the inner ring of the bearing will be deflecting, and the roller will be stressed unevenly along the generatrix bar direction.

When the roller is divided into k slices along the generatrix direction, the deformation on each slice λ consists of three parts: deformation δ_j caused by radial load at position j, deformation c_λ caused by the convexity of the roller at λ, and deformation of the roller at position j caused by inner ring deflection, as shown in Figure 6a. The contact force between the roller slice and the inner and outer raceway is calculated using the Hertz contact theory, as shown in Figure 6b.

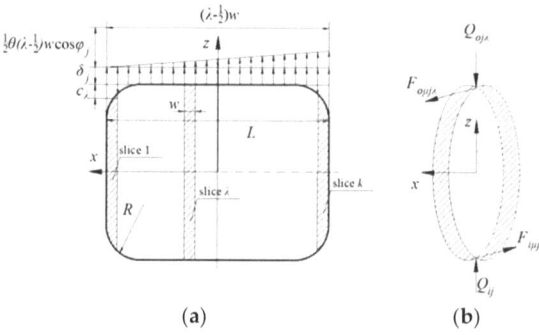

Figure 6. Schematic diagram of deformation and stress on roller slices. (**a**) The roller slice is deformed. (**b**) Roller slices are stressed.

In Figure 6, θ is the deflection angle of the bearing inner ring, L is the length of the roller, w is the slice width, $w = L/k$, R is the radius of roller repair, $Q_{ij\lambda}$ is the contact force between the roller slice and the inner raceway, $Q_{oj\lambda}$ is the contact force between the roller slice and the outer raceway, $F_{i\mu j\lambda}$ is the drag force of the inner raceway on the roller slice, and $F_{o\mu j\lambda}$ is the drag force of the outer raceway on the roller slice.

According to the analysis in Figure 6, the contact force between a single roller and the inner raceway is as follows:

$$Q_{ij} = \sum_{\lambda=1}^{k} Q_{ij\lambda} \tag{5}$$

The contact force between a single roller and the outer raceway is

$$Q_{oj} = \sum_{\lambda=1}^{k} Q_{oj\lambda} \tag{6}$$

Due to the uneven force on the roller along the generatrix bar direction, the resulting torques around the y-axis of the roller are

$$\begin{aligned} M_{iry} &= \sum_{\lambda=1}^{k} \frac{Q_{ij\lambda}}{w}\left[-\frac{L}{2}+\left(\lambda-\frac{1}{2}\right)w\right] \\ M_{ory} &= \sum_{\lambda=1}^{k} \frac{Q_{oj\lambda}}{w}\left[-\frac{L}{2}+\left(\lambda-\frac{1}{2}\right)w\right] \end{aligned} \tag{7}$$

According to the Hertz contact theory, the contact force between the roller slice and the inner raceway is [23]

$$Q_{ij\lambda} = \frac{w\delta_{ij\lambda}^{1.11}}{A^{1.11}(kw)^{0.11}} \tag{8}$$

The contact force between the roller slice and the outer raceway is

$$Q_{oj\lambda} = \frac{w\delta_{oj\lambda}^{1.11}}{A^{1.11}(kw)^{0.11}} \tag{9}$$

where $\delta_{ij\lambda}$ and $\delta_{oj\lambda}$ are the contact deformations between the roller slice and the inner and outer raceways. And $A = 1.36\eta^{0.9}$, η is the combined elastic constant of the two bodies.

$$\delta_{oj\lambda} = \delta_{oj} + c_{xo\lambda} \tag{10}$$

where $c_{xo\lambda}$ refers to the contact deformation between roller and outer raceway caused by ring deflection and roller modification.

$$\delta_{ij\lambda} = \delta_r \cos\varphi_j - \frac{P_d}{2} - \delta_{oj} + c_{xi\lambda} \tag{11}$$

where P_d is the radial clearance of the bearing and $c_{xi\lambda}$ refers to the contact deformation between the roller and the inner raceway caused by the deflection of the ring and the modification of the roller [23].

When Equations (5) and (11) are substituted into Equations (3) and (4), a system of static equilibrium equations for 2Z rollers is obtained. This system comprises a total of 2Z + 2 unknowns, including δ_r, θ, δ_{oj}, and β_j. To facilitate the numerical solution of this system of equations, the inner race force balance equation is introduced.

$$F_Z - \sum_{j=1}^{Z} Q_{ij}\cos\varphi_j = 0 \tag{12}$$

$$M - \sum_{j=1}^{Z} Q_{ij} e_{ij} \cos \varphi_j = 0 \tag{13}$$

where Z is the number of rollers, φ_j is the position angle of rollers, and e_{ij} is the eccentricity of the inner raceway of rollers.

The Newton–Raphson method is employed to solve the nonlinear equations, and the contact forces $Q_{ij\lambda}$ and $Q_{oj\lambda}$ between each slice of a roller and the inner and outer raceways were obtained.

Due to the role of lubricating oil, the drag force of the inner and outer raceway on the roller slice is

$$F_{i\mu j\lambda} = \mu_{ij\lambda} Q_{ij\lambda} \tag{14}$$

$$F_{o\mu j\lambda} = \mu_{oj\lambda} Q_{oj\lambda} \tag{15}$$

where $\mu_{ij\lambda}$ and $\mu_{oj\lambda}$ are the drag coefficients between roller slices and inner and outer raceways, respectively.

The drag coefficient is determined by the Hertzian contact stress p and slip velocity V at the contact [24]:

$$\mu_{ij\lambda} = AV_{ij\lambda} \exp\left(\frac{-BCV_{ij\lambda}}{p_{ij\lambda}}\right)[1 - \exp\left(\frac{B - p_{ij\lambda}}{B}\right)] + D[1 - \exp(-CV_{ij\lambda})] \tag{16}$$

$$\mu_{oj\lambda} = AV_{oj\lambda} \exp\left(\frac{-BCV_{oj\lambda}}{p_{oj\lambda}}\right)[1 - \exp\left(\frac{B - p_{oj\lambda}}{B}\right)] + D[1 - \exp(-CV_{oj\lambda})] \tag{17}$$

where A, B, C, D are determined for lubricant properties, $A = 0.27$, $B = 0.43\ GPa$, $C = 6.93$, $D = 0.022$.

The slip velocity V is determined by the relationship shown in Figure 7, in which ω_i is the inner circle velocity, ω_m is the roller's revolution velocity, and ω_{rj} is the roller's rotation velocity.

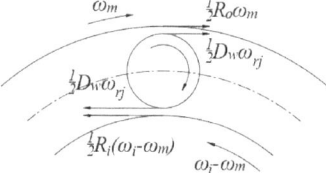

Figure 7. Roller motion state.

The relative sliding speed of the roller with the inner and outer raceway is

$$V_{ij} = (\omega_i - \omega_m)R_i - \omega_{rj} r \tag{18}$$

$$V_{oj} = \omega_m R_o - \omega_{rj} r \tag{19}$$

where R_i is the radius of the inner raceway, r is the radius of the roller, and R_o is the radius of the outer raceway.

The sliding speed [25] of the λ slice and ring at angular position j from the center of the roller is

$$V_{ij\lambda} = \frac{1}{2}(d_m - D_w)(\omega_i - \omega_m) - \frac{1}{2} D_w \omega_{rj}[1 + (\frac{L}{2} - \lambda w) \tan \theta_{zj}] \tag{20}$$

$$V_{oj\lambda} = \frac{1}{2}(d_m D_w)\omega_m - \frac{1}{2}D_w\omega_{rj}[1 + (\frac{L}{2} - \lambda w)\tan\theta_{zj}] \qquad (21)$$

where θ_{zj} is the roller skew angle about the z-axis.

(2) Roller and cage interaction

The roller pushes the cage forward in the bearing zone and the roller pushes forward in the non-bearing zone cage. The interaction between the roller and the cage is equivalent to a spring. Due to the gap in the pocket holes, the roller can only contact one side of the cage, which defines that the spring can only be compressed but not stretched. The compression amount of the spring is shown in Figure 8, where the dotted line represents the position of the roller when it is not in contact with the cage, and the solid line represents the position when the roller is in contact with the cage. In the figure, θ_{Xj} is the rotation angle of the roller and θ_c is the rotation angle of the cage.

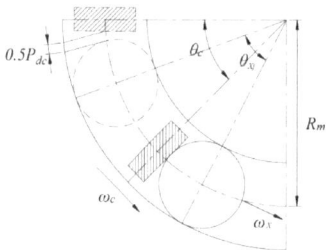

Figure 8. Spring compression amount.

Due to the deviation of the inner ring, the roller will be skewed due to the different drag force at both ends of the roller. The position relationship between the roller and the cage after the skew is shown in Figure 9.

$$F_{acrj} = \begin{cases} K_a[(\theta_c - \theta_{Xj})R_m - \frac{L}{2}\theta_{zj} - \frac{Pd_c}{2}] & (\theta_c - \theta_{Xj})R_m - \frac{L}{2}\theta_{zj} > \frac{Pd_c}{2} \\ -K_a[(\theta_{Xj} - \theta_c)R_m + \frac{L}{2}\theta_{zj} - \frac{Pd_c}{2}] & (\theta_{Xj} - \theta_c)R_m + \frac{L}{2}\theta_{zj} < \frac{Pd_c}{2} \\ 0 & \text{else} \end{cases} \quad (22)$$

$$F_{bcrj} = \begin{cases} K_b[(\theta_c - \theta_{Xj})R_m + \frac{L}{2}\theta_{zj} - \frac{Pd_c}{2}] & (\theta_c - \theta_{Xj})R_m + \frac{L}{2}\theta_{zj} > \frac{Pd_c}{2} \\ -K_b[(\theta_{Xj} - \theta_c)R_m - \frac{L}{2}\theta_{zj} - \frac{Pd_c}{2}] & (\theta_{Xj} - \theta_c)R_m - \frac{L}{2}\theta_{zj} > \frac{Pd_c}{2} \\ 0 & \text{else} \end{cases} \quad (23)$$

where K_a and K_b are the spring stiffness of the roller in contact with the cage on the a side and b side, respectively.

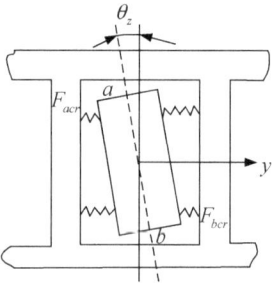

Figure 9. Position relationship between the roller and the cage after the skew.

The normal force exerted by the cage on the roller can be determined through the aforementioned analysis:

$$F_{crjy} = F_{acrj} + F_{bcrj} \quad (24)$$

The torque exerted by the cage on the roller is found as follows:

$$M_{crjz} = \frac{1}{2}L(F_{acrj} - F_{bcrj}) \quad (25)$$

The tangential friction force between the roller and the cage pocket hole is modeled as Coulomb friction force [26], which can be mathematically expressed as follows:

$$F_{\mu crj} = \mu_c (F_{acrj} + F_{bcrj}) \tag{26}$$

where μ_c is the friction coefficient between the roller and the cage, $\mu_c = 0.002$.

M_{crjx} is the friction torque of the cage on the roller:

$$M_{crjx} = \frac{1}{2} D_w F_{\mu crj} \tag{27}$$

2.4. Calculation Process

The specific calculation procedure of the dynamic model for cylindrical roller bearings in MATLAB is shown in Figure 10.

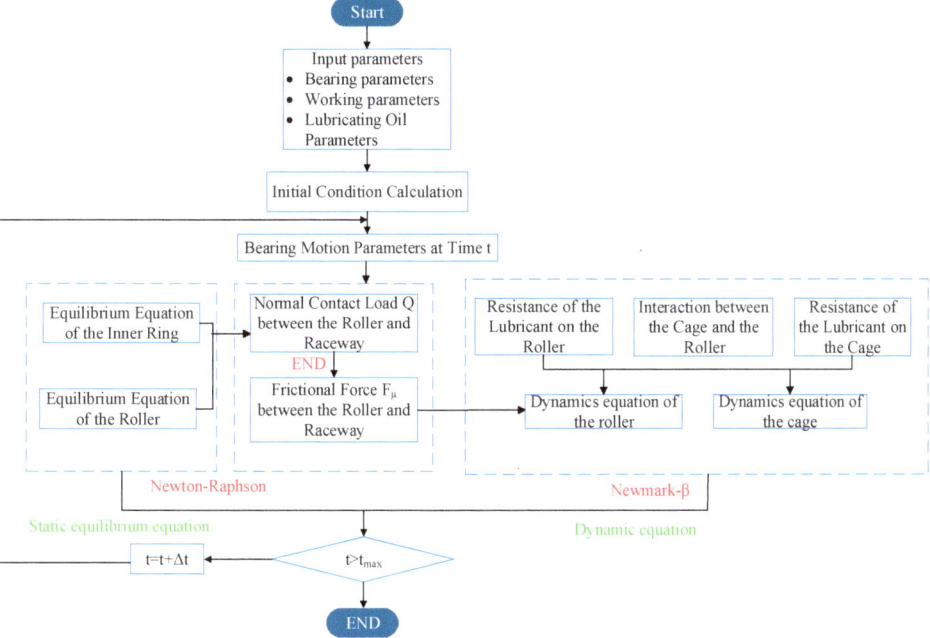

Figure 10. Calculation process.

(1) The initial motion parameter of the bearing is determined by calculating the initial value of the dynamic equation.
(2) The Newton–Raphson method is employed to solve the nonlinear equilibrium equations of the inner ring and the roller, thereby obtaining the contact force between the roller slice and the raceway.
(3) Through the interaction of bearing parts, the drag force of the raceway on the roller, the interaction force between the roller and the cage, and the resistance of the lubricant are obtained.
(4) The aforementioned results are then utilized as initial values in solving dynamics equations for both roller and cage using the Newmark-β method.

2.5. Model Verification

The bearing parameters specified in the Harris model [27] are incorporated into the aforementioned established model, and these parameters are presented in Table 1. By setting the inner ring speed of the bearing to 5000 r/min, we calculate the stable speed of the cage as the radial load varies from 0 to 12,000 N. The corresponding results are illustrated in Figure 11. It is observed that with an increase in radial load, there is a corresponding increase in cage speed until it reaches its maximum value, at which point it remains constant. These findings demonstrate a strong agreement between our proposed model and that of Harris, thereby validating the accuracy of our established model.

Table 1. Harris model bearing geometric and material parameters.

Parameter	Numerical Value
Number of rollers	36
Bearing outer ring diameter D (mm)	220
Bearing inner ring diameter d (mm)	146
Roller diameter D_w (mm)	14
Total length of roller L (mm)	20
Effective roller length L_s (mm)	17
Roller trim radius R_y (mm)	40
Radial clearance Pd (mm)	0.064
Elastic modulus of roller E_b (GPa)	207
Elastic modulus of inner and outer raceway E_r (GPa)	207
Poisson ratio of rollers ν_b	0.3
Poisson ratio of inside and outside raceway ν_r	0.3
Roller density ρ_b (kg·m^{-3})	7850
Inner and outer ring density ρ_r (kg·m^{-3})	7850

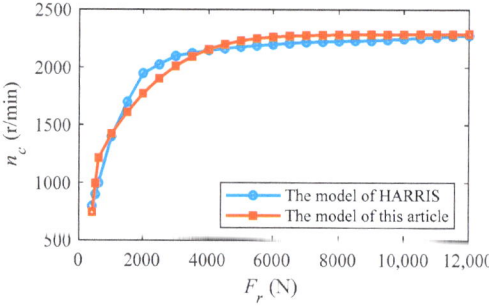

Figure 11. Cage speed under different loads.

3. Dynamic Characteristics Analysis of Cylindrical Roller Bearing

The analysis of roller slipping and skewing characteristics under eccentric loading conditions necessitates the application of a specific moment load to the bearing while it is subjected to radial load. Hence, in the dynamic model, the radial load F_Z is set to 800 N, the moment load M is set to 3 N·m, and the inner race speed n is set at 2000 r/min.

When the roller rotates once around the X-axis, the variation of contact force between each slice and raceway is shown in Figure 12, in which θ_X represents the rotation angle of the roller and λ denotes the section number of the roller along the direction of the generatrix. It is observed that the contact force between the roller and raceway changes with respect to the rotation angle. The maximum contact force occurs when the roller is positioned at the center of the bearing zone ($\theta_X = 360°$ or $0°$). Specifically, this maximum contact force is present on slice 10 during rotations ranging from 0 to 90° and 270 to 360°. This is due to the linear contact between the roller and raceway, which results in a skewed ring when an unbalanced load is present. As a result, the contact force between the roller and raceway changes along the generatrix direction, leading to an asymmetrical load distribution that causes a large contact force at the heavy end and a small contact force at the light end. The maximum load of the roller along the generatrix bar occurs in slices located at both ends of the roller, with slice 10 experiencing maximum contact force between roller and raceway.

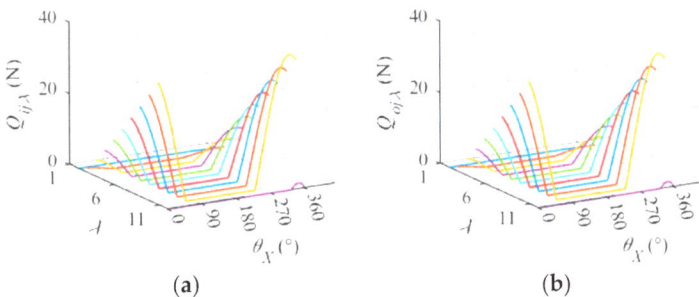

Figure 12. Contact force between each slice of the roller and the raceway. (**a**) Contact force between the roller slice and the inner raceway. (**b**) Contact force between the roller slice and the outer raceway.

The variations in rotation speed and revolution speed of the rollers are depicted in Figure 13a,b. It can be observed that as the roller moves from the bearing region ($\theta_X = 0°$) towards the non-bearing region ($\theta_X = 90°$), both rotation and revolution speeds initially remain stable before experiencing a sharp decline. Upon entering the non-bearing region, the rotational velocity and revolution velocity gradually increase to approach the theoretical level, exhibiting fluctuations within a certain range. During the transition from the non-bearing region ($\theta_X = 180°$) to the bearing region ($\theta_X = 270°$), rotational velocity exhibits steady oscillations followed by a sudden decrease. Once inside the bearing zone, rotation speed rapidly rises to approximate and fluctuate around its theoretical value with a wide range of variation.

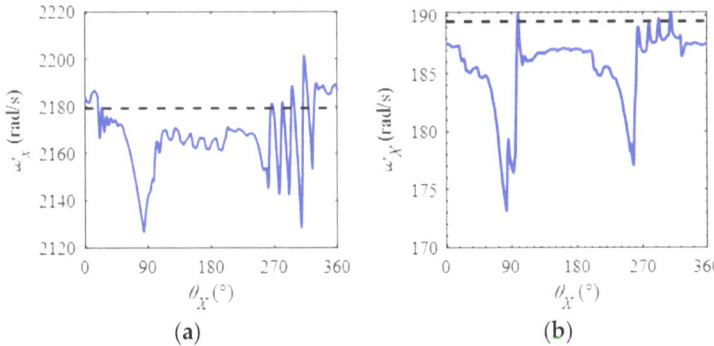

Figure 13. Roller rotation and revolution speed. (**a**) Roller rotation speed. (**b**) Roller revolution speed.

The variation of the roller's skew angle around the z-axis is illustrated in Figure 14. The positive skew angle of the roller, defined as its counterclockwise rotation about the z-axis, exhibits a dependency on its revolution position. As the roller transitions from the loaded region to the non-loaded region, its skew angle remains minimal. However, upon approaching the non-loaded region, clockwise rotation around the z-axis leads to an increase in skew angle, whereas counterclockwise rotation results in a decrease after entering this region. Within the non-load region, clockwise rotation persists until reaching maximum value and stabilizing. Conversely, upon entering the loaded region again, counterclockwise rotation causes an increase in skew angle.

To further investigate the causes of changes in roller rotation and revolution speed, a plot is presented in Figure 15 illustrating the relative sliding speed of each slice of a single roller over a rotation cycle with the inner and outer raceways. Overall, as the roller transitions from the loaded region ($\theta_X = 0°$) to the non-loaded region ($\theta_X = 90°$), its relative sliding speed with both raceways initially exhibits steady fluctuations followed by gradual increments. Upon entering the non-loaded region, this relative sliding speed decreases and fluctuates within a specific range.

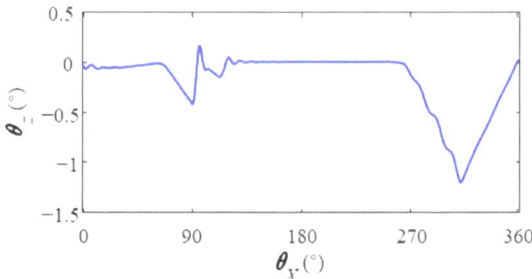

Figure 14. Roll skew angle.

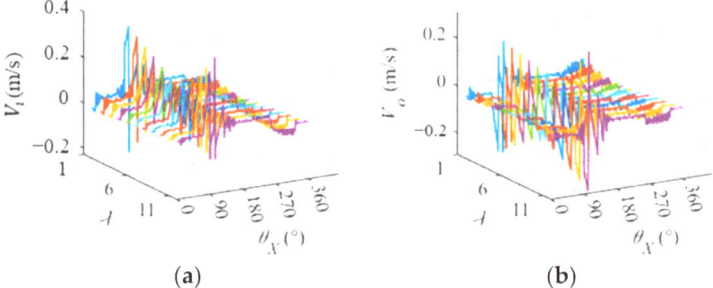

Figure 15. Relative sliding speed of roller slice and raceway. (**a**) Relative sliding speed between the roller slice and the inner raceway. (**b**) Relative sliding speed between the roller slice and the outer raceway.

In the process of transitioning from the non-loaded region ($\theta_X = 180°$) to the bearing region ($\theta_X = 270°$), there is an initial steady fluctuation followed by a sharp decrease in relative sliding velocities between the inner and outer raceways. Once in the bearing zone ($\theta_X = 270°$), the relative sliding speed between the roller and raceway decreases and fluctuates sharply within a specific range. The rotation and revolution speed of the roller correspond to changes in relative sliding speed with both raceways, indicating that variations in rotation and revolution speed are caused by changes in relative sliding speed. Relative sliding between the roller and raceway primarily occurs in the transitional region between the bearing region and non-loaded region, where the rotation angle of the roller is close to $\theta_X = 90°$ and $\theta_X = 270°$. From different slices of rollers, it can be observed that end slices exhibit higher relative sliding speeds with both inner and outer raceways compared to central slices.

In the process of bearing operation, rollers are influenced by both the raceway and cage. To further investigate the factors contributing to changes in the relative sliding speed of rollers, the drag forces exerted by the raceway on roller slices and the normal forces applied by the cage on rollers during rotation are illustrated in Figures 16 and 17. As can be seen from the figure, the roller is located in the center of the bearing zone ($\theta_X = 0°$) towards the non-bearing zone ($\theta_X = 90°$). During revolution, both the drag force exerted by the raceway on the roller and the force applied by the cage onto it experience an increase.

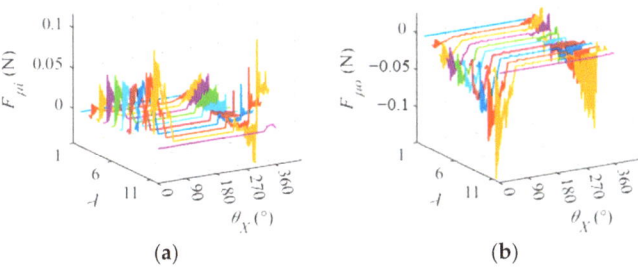

Figure 16. Drag force of the raceway on the roller slice. (**a**) Drag force of the inner raceway on the roller slice. (**b**) Drag force of the outer raceway on the roller slice.

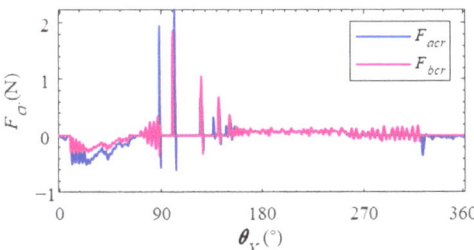

Figure 17. Normal force applied by the cage on the rollers.

The force exerted by the cage on the roller is opposite to the direction of the roller's revolution and greater than the drag force of the raceway on the roller. As a result, there is an increase in relative sliding speed between the roller and raceway, leading to a decrease in both the rotation and revolution speed of the roller. Upon entering the non-bearing zone, there is a rapid increase in normal force exerted by the cage on the roller, resulting in an acceleration of both rotation and revolution speed.

During the revolution from the non-bearing zone ($\theta_X = 180°$) to the bearing zone ($\theta_X = 270°$), the drag force of the raceway on the rollers and the normal force of the cage on the rollers change steadily first. However, as it approaches the boundary of the bearing zone ($\theta_X = 270°$), there is a sharp increase in the normal force exerted by the cage on the rollers, which simultaneously experience a significant decrease in the drag force exerted on them by the raceway. The relative sliding speed between the roller and the raceway increases, resulting in a decrease in the rotation and revolution speed of the roller. Upon entering the bearing zone ($\theta_X = 270°$), there is an increase in both the amplitude and fluctuation range of the drag force on the raceway by the rollers, as well as an increase in the normal force exerted by the cage on the rollers. Consequently, there is an increase in both the amplitude and fluctuation range of the relative sliding velocity between the roller and raceway, as well as the rotation and revolution velocity amplitude and fluctuation range of the roller.

The skew of the roller around the z-axis is influenced by the combined effect of the raceway and the cage. As depicted in Figure 16, it can be observed that there is an unequal drag force exerted by the raceway on each section of the roller, resulting in a torque generated by the raceway on the roller around the z-axis. The normal force applied by the cage on the rollers is illustrated in Figure 17. It is evident that there exists a discrepancy in the normal force exerted by both ends of the cage on the rollers, leading to a torque exerted by the cage on the rollers along with their z-axis direction. Consequently, due to this joint action between the raceway and cage, a skew angle about the z-axis direction is induced in the roller.

4. Dynamic Characteristics of Cylindrical Roller Bearing under Unbalanced Load Conditions

Cylindrical roller bearings are mainly used to bear radial load, and under certain conditions there are eccentric load conditions, which will have a direct impact on the dynamic characteristics of the bearing; based on the analysis of the influence of rotational speed and radial load, the influence of the moment load on the dynamic characteristics of the bearing under eccentric load is emphatically analyzed. The bearing structure and material parameters are presented in Table 2.

Table 2. Structural and material parameters of cylindrical roller bearings.

Parameter	Numerical Value
Number of rollers	20
Bearing outer ring diameter D (mm)	60
Bearing inner ring diameter d (mm)	30
Roller diameter D_w (mm)	4
Total length of roller L (mm)	25.5
Effective roller length L_s (mm)	22.95
Roller trim radius R_y (mm)	20
Radial clearance Pd (mm)	0.01
Elastic modulus of roller E_b (GPa)	207
Elastic modulus of inner and outer raceway E_r (GPa)	207
Poisson ratio of rollers v_b	0.3
Poisson ratio of inside and outside raceway v_r	0.3
Roller density ρ_b (kg·m^{-3})	7850
Inner and outer ring density ρ_r (kg·m^{-3})	7850

4.1. Effect of Rotational Speed on Dynamic Characteristics of Cylindrical Roller Bearing

The inner ring speed n was set to 2000 r/min, 6000 r/min, 10,000 r/min, and 14,000 r/min in order to investigate the influence of speed on bearing dynamic characteristics and analyze the dynamic performance of bearings under different inner ring speeds. The radial load was fixed at 800 N, while the torque load remained constant at 3 N·m. The bearing structure and material parameters are presented in Table 2.

Figure 18 shows the change in the skew angle of the roller during one revolution under different inner ring speeds. The higher the speed of the bearing inner ring, the greater the skew angle amplitude of the roller. This is because the inner ring speed increases, the rotational speed of the roller and the cage increases, the relative speed increases, and the normal force of the cage on the roller (as shown in Figure 19) increases, making the skew angle of the roller larger.

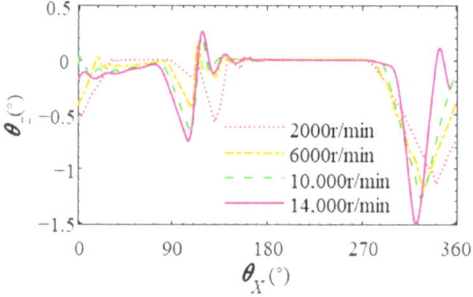

Figure 18. Deflection angle of the roller at different rotational speeds.

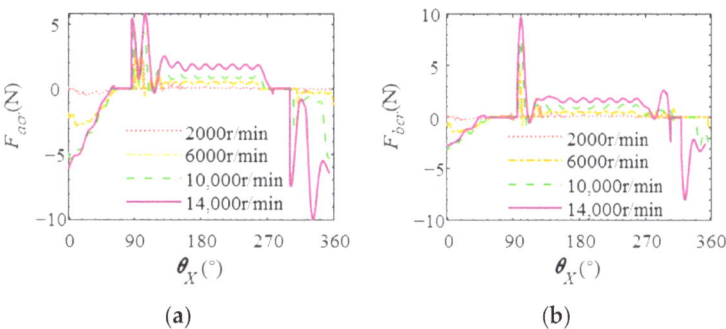

Figure 19. Normal force of the cage on the roller at different rotational speeds. (**a**) Normal force of the cage on roller point a. (**b**) Normal force of the cage on roller b.

The relative sliding speed between the roller slice and the inner raceway at different inner ring speeds is shown in Figure 20. The different colored curves in the figure represent the different slices of the roller. Overall, the higher the inner ring speed, the greater the relative sliding speed between the roller and the inner raceway. From the different slices of rollers, the higher the inner ring speed, the greater the distance from the center of the roller and the relative sliding speed between the slice and the inner raceway.

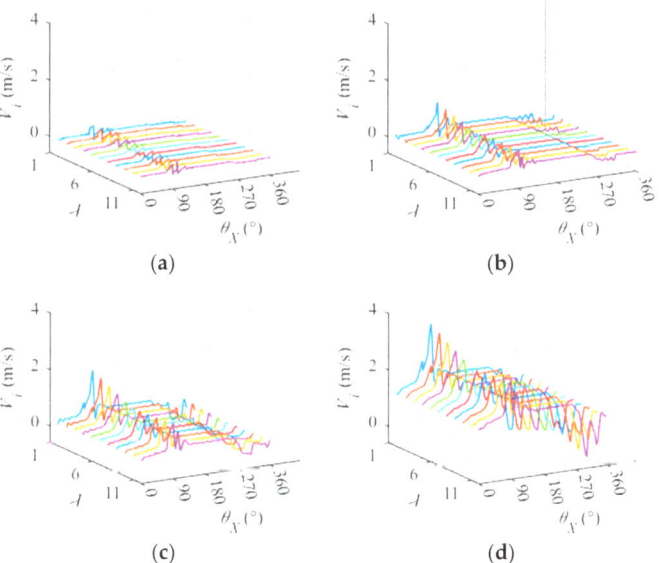

Figure 20. Relative sliding speed between roller slice and inner raceway at different rotational speeds. (**a**) Inner ring speed 2000 r/min. (**b**) Inner ring speed 6000 r/min. (**c**) Inner ring speed 10,000 r/min. (**d**) Inner ring speed 14,000 r/min.

4.2. Effect of Torque Load on Dynamic Characteristics of Cylindrical Roller Bearing

The torque load on the bearing was varied to investigate its influence on the dynamic characteristics and analyze the dynamic performance under different loads. Specifically, four torque loads M were applied: 0 N·m, 3 N·m, 5 N·m, and 10 N·m. The inner ring speed was set at 6000 r/min, and a radial load of $F_Z = 400$ N was applied.

The rotation and revolution speed of rollers under different torque loads are shown in Figure 21. It can be seen that the rotation and revolution velocities of rollers with larger torque loads are larger, and the fluctuation range is smaller and the frequency is larger. This is because the greater the torque load, the greater the contact load between the roller and the raceway, and the smaller the degree of slip between the raceway.

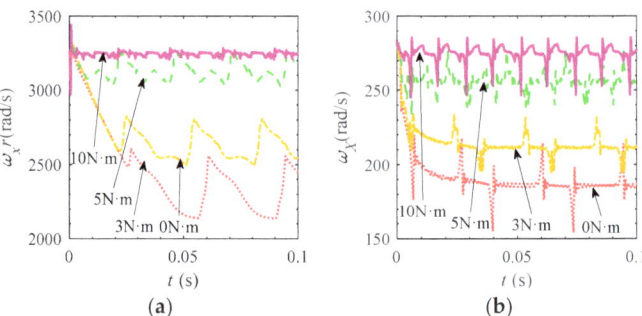

Figure 21. Roller rotation and revolution speed under different torque loads. (**a**) Roller rotation speed. (**b**) Roller revolution speed.

Figure 22 shows the influence of different torque loads on the skew angle of rollers. It can be seen that the greater the torque load, the greater the roller skew angle and the greater the skew range. Figure 23 shows the normal force of the cage on the roller under different torque loads. It can be seen that the greater the torque load, the greater the normal force of the cage on the roller, and the smaller the frequency of action. This is because the increase in torque load and the increase in the contact load between the roller and the raceway aggravate the biased load effect, and as the difference between the drag force of the inner raceway and the different slices of the roller increases, the larger the drag force of the slice with the larger distance from the center of the roller, the larger the torque of the raceway to the roller in the z-axis, and the larger the skew angle of the roller, and because the normal force of the cage on the roller decreases. The frequency of action is reduced, and the frequency of the roller skew angle fluctuation is reduced.

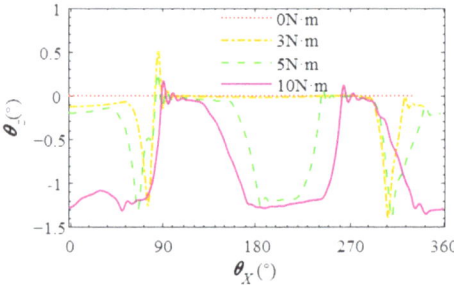

Figure 22. Deflection angle of rollers under different torque loads.

Figure 24 illustrates the variation in relative sliding speed between the roller slice and the inner raceway under different torque loads. The distinct curves of various colors represent different slices of the roller. On the whole, the greater the torque load, the smaller the relative sliding speed between the roller and the inner raceway. This can be attributed to the elevated contact load between them, which hinders roller sliding. Considering different slices of rollers, higher torque loads result in greater disparities among these sections and their respective relative sliding speeds with respect to the inner raceway. This is due to an augmented skew angle of the roller caused by increased torque load, leading to amplified differences in relative sliding speeds between various slices of the roller and its corresponding inner raceway.

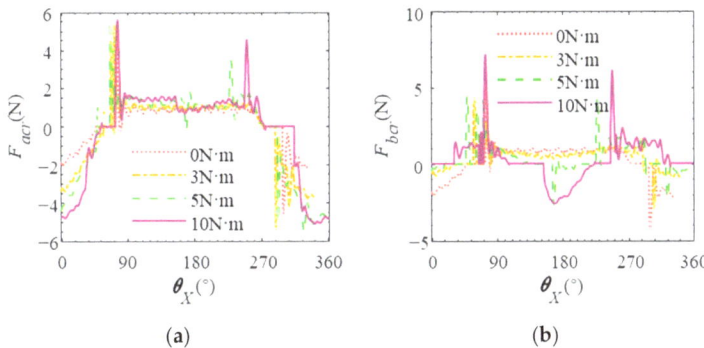

Figure 23. Normal force of cage on roller under different torque loads. (**a**) Roller rotation speed. (**b**) Roller revolution speed.

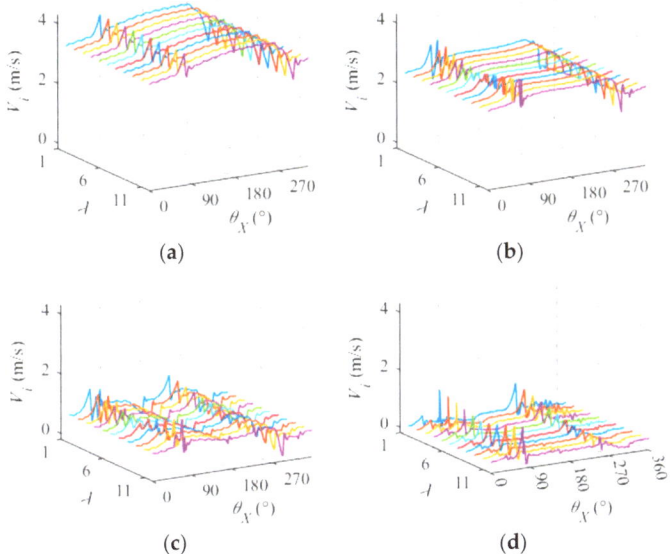

Figure 24. Relative sliding velocity between the roller slice and inner raceway under different torque loads. (**a**) Torque load 0 N·m. (**b**) Torque load 3 N·m. (**c**) Torque load 5 N·m. (**d**) Torque load 10 N·m.

5. Conclusions

To investigate the dynamic behavior of cylindrical roller bearings with roller skew under eccentric load conditions, a dynamic analysis model was developed that takes into account roller skew and cage interaction. The model was used to investigate the impact of rotational speed, radial load, and torque load on the dynamic characteristics of the bearings. Based on the analysis conducted, the following conclusions can be drawn:

(1) The increase in rotational speed leads to a rise in the relative sliding speed between the roller and the raceway, resulting in a higher skew angle and an augmented normal force exerted by the cage on the roller. Moreover, this also amplifies the disparity in relative sliding speeds between different slices of rollers and raceway, as well as intensifies the variation in drag forces experienced by different sections of rollers on the raceway.

(2) The relative sliding speed between the roller and raceway decreases as the radial load increases, leading to a decrease in skew angle and the normal force of the

cage on the roller. Additionally, there is a reduction in the relative sliding speed difference between different roller slices and raceway, as well as a decrease in drag force difference between the raceway and different roller slices.

(3) When the torque load increases, the relative sliding speed between the roller and the raceway decreases, resulting in an increase in skew angle and a decrease in the normal force of the cage on the roller. Additionally, there is an increased difference in relative sliding speed between different slices of rollers and the raceway, as well as an increased difference in drag force between different slices of rollers on the raceway.

Author Contributions: Conceptualization, Y.Y.; methodology, Y.Y.; software, J.W.; validation, B.W.; formal analysis, Y.Y.; investigation, J.W.; resources, Y.Y.; data curation, B.W.; writing—original draft preparation, M.W.; writing—review and editing, J.W.; visualization, J.W.; supervision, B.W.; project administration, B.W.; funding acquisition, B.W. All authors have read and agreed to the published version of the manuscript.

Funding: This work was financially supported by the Natural Science Foundation of Liaoning Province (2023-MS-280) and the Scientific research project of Education Department of Liaoning Province (LJKMZ20220864).

Data Availability Statement: Restrictions apply to the datasets.

Conflicts of Interest: Author Yang Yang was employed by the China North Vehicle Research Institute. The remaining authors declare that the research was conducted in the absence of any commercial or financial relationships that could be construed as a potential conflict of interest.

References

1. Walters, C.T. The Dynamics of Ball Bearings. *J. Lubr. Technol.* **1971**, *93*, 1–10. [CrossRef]
2. Gupta, P.K. Dynamics of Rolling-Element Bearings—Part I: Cylindrical Roller Bearing Analysis. *J. Lubr. Technol.* **1979**, *101*, 293–302. [CrossRef]
3. Meeks, C.R. The dynamics of ball separators in ball bearings—Part II: Results of optimization study. *ASLE Trans.* **1985**, *28*, 288–295. [CrossRef]
4. Meeks, C.R.; Ng, K.O. The dynamics of ball separators in ball bearings—Part I: Analysis. *ASLE Trans.* **1985**, *28*, 277–287. [CrossRef]
5. Meeks, C.R.; Tran, L. Ball Bearing Dynamic Analysis Using Computer Methods—Part I: Analysis. *J. Tribol.* **1996**, *118*, 52–58. [CrossRef]
6. Xi, H.; Wang, H.Y.; Han, W.; Le, Y.; Xu, H.; Chen, W.; Xu, S.N.; Wang, F.C. Contact trajectory of angular contact ball bearings under dynamic operating condition. *Tribol. Int.* **2016**, *104*, 247–262. [CrossRef]
7. Ye, Z.; Wang, L. Effect of external loads on cage stability of high-speed ball bearings. *Proc. Inst. Mech. Eng. Part J J. Eng. Tribol.* **2015**, *229*, 1300–1318. [CrossRef]
8. Shan, W.; Chen, Y.; Wang, X.; Yu, C.; Wu, K.; Han, Z. Nonlinear Dynamic Characteristics of Deep Groove Ball Bearings with an Improved Contact Model. *Machines* **2023**, *11*, 340. [CrossRef]
9. Zhao, D.; Hong, J.; Yan, K.; Zhao, Q.; Fang, B. Dynamic interaction between the rolling element and cage of rolling bearing considering cage flexibility and clearance. *Mech. Mach. Theory* **2022**, *174*, 104905. [CrossRef]
10. Ye, Z.H.; Wang, L.Q. Cage Instabilities in High-Speed Ball Bearings. *Appl. Mech. Mater.* **2013**, *278–280*, 3–6. [CrossRef]
11. Wang, Z.; Song, J.; Li, X.; Yu, Q. Modeling and Dynamic Analysis of Cylindrical Roller Bearings Under Combined Radial and Axial Loads. *J. Tribol.* **2022**, *144*, 121203. [CrossRef]
12. Han, Q.; Li, X.; Chu, F. Skidding behavior of cylindrical roller bearings under time-variable load conditions. *Int. J. Mech. Sci.* **2018**, *135*, 203–214. [CrossRef]
13. Chen, S.; Chen, X.; Li, Q.; Gu, J. Experimental Study on Cage Dynamic Characteristics of Angular Contact Ball Bearing in Acceleration and Deceleration Process. *Tribol. Trans.* **2021**, *64*, 42–52. [CrossRef]
14. Liu, Y.; Chen, Z.; Wang, K.; Zhai, W. Non-uniform roller-race contact performance of bearings along width in the rotor-bearing system under dynamic loads. *J. Sound Vib.* **2022**, *538*, 117251. [CrossRef]
15. Gao, S.; Wang, L.; Zhang, Y. Modeling and dynamic characteristic analysis of high speed angular contact ball bearing with variable clearance. *Tribol. Int.* **2023**, *182*, 108330. [CrossRef]
16. Gao, S.; Han, Q.; Zhou, N.; Pennacchi, P.; Chatterton, S.; Qing, T.; Zhang, J.; Chu, F. Experimental and theoretical approaches for determining cage motion dynamic characteristics of angular contact ball bearings considering whirling and overall skidding behaviors. *Mech. Syst. Signal Process.* **2022**, *168*, 108704. [CrossRef]
17. Gao, S.; Chatterton, S.; Pennacchi, P.; Han, Q.; Chu, F. Skidding and cage whirling of angular contact ball bearings: Kinematic-hertzian contact-thermal-elasto-hydrodynamic model with thermal expansion and experimental validation. *Mech. Syst. Signal Process.* **2022**, *166*, 108427. [CrossRef]

18. Cao, W.; Wang, J.; Pu, W.; Zhang, Y.; Wu, J.; Chu, K.; Wu, H. A study on the effect of acceleration on slip velocity and lubrication performance in cylindrical roller bearings. *Proc. Inst. Mech. Eng. Part J J. Eng. Tribol.* **2016**, *230*, 1231–1243. [CrossRef]
19. Prisacaru, G.; Bercea, I.; Cretu, S.; Mitu, N. Load distribution in a cylindrical roller bearing in high speed and combined load conditions. *Eur. J. Mech. Eng.* **1995**, *40*, 19–25.
20. Kabus, S.; Hansen, M.R.; Mouritsen, O.Ø. A New Quasi-Static Cylindrical Roller Bearing Model to Accurately Consider Non-Hertzian Contact Pressure in Time Domain Simulations. *J. Tribol.* **2012**, *134*, 041401. [CrossRef]
21. Chen, H.; Zhang, H.; Liang, H.; Wang, W. The collision and cage stability of cylindrical roller bearing considering cage flexibility. *Tribol. Int.* **2024**, *192*, 109219. [CrossRef]
22. de Nul, J.M.; Vree, J.M.; Maas, D.A. Equilibrium and Associated Load Distribution in Ball and Roller Bearings Loaded in Five Degrees of Freedom While Neglecting Friction—Part II: Application to Roller Bearings and Experimental Verification. *J. Tribol.* **1989**, *111*, 149–155. [CrossRef]
23. Harris, T.A.; Kotzalas, M.N. *Rolling Bearing Analysis—2 Volume Set*; CRC Press: Boca Raton, FL, USA, 2006.
24. Wang, Y.; Zhang, G.; Wang, E. Traction Behavior of No. 4129 Synthetic Oil for Space Lubrication. *J. Fail. Anal. Prev.* **2019**, *19*, 138–143. [CrossRef]
25. Cui, Y.; Deng, S.; Yang, H.; Zhang, W.; Niu, R. Effect of cage dynamic unbalance on the cage's dynamic characteristics in high-speed cylindrical roller bearings. *Ind. Lubr. Tribol.* **2019**, *71*, 1125–1135. [CrossRef]
26. Tu, W.; Liang, J.; Yu, W.; Shi, Z.; Liu, C. Motion stability analysis of cage of rolling bearing under the variable-speed condition. *Nonlinear Dyn.* **2023**, *111*, 11045–11063. [CrossRef]
27. Harris, T.A. An Analytical Method to Predict Skidding in High Speed Roller Bearings. *ASLE Trans.* **1966**, *9*, 229–241. [CrossRef]

Disclaimer/Publisher's Note: The statements, opinions and data contained in all publications are solely those of the individual author(s) and contributor(s) and not of MDPI and/or the editor(s). MDPI and/or the editor(s) disclaim responsibility for any injury to people or property resulting from any ideas, methods, instructions or products referred to in the content.

Article

Combining Artificial Neural Networks and Mathematical Models for Unbalance Estimation in a Rotating System under the Nonlinear Journal Bearing Approach

Ioannis Tselios and Pantelis Nikolakopoulos *

Machine Design Laboratory, Department of Mechanical Engineering and Aeronautics, University of Patras, 26504 Patras, Greece; itselios@ac.upatras.gr
* Correspondence: pnikolakop@upatras.gr; Tel.: +30-261-096-9421

Abstract: Rotating systems are essential components and play a critical role in many industrial sectors. Unbalance is a very common and serious fault that can cause machine downtime, unplanned maintenance, and potential damage to vital rotating machines. Accurately estimating unbalance in rotor–bearing systems is crucial for ensuring the reliable and efficient operation of machinery. This research paper presents a novel approach utilizing artificial neural networks (ANNs) to estimate the unbalance masses in a multidisk system based on simulation data from a nonlinear rotor–bearing system. Additionally, this study explores the effect of various operating parameters on oil film stability and vibration response through a combination of bifurcation diagrams, spectrum cascades, Poincare maps, and orbit and FFT plots. This study demonstrates the effectiveness of ANNs for unbalance estimation in a fast and accurate way and discusses the potential of ANNs in smart online condition monitoring systems.

Keywords: artificial neural network; rotordynamics; unbalance estimation; nonlinear vibration; stability threshold; lubricant temperature; journal bearing

1. Introduction

Rotating machines are fundamental components of mechanical systems and are extensively utilized in various industries. The high vibration levels that can occur may cause significant damage to the machine or even lead to complete failure of the system [1]. Rotating machinery, including turbines and compressors, is crucial equipment in aeronautics, power generation, and the shipping industry. Defects and malfunctions in these machines can lead to substantial economic losses and, therefore, must be under constant surveillance and monitoring [2].

Rotor unbalance is a common fault that results in increased vibration levels and decreases the expected life of the rotating machines. The widespread use of rotating machinery in energy generation, aviation, and the marine industry indicates the importance of monitoring its operational performance. Fault diagnosis methods allow for the detection and estimation of any deviation of the operating parameters from their normal values. Rotor unbalance is a state in which the mass center of the shaft and its components is not coincident with the rotation center. In practice, achieving perfect balance in rotors is impossible due to inherent imperfections from manufacturing errors, despite the use of sophisticated machinery and strict tolerances. Therefore, rotors often have some degree of mass imbalance, leading to a synchronous excitation force that rotates with the rotor system at a frequency equal to the rotor's spin speed [3,4]. An effective maintenance program could provide early fault detection of rotors which would reduce unplanned and costly maintenance actions. The implementation of an effective online condition monitoring system enables accurate, real-time fault prediction within the machinery [5,6].

Citation: Tselios, I.; Nikolakopoulos, P. Combining Artificial Neural Networks and Mathematical Models for Unbalance Estimation in a Rotating System under the Nonlinear Journal Bearing Approach. *Lubricants* 2024, 12, 344. https://doi.org/10.3390/lubricants12100344

Received: 20 August 2024
Revised: 2 October 2024
Accepted: 4 October 2024
Published: 6 October 2024

Copyright: © 2024 by the authors. Licensee MDPI, Basel, Switzerland. This article is an open access article distributed under the terms and conditions of the Creative Commons Attribution (CC BY) license (https://creativecommons.org/licenses/by/4.0/).

Unbalance is a very common fault in rotating machinery, and it has been investigated thoroughly using mathematical models and signal processing techniques, such as spectrum analysis, and time–frequency analysis [7]. However, it is still very difficult to fully estimate the severity of the unbalance of the rotor system under real working conditions [8].

In the Industry 4.0 era, accurate and fast diagnosing can be achieved by leveraging machine learning techniques alongside model-based monitoring systems. Compared to signal-based systems, model-based monitoring provides more precise and quicker information, as it incorporates prior knowledge of the vibration system into the fault identification process [9].

Journal bearings are widely used to support rotors in a variety of applications and industries. Hydrodynamic journal bearings are very common and have been used in various industries, mainly due to their reliability and high load capacity. However, due to the complex dynamic coupling between the rotating shaft and the bearing (fluid media), it is very important to study the vibration behavior under various operating conditions [10,11]. The instability of rotor–bearing systems is a critical issue, as it can lead to performance degradation, a high vibration level, and potential failure. Understanding and analyzing the parameters contributing to rotor–bearing system instability is essential for ensuring the reliable and safe operation of rotating machinery [12,13].

It would be very beneficial to consider the nonlinear aspects of vibration in unbalance analysis, since this approach can improve the efficiency of balancing process. Unfortunately, there are not a lot of research works that have developed balancing procedures considering the nonlinear vibration conditions and have practical applications, as reported in [14]. Linear models and synchronous response are generally adequate to analyze rotors supported by hydrodynamic bearings for certain operating conditions. Hence, stiffness and damping coefficients can provide an accurate model for a wide range of situations. However, in certain instances, this approach proves inadequate in capturing the dynamic behavior of the rotor–bearing system [15]. At specific rotation speeds, the perturbed motion of the rotor is no longer small, and the linear model is insufficient for analyzing the nonlinear dynamic behavior of the rotor–bearing system.

When the shaft rotates at a low rotating speed, only synchronous vibrations with minor amplitudes occur. These synchronous vibrations are caused by the unbalance forces acting on the rotor. As the rotation speed increases, the unbalance-induced vibration is not the only regime of motion. Along with synchronous vibrations, sub-synchronous vibration occurs due to oil whirl and oil whip phenomena. Oil whirl is the rotor lateral forward precessional subharmonic vibration around the center of the bearing. The amplitudes of oil whirl are higher compared to those of synchronous vibrations. As the rotation speed approaches double the value of the rotor's first critical speed, the oil whirl pattern becomes replaced by an oil whip—a lateral forward precessional subharmonic vibration of the rotor. Oil whip maintains a constant frequency independent of rotation speed, close to the first natural frequency of the rotor, with amplitudes exceeding those of both synchronous vibrations and oil whirl [16].

Oil film instability induces high vibration in large rotating machinery mounted on hydrodynamic bearings, constraining the reliable and efficient operation of rotating systems. Consequently, it becomes a substantial issue for industries, and it is rather necessary to obtain a comprehensive understanding of the mechanism to offer effective technical support in diagnosing system malfunctions related to oil whirl/whip [17]. As a result, representative mathematical models, including the use of nonlinear forces in bearing modeling, have been developed to simulate the vibration response for the specific working conditions of such systems [18]. In recent years, there has been a growing research interest in multidisk rotating systems under the nonlinear bearing approach. Hui Ma et al. [19] investigated how the eccentric phase difference between two disks influences oil film instability in a rotor–bearing system. They employed a model using lumped masses for the rotor system, with the journal bearings simulated using a nonlinear oil film force model based on the assumption of short bearings. The study revealed that the threshold for oil film instability increased with the growth of the eccentric phase.

Amaroju Kartheek et al. [20] propose a design strategy to reduce the response of a flexible shaft supported on hydrodynamic bearings. The control strategy proposed adjusts the relative position of the unbalance masses to decrease the excitation level of the first mode. Luo et al. [21] developed a Dynamic Recurrence Index (DRI) to detect the occurrence and estimate the evolution of oil whirl in a rotating system by measuring the dynamic similarity of normal working condition and unstable operating states. Sayed H. et al. [22] study the bearing coefficients as a function of the bearing parameters based on a novel method technique of polynomial fitting.

In modern industries, maintenance is considered one of the most crucial activities, making it critical to develop improved methods for fault identification and maintenance planning. In the Industry 4.0 era, traditional fault detection methods have been sidelined in favor of online monitoring systems, which allow real-time fault estimation. This paper explores this perspective, focusing on the application of artificial neural networks. These smart systems can learn the characteristics of complex systems and give results faster than traditional methods. Artificial neural networks (ANNs) imitate biological nervous systems. The neurons (elements) of the ANN work in parallel, and they can accomplish several different functions depending on the tuning of the connection weights. Typically, ANNs are submitted to a process of adjusting connection weights known as training, with the expectation that specific inputs should lead to specific target values. The network's training is based on the comparison of the output and the target values and is deployed until convergence is achieved. The artificial neural network is a machine learning method that has already given sufficient results not only for fault identification but also for the fast and accurate modeling of complex systems.

Reddy M.C.S. and Sekhar A.S. [23] utilize ANNs to identify unbalance and looseness in rotor–bearing systems. Mohammad Gohari and Ahmad Kord [24] investigate a model that can estimate the radius, mass, and the location of unbalance based on an ANN approach. Katsaros and Nikolakopoulos [25] combine numerical and machine learning techniques to develop a predictive tool for estimating the operating parameters of tilting pivoted pad thrust bearings. They employ the finite difference method to solve the Reynolds equation and retrieve the hydrodynamic lubrication characteristics of a pad. The innovation in this work lies in the development of a machine learning system capable of providing accurate estimation in considerably less time compared to traditional complex analyses.

The digitization of information, coupled with wide networks capable of high-speed data exchange and the availability of massive computational power, has become a crucial aspect of Rotordynamics and Mechanical Engineering as a whole. Condition monitoring of rotating machinery typically involves maintenance strategies aimed at reducing operating and maintenance costs, while ensuring maximum operating time and achieving the highest possible production rate.

The main novelty of the current work is the development of an online unbalance monitoring system based on an artificial neural network and nonlinear mathematical models that offers higher accuracy and significantly increases the performance and the maintenance activities of rotating systems.

The key novelty points of this work are the following:

- Vibration Response Prediction: Utilization of coupled mathematical models to predict the vibration response of a rotor supported by nonlinear journal bearings.
- Investigation of Critical Parameters: Investigation of the effect of essential factors, such as lubricant temperature and rotation speed, on the rotor's vibration response.
- ANN Development: Development of an artificial neural network (ANN) to estimate the unbalance masses, considering the effect of oil temperature on the bearing.

This paper is organized into six sections. Following this introduction, Section 2 focuses on developing the nonlinear mathematical models of a rotating system with two disks and presents the numerical integration methods. In Section 3, using the nonlinear models, the influence of several factors on the vibration response is investigated using the appropriate nonlinear metrics like Poincare maps, bifurcation diagrams, waterfall plots, and orbit and FFT plots. In Section 4, the ability of ANNs to perform unbalance estimation prediction under various conditions is evaluated using simulation data. Lastly, Sections 5 and 6 are the Discussion and Conclusions Sections.

2. Mathematical Analysis of Nonlinear Rotor–Bearing System

2.1. Presentation of Rotor–Bearing System

In a fully balanced rotating system and at rotation speeds below critical speed, the only mode of motion is the rotor spinning at its operating speed. An unbalance force will push the shaft to orbit around its equilibrium point [26]. The unbalance force is a centrifugal force: $F_u = m_u r_u \omega^2 e^{j(\omega t + \varphi)}$, where m is the unbalance mass, r is the unbalance radius, ω the rotation speed, and φ the angle of unbalance mass. In Figure 1, the rotor-bearing system that will be analyzed is depicted and in Table 1, the properties of this rotating system are presented.

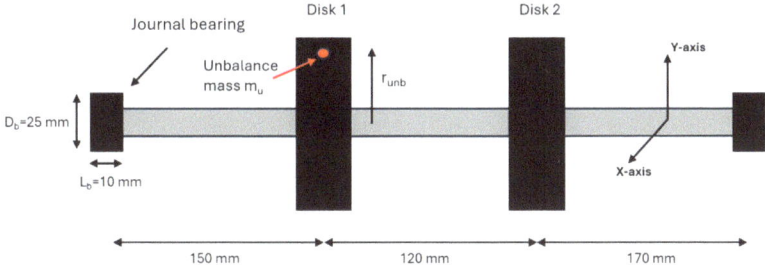

Figure 1. Rotating system supported by hydrodynamic journal bearings.

Table 1. Properties of the rotating system.

Properties of the Rotating System	
Young's Modulus	E = 210 GPa
Density	ρ = 7810 kg/m^3
Diameter of shaft	d_s = 10 mm
Diameter of disk	d_d = 75 mm
Thickness of disk	t_h = 25 mm
Length of shaft	L_s = 440 mm

The properties of the rotor–bearing system are presented in the following table.

Hydrodynamic journal bearings are widely used in industry due to their simplicity, efficiency, and cost-effectiveness. The geometry of the journal bearing is shown in Figure 2, where O_b is the center of the bearing, O_j is the journal center, R_b is the bearing radius, R_j is the journal radius, and e is the bearing eccentricity. The eccentricity ratio is defined as $\varepsilon = \frac{e}{C}$, where C is the radial clearance.

The instability of rotor–bearing systems, due to the complex dynamic coupling between the rotating shaft and the oil film of the bearing, can lead to performance degradation, a high vibration level, and potential failure. Therefore, it is very important to study the vibration behavior under various operating conditions to ensure the reliable and safe operation of the system.

As mentioned earlier, stiffness and damping coefficients, under the assumption of a linear bearing model, offer accurate predictions for many situations. However, at certain rotational speeds, the rotor's vibration amplitude becomes significant at multiple frequencies, making the linear model inadequate for capturing the system's behavior. Nonlinear models become necessary in regions of instability (oil whirl and whip regions), near critical speeds, and when there is a high level of unbalance in the system. Mathematical models using nonlinear bearing forces must be developed to simulate the vibration response at the above-mentioned regions.

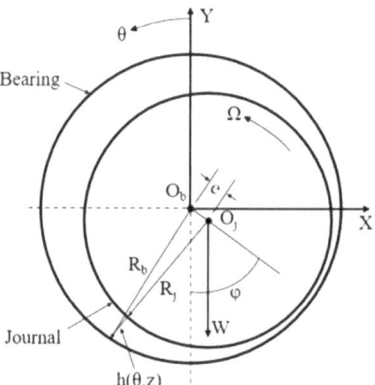

Figure 2. Geometry of the journal bearing.

In this study, the effect of lubricant temperature on the vibration response is also investigated, simulating realistic operating conditions of rotating machinery, as the oil film temperature increases during operation.

The hydrodynamic bearings are modeled using a nonlinear oil film force model based on short bearing theory, rather than relying on linear dynamic coefficients. This nonlinear bearing model was developed by Capone under the assumption of a short journal bearing (pressure distribution in circumference is negligible). Although this model has been in use for decades, it remains widely adopted in recent research due to its ability to provide accurate results for short journal bearings [11,20,27]. An alternative approach is solving the Reynolds equation at each time step, which offers greater precision but comes with a significantly higher computational cost.

The nonlinear journal bearing forces are produced using the Reynolds equation [27]. The dimensionless approach of Reynolds equation is as follows:

$$\frac{\partial}{\partial \theta}\left(h^3 \frac{\partial P}{\partial \theta}\right) + \left(\frac{R_b}{L_b}\right)^2 \frac{\partial}{\partial z}\left(h^3 \frac{\partial P}{\partial z}\right) = \frac{\partial h}{\partial \theta} + 2\frac{\partial h}{\partial (\Omega t)} \quad (1)$$

where $z = Z/L_b$, $h = H/C$ is the oil film thickness, P is the nondimensional pressure distribution $P = \frac{P}{6\mu\Omega(\frac{R_b}{C})}$, R_b is the bearing radius, and L_b is the bearing length.

The oil film thickness H is given from the following equation:

$$H = C(1 + \varepsilon \cos \theta) \quad (2)$$

In the short journal bearing, the ratio of length to diameter is lower than 0.5. Ocvirk developed a solution for Equation (1) considering the short bearing approach. In this case, the pressure gradient on the circumferential direction is negligible, compared to the pressure gradient on the axial direction.

Sommerfeld half-bearing conditions are assumed under the short bearing approach and the following expression for pressure distribution arises from the Reynolds equation.

$$p(\theta, z) = \frac{1}{2}(4z^2 - 1)\left(\frac{L_b}{D_b}\right)^2 \left[\frac{(x - 2\dot{y})\sin(\theta) - (y - 2\dot{x})\cos(\theta)}{(1 - x\cos\theta - y\sin\theta)^3}\right] \quad (3)$$

It is necessary to present the nondimensionalization process of the displacements and velocities of the shaft vibration. The nondimensional displacements x and y are given by the following equation:

$$x = \frac{X}{C}, \quad y = \frac{Y}{C} \qquad (4)$$

The nondimensional velocities $\dot{x}$ and $\dot{y}$ are given by the following equation:

$$\dot{x} = \frac{\dot{X}}{\Omega C}, \quad \dot{y} = \frac{\dot{Y}}{\Omega C} \qquad (5)$$

Capone [28,29] developed a nonlinear journal bearing model for the short bearing hypothesis, which was developed by Ocvirk [30]. Mathematical expressions for reaction forces can be determined and added to the motion equation of a rotating system as an external force. By integrating Equation (1) on z from $-L_b/2$ to $L_b/2$, the oil film force by length unit in the circumferential direction is obtained. Integrating this oil film force on the circumferential direction θ from 0 to π, the vertical and horizontal forces acting on the shaft are obtained.

The oil film forces which are obtained from the short journal bearing theory can be expressed as follows:

$$\begin{Bmatrix} Fb_x \\ Fb_y \end{Bmatrix} = \mu \Omega R_b L_b \left(\frac{R_b}{C}\right)^2 \left(\frac{L_b}{D_b}\right)^2 \begin{Bmatrix} fb_x \\ fb_y \end{Bmatrix} \qquad (6)$$

$$\begin{Bmatrix} fb_x \\ fb_y \end{Bmatrix} = -\frac{\left[(x-2\dot{y})^2 + (y+2\dot{x})^2\right]^{0.5}}{1 - x^2 - y^2} \begin{Bmatrix} 3xV(x,y,\alpha) - \sin\alpha\, G(x,y,\alpha) - 2\cos\alpha\, S(x,y,\alpha) \\ 3yV(x,y,\alpha) + \cos\alpha\, G(x,y,\alpha) - 2\sin\alpha\, S(x,y,\alpha) \end{Bmatrix} \qquad (7)$$

$$V(x,y,\alpha) = \frac{2 + (y\cos\alpha - x\sin\alpha)G(x,y,\alpha)}{1 - x^2 - y^2} \qquad (8)$$

$$S(x,y,\alpha) = \frac{x\cos\alpha + y\sin\alpha}{1 - (x\cos\alpha + y\sin\alpha)^2} \qquad (9)$$

$$G(x,y,\alpha) = \frac{2}{(1 - x^2 - y^2)^{0.5}} \left[\frac{\pi}{2} + \arctan\left(\frac{y\cos\alpha - x\sin\alpha}{(1 - x^2 - y^2)^{0.5}}\right)\right] \qquad (10)$$

$$\alpha = \arctan\left(\frac{y + 2\dot{x}}{x - 2\dot{y}}\right) - \frac{\pi}{2}\text{sign}\left(\frac{y + 2\dot{x}}{x - 2\dot{y}}\right) - \frac{\pi}{2}\text{sign}(y + 2\dot{x}) \qquad (11)$$

The mathematical model of the rotor bearing system is divided into two parts: the finite element model of the shaft and the disks, and the nonlinear hydrodynamic forces from the bearings, which are obtained by solving the Reynolds equation under the short bearing theory. The differential equation of motion for the nonlinear analysis of the system is given by Equation (12), which considers the hydrodynamic force, the unbalance forces, and the rotor weight.

$$[M]\{\ddot{u}\} + ([C] + \Omega[G] + \beta_{rid}[K])\{\dot{u}\} + ([K] + \beta_{rid}\Omega[K_{rid}])\{u\} = \{F_b\} + \{F_u\} + \{W\} \qquad (12)$$

[M], Mass matrix;
[C], Damping matrix;
[G], Gyroscopic matrix;
Ω, the rotation speed of the rotor;
[K], Stiffness matrix;
[K_{rid}], Stiffness matrix for Rotor Internal Damping contribution;
β_{rid}, Rotor Internal Damping coefficient;
$\{u\} = \{x\ y\ \theta\ \psi\}$, Displacement vector;
$\{F_b\}$, Nonlinear bearing force vector;

{F_u}, Unbalance force vector;
{W}, Weight.

The full matrices of the system consist of the sum of the matrices of the shaft, the disks, and the bearings. It is well known that Rotor Internal Damping produces non-conservative forces and can be destabilizing at some rotation speeds since it inserts power into the rotor during each revolution [31,32]. The rotor is simulated as three Timoshenko beams; each Timoshenko beam has two nodes and each node has four dofs, translation, and rotation in the horizontal and vertical directions, respectively. Disks are modeled as concentrated parameters. Bearings are introduced using the forces obtained from nonlinear bearing equations as explained earlier.

Figure 3 illustrates the difference between the linear and nonlinear approaches. As previously mentioned, linear coefficients yield accurate results for small-amplitude vibrations near the equilibrium position. In contrast, nonlinear bearing forces provide accurate results for the entire operating range. As shown in the left orbit plot, the linear dynamic bearing coefficients (K, C) accurately predict small vibration amplitudes around equilibrium position (see green line). On the other hand, the nonlinear bearing forces (see orange line in the right orbit plot) are capable of predicting fully the rotor motion even for large vibration amplitude.

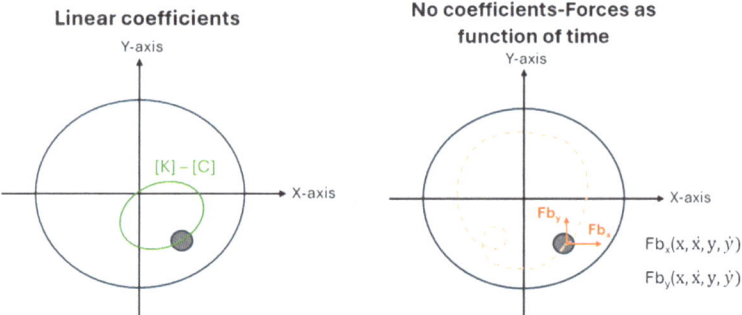

Figure 3. Linear and nonlinear bearing approaches.

2.2. Mathematical Analysis of Nonlinear Rotating System

The hydrodynamic bearings are simulated using a nonlinear oil film force model under the assumption of short bearing theory, instead of linear dynamic coefficients. Numerical integrations must be utilized to find the vibration response since the oil film force is nonlinear. Two different integration methods for the numerical solution of the ODEs were implemented, the Newmark method and the ODE23S stiff solver of MATLAB, which is based on a modified Rosenbrock formula [33]. In Section 3.2, the two methods are compared and yield similar results, demonstrating that both are robust and accurate for the vibration analysis of nonlinear systems.

In order to apply the numerical integration methods, the equations of the nonlinear system must be integrated from a selected initial state. The steady-state vibration responses of the rotor on the position of bearing 1 (left) in horizontal and vertical directions are assumed to be achieved after neglecting the first 450 revolutions. It should be noted that, in the frequency spectrum, $1\times$ represents the synchronous vibration component corresponding to the speed equal to the speed of the rotor and $0.5\times$ denotes the whirl speed equal to half of the rotor speed [31].

2.2.1. Newmark Method

Newmark's robust numerical time integrator is widely used to find the displacement, velocity, and acceleration values for each selected time step [34,35]. To reduce the computational time and avoid truncation errors, the dimensionless transformations are adopted as follows:

$$\tau = \Omega t, \tilde{u} = \frac{u}{C} \tag{13}$$

The dynamic equations of the system are expressed in dimensionless form:

$$\Omega^2 M\ddot{\tilde{u}} + \Omega(G+C)\dot{\tilde{u}} + K\tilde{u} = \frac{F_b + F_u + W}{C} \tag{14}$$

substituting $M' = \Omega^2 M$, $C' = \Omega(G+C)$, $K' = K$, $w = \tilde{u}$, and $F' = \frac{F_b + F_u + W}{C}$.

The Newmark integration method is a robust algorithm used to solve nonlinear equations in the time domain. The difference format of Newmark is as follows:

$$\dot{w}_{t+\Delta t} = \dot{w}_t + \left[(1-\delta)\ddot{w}_t + \delta\ddot{w}_{t+\Delta t}\right]\Delta t \tag{15}$$

$$w_{t+\Delta t} = w_t + \dot{w}_t \Delta t + \left[(0.5-a)\ddot{w}_t + a\ddot{w}_{t+\Delta t}\right]\Delta t^2 \tag{16}$$

where $\delta = 0.5$ and $a = 0.25$ are parameters depending on the precision and stability of the numerical integration and $\Delta \tau = 0.0001$ sec is the integration time step.

The following formula can be derived from the above equations:

$$\ddot{w}_{t+\Delta t} = \frac{1}{a\Delta t^2}(w_{t+\Delta t} - w_t) - \frac{1}{a\Delta t}\dot{w}_t - \left(\frac{1}{2a} - 1\right)\ddot{w}_t \tag{17}$$

$$\dot{w}_{t+\Delta t} = \frac{\delta}{a\Delta t}(w_{t+\Delta t} - w_t) + \left(1 - \frac{\delta}{a}\right)\dot{w}_t + \Delta t\left(1 - \frac{\delta}{2a}\right)\ddot{w}_t \tag{18}$$

The final equations of Newmark method can be derived from all the above:

$$M'\ddot{w}_{t+\Delta t} + C'\dot{w}_{t+\Delta t} + K'w_{t+\Delta t} = F'_{t+\Delta t} \tag{19}$$

$$K^* w_{t+\Delta t} = F^*_{t+\Delta t} \tag{20}$$

where

$$K^* = \frac{1}{a\Delta t^2}M' + \frac{\delta}{a\Delta t}C' + K' \tag{21}$$

$$F^* = F'_{t+\Delta t} + \left[\frac{1}{a\Delta t^2}w_t + \frac{1}{a\Delta t}\dot{w}_t + \left(\frac{1}{2a} - 1\right)\ddot{w}_t\right]M' + \left[\frac{\delta}{a\Delta t}w_t + \left(\frac{\delta}{a} - 1\right)\dot{w}_t + \left(\frac{\delta}{2a} - 1\right)\Delta t \ddot{w}_t\right]C' \tag{22}$$

2.2.2. State Space Representation

The ODE23S stiff solver in MATLAB R2023b is widely used due to its adaptive time stepping capability, offering advantages in both accuracy and computational efficiency. The same transformations described above are also used for ODE23S. The time response of the system is obtained for each rotor speed using the inbuilt ordinary differential equation solver in MATLAB, which solves the equations in state space form. To apply the ODE23S solver, it is necessary to represent the system in state space form as follows:

$$\dot{q} = Aq + Bf \tag{23}$$

$$\begin{Bmatrix} \dot{u} \\ \ddot{u} \end{Bmatrix} = \begin{bmatrix} 0 & I \\ -M^{-1}K & -M^{-1}(C + \Omega G) \end{bmatrix} \begin{Bmatrix} u \\ \dot{u} \end{Bmatrix} + \begin{bmatrix} 0 & 0 \\ 0 & M^{-1} \end{bmatrix} \begin{Bmatrix} 0 \\ W + f_{unb} + f_{bear} \end{Bmatrix} \tag{24}$$

3. Simulation Results of Nonlinear Rotor–Bearing System

3.1. Nonlinear Analysis of Rotor Vibrations

This section focuses on the effect of various parameters, like rotation speed and lubricant temperature, on the vibration response of the rotor by using bifurcation diagrams, waterfall plots, orbit plots, frequency domain plots, and Poincare maps. The rotor model for this section has an unbalance of 1 g at the left disk at a radius r = 30 mm and Rotor Internal Damping is not taken into account ($\beta_{rid} = 0$). The bearing length is L_b = 10 mm, bearing diameter D_b = 25 mm, viscosity μ = 0.1 Pas, and bearing radial clearance C = 100 µm.

The waterfall plot of the rotor system is shown in Figure 4 for unbalance m = 1 g at disk 1. The high vibration response is obtained at the rotor speeds corresponding to the (first) critical speed and on oil whirl–whip regions. The critical speed is the rotational speed at which the rotor system experiences resonance, resulting in a large vibration response. This occurs when the excitation frequency (in this case the rotor speed) matches one of the natural frequencies of the system. At around 3275 rpm, oil whirl starts, and sub-synchronous vibration exists together with the synchronous unbalance response. At 3350 rpm, the rotor inserts in the oil whip region with very high vibration amplitudes at a subharmonic frequency (oil whip locks at the resonance frequency of the system).

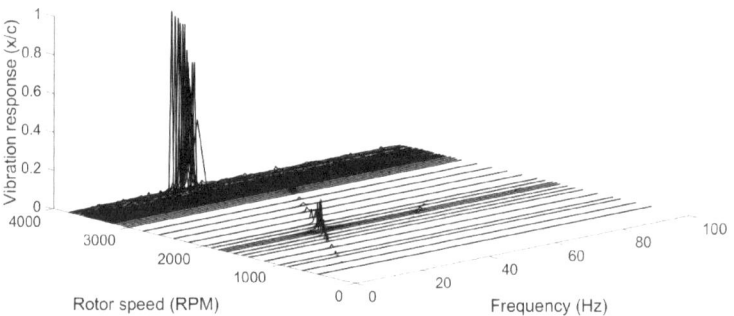

Figure 4. Waterfall plot for horizontal vibration displacement from 500 to 3800 rpm.

In Figure 4, the waterfall plot for the vibration response is depicted from 500 rpm to the rotation speed of 3800 rpm.

In Figure 5, the waterfall plot for the vibration response is illustrated from 3000 rpm to the rotation speed of 3800 rpm.

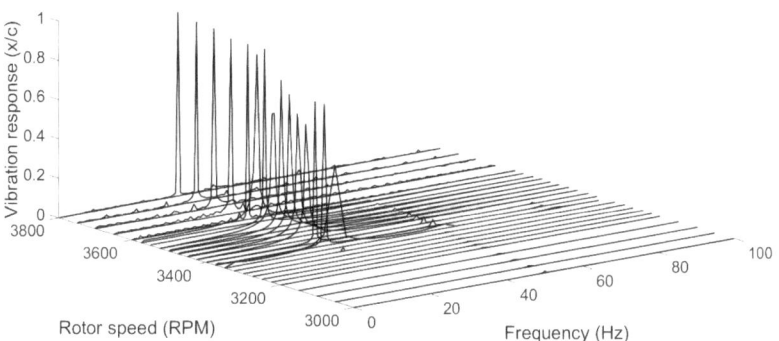

Figure 5. Waterfall plot for horizontal vibration displacement from 3000 to 3800 rpm.

3.1.1. Poincare Map

The Poincare map is a stroboscopic picture of motion in a phase plane, and it consists of a time series at a constant interval of T (T = $2\pi/\omega$) and is used to indicate the nature of the system motion [19]. The periodicity of the motion is employed in the stroboscopic method and the "map points" can be obtained directly from the time waveform results. Thus, the trajectories will return to the suitably selected Poincare section by a time interval corresponding to the period of unbalance excitation, i.e., the rotor speed. So, the values of state parameters are extracted from the steady-state response at the corresponding time points. All the extracted points are plotted in the plane (in this paper, the x-y plane is chosen) and the Poincare maps are obtained. A change in the number and topology of map points often indicates a change in the motion mode [36]. In the case of pure synchronous motion (period-1 motion), the points in the plane points coincide with each other and only one map point exists in the Poincare map.

In Figures 6–8, Poincare maps for speeds at oil whirl–whip regions are illustrated. In Figure 6, the Poincare map shows the start of oil whirl since there is a specific topology of map points which indicates the subharmonic vibration aspect. At lower speeds where only synchronous vibration occurs, the Poincare map would consist of only one point. The period-1 motion becomes unstable when oil whirl occurs because the synchronous motion coexists with the sub-synchronous motion.

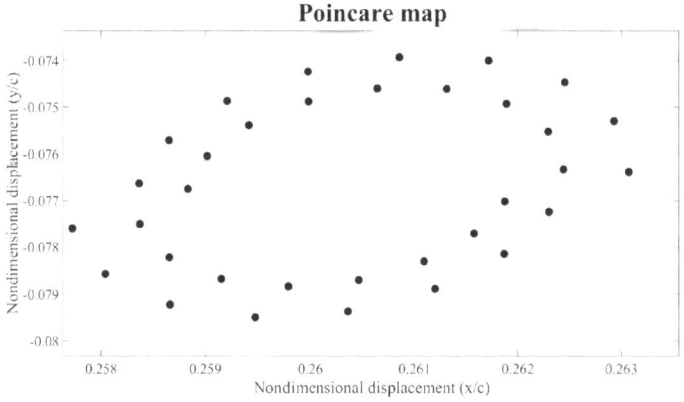

Figure 6. Poincare map of vibration response at 3275 rpm.

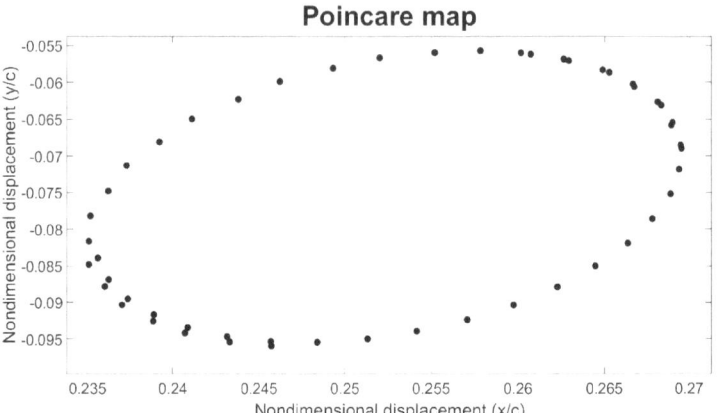

Figure 7. Poincare map of vibration response at 3325 rpm.

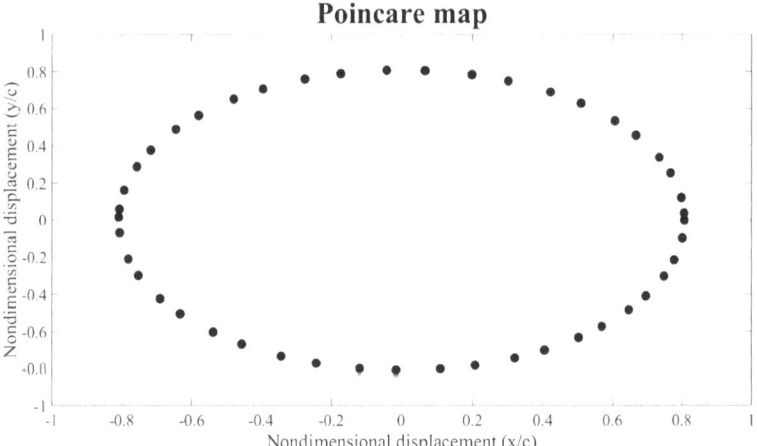

Figure 8. Poincare map of vibration response at 3800 rpm.

The motion between 3275 rpm and 3350 rpm is the transition motion from oil whirl (P2-motion) to oil whip motion (quasi-periodic motion). In Figure 7, the Poincare map is illustrated for 3325 rpm and the specific topology of the map points indicates the transition from P2 to quasi-periodic motion.

The mapping points of 3800 rpm form a closed loop in the Poincare map (Figure 8). This means the rotor motion is a pure quasi-periodic motion, since the rotor is in the oil whip region.

3.1.2. Bifurcation Analysis

The map bifurcation diagram is obtained from Poincare map points plot at each rotating speed [36]. The bifurcation of the Poincare map can also reveal the transition from period-1 motion to quasi-periodic motion. The bifurcation map of the rotor vibration response (see Figure 9) shows that the response is a synchronous vibration with period-1 (P1 motion) from 0 to 3275 rpm where oil whirl starts to appear. At 3350–3400 rpm, oil whip (quasi periodic motion) starts to dominate the vibration and gives a high subharmonic vibration amplitude.

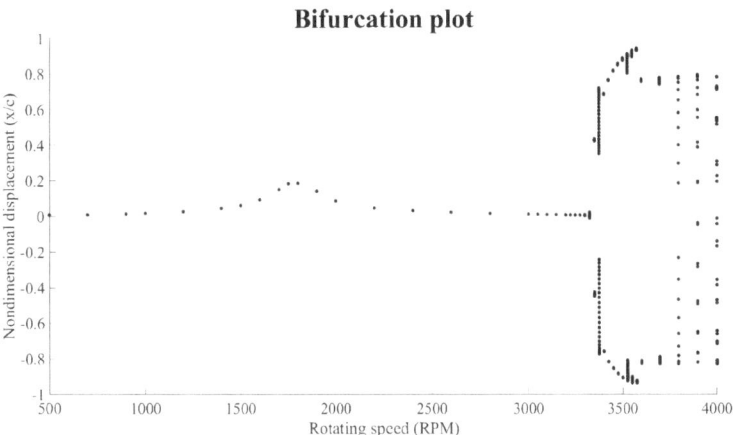

Figure 9. Bifurcation diagram for horizontal vibration response of the rotor.

3.1.3. Frequency Domain Analysis

To examine in more detail the vibration response, it is appropriate to study the vibration response in the frequency domain using the Fast Fourier Transform (FFT) analysis). In Figure 10, the vibration response at the frequency domain is illustrated for a rotation speed of 3200 rpm. It is clearly shown that the only source of vibration is due to unbalance for a frequency equal to the rotor speed (synchronous vibration).

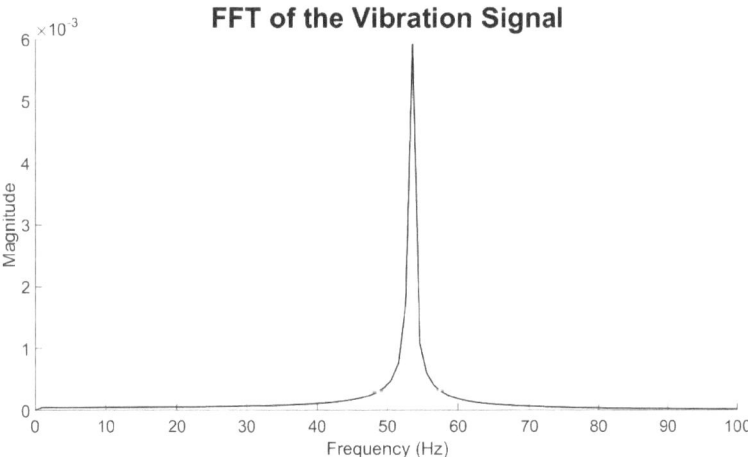

Figure 10. FFT of vibration signal at 3200 rpm.

In the next diagrams, the FFT plots are illustrated for various speeds in the oil whirl–whip region to have a better understanding of the dynamics of the system. In Figure 11, it is shown that oil whirl starts at around 3275 rpm, but the subharmonic vibration magnitude is small compared to oil whip and critical speed regions. At 3275 rpm, unbalance still dominates the response, but at 3325 rpm sub-synchronous vibration starts to dominate as expected.

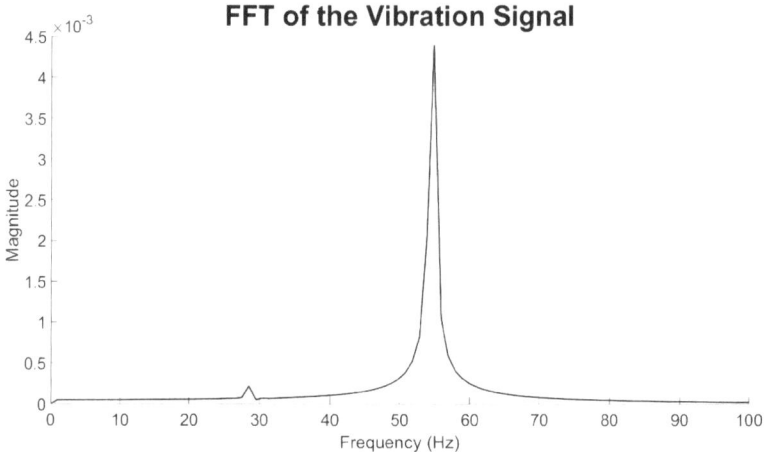

Figure 11. FFT of vibration signal at 3275 rpm.

As shown in Figure 12, at 3325 rpm sub-synchronous vibration starts to dominate, which indicates the occurrence of oil whirl instability.

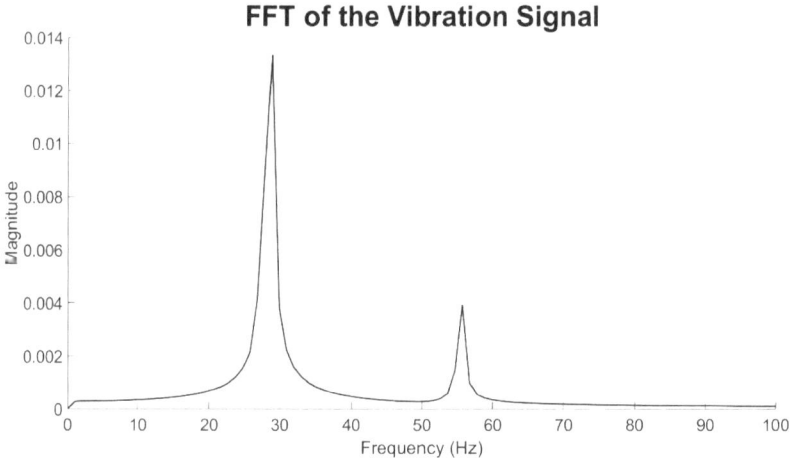

Figure 12. FFT of vibration signal at 3325 rpm.

At 3400 rpm (Figure 13), the system is purely in instability—oil whip region as illustrated in the diagram below. The characteristic of oil whip is that it does not change the frequency component with rotation speed. Oil whip "locks" at the first critical speed and does not change its frequency. The amplitude of oil whip vibration is limited by the bearing clearance, but the shaft vibration may become very high at different locations (i.e., at disks). Therefore, it is important to study these aspects in order to have a better understanding of the vibration behavior of the system.

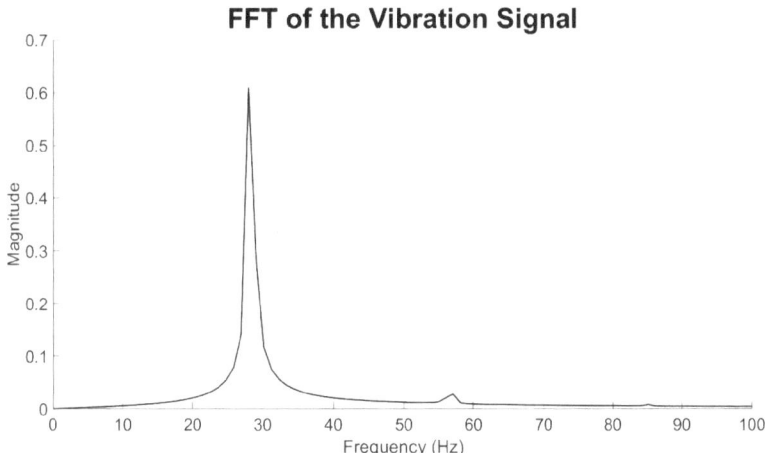

Figure 13. FFT of vibration signal at 3400 rpm.

3.1.4. Orbit Plots

The following figures illustrate the orbits of the rotor at the left bearing position for various rotational speeds. These orbit plots provide valuable insights into the system's vibration for evaluating the nonlinear behavior of the system.

In Figure 14, the orbit plot for the speed of 3275 rpm is depicted. This orbit scheme indicates the simultaneous presence of synchronous and sub-synchronous vibration.

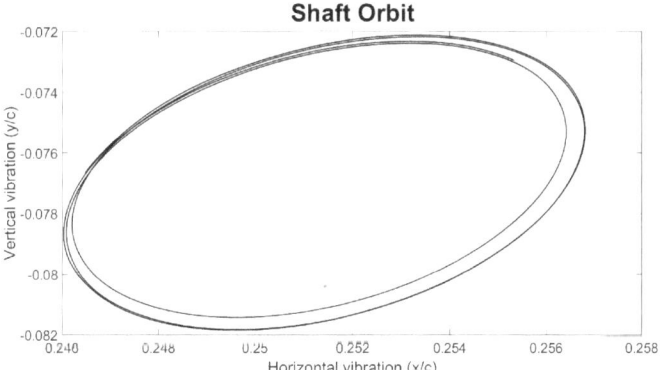

Figure 14. Orbit plot for 3275 rpm.

In Figure 15, the orbit plot for 3300 rpm illustrates the transition from low sub-synchronous vibration (Figure 14) to a higher vibration level, as expected from oil whirl–whip theory.

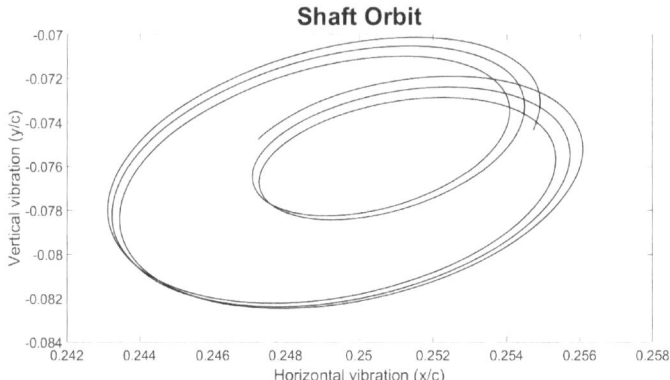

Figure 15. Orbit plot for 3300 rpm.

In Figure 16, the orbit plot for 3700 rpm illustrates the quasi-periodic motion (oil whip region) with a very high vibration response (subharmonic vibration), close to the radial clearance of the bearing.

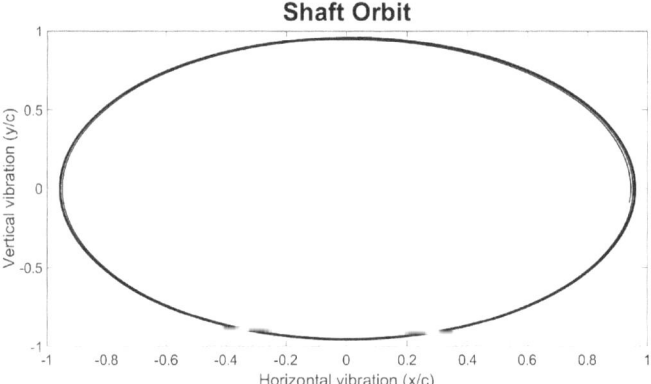

Figure 16. Orbit plot for 3700 rpm.

3.1.5. Validation of Mathematical Model

As mentioned earlier, with increasing rotational speed, unbalance-induced vibration is not the only source of vibration. In addition to synchronous motion, sub-synchronous vibrations arise due to oil whirl and oil whip phenomena. The models developed in this paper are robust and have been validated through multiple approaches. Firstly, the system's behavior within the instability region aligns with existing research on vibration responses in the oil whirl and oil whip regions. Moreover, the system's critical speed, approximately 1780 rpm (as indicated by the bifurcation plot), closely corresponds to the value reported by Hui Ma et al. [19] for a similar rotating system.

3.2. Comparison of Numerical Integration Methods

In Section 2, the numerical integration methods of Newmark and the ODE23S solver of MATLAB were presented and it was mentioned that both are considered very robust and accurate for nonlinear vibration response prediction. In this section, a comparative analysis of the two methods is illustrated to gain a better understanding of these numerical methodologies.

As depicted in Figures 17 and 18, the two methods produce similar results even for speeds at the instability region, which indicates the capabilities and the applicability of both methods in nonlinear vibration analysis. The results of the comparative analysis indicate that the Newmark method gives a slightly lower value for oil whirl initiation compared to the ODE23S solver. In Figure 17, the FFT comparative diagram at 3400 rpm reveals that the two methodologies differ in the magnitude of subharmonic vibration. The important point of this diagram is that both methods capture the subharmonic vibration level of the system.

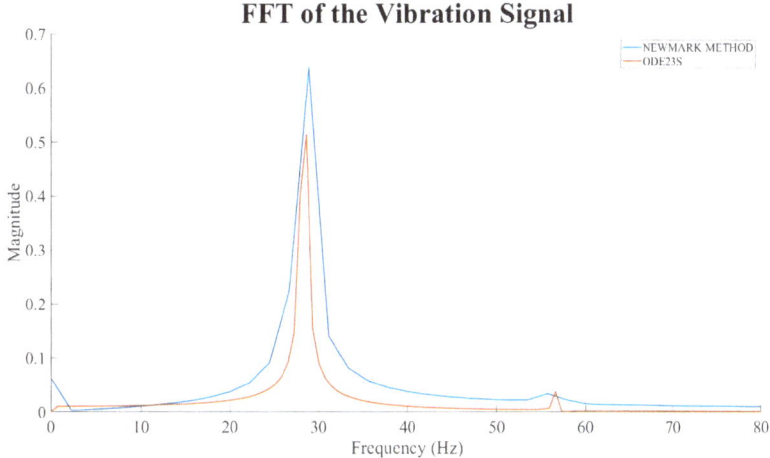

Figure 17. Comparison of solvers for horizontal vibration response at left bearing at 3400 rpm.

In Figure 18, the rotor is in the oil whip region, with a large vibration amplitude at a subharmonic resonance frequency (as explained earlier). The differences between the two numerical schemes are minimal, which indicates that both methodologies can efficiently represent the nonlinear effects of the rotor–bearing system.

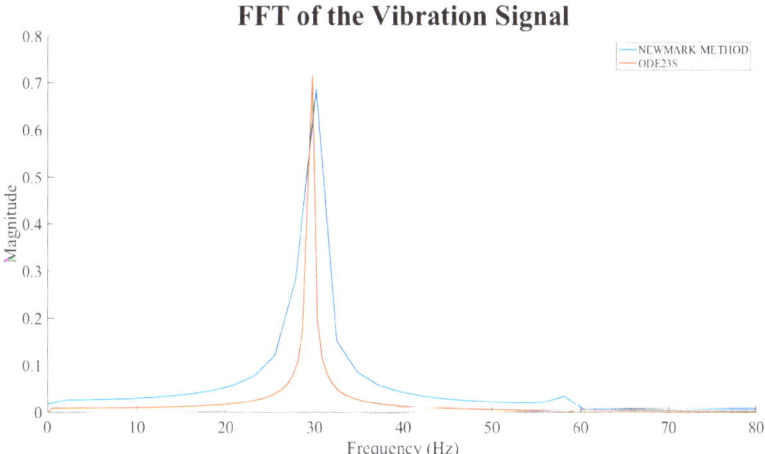

Figure 18. Comparison of solvers for horizontal vibration response at left bearing at 3550 rpm.

3.3. Effect of Lubricant Temperature on Vibration Response and Necessity of Nonlinear Model

It is known that the lubricant temperature is increased for increased operation speed and time. The lubricant viscosity changes with the temperature, as depicted in Figure 19. It would be very beneficial to consider temperature effects in the vibration analysis of the rotating system, since overheating is a very common issue in rotating systems [37,38].

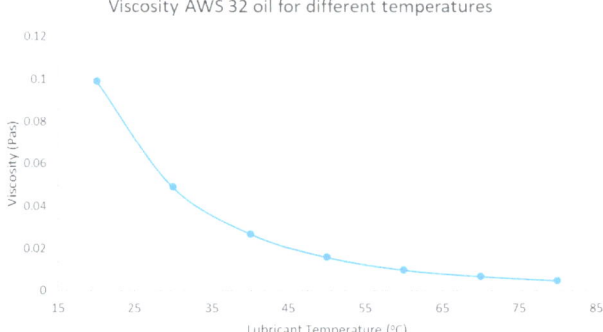

Figure 19. Viscosity as function of the temperature for AWS 32 oil.

The lubricant is AWS 32 oil, for which there exists the following equation that describes the change in viscosity for different temperatures [39].

$$\log[\log(\mu) + 4.2] = -S_o \log(1 + \frac{T}{135}) + \log(G_o) \qquad (25)$$

T is the temperature of the lubricant and So and Go are constants of the lubricant. For AWS 32 oil, So = 1.58 and Go = 3.98, according to Larson [40].

As mentioned earlier, nonlinear vibration models for rotor–bearing systems are necessary to study the vibration response in the oil whirl–whip regions, near critical speeds and under conditions of significant unbalance in the system. While linear dynamic coefficients of the bearing can provide accurate results, this is typically limited to cases where the vibration amplitude remains below 20% of the radial bearing clearance. Beyond this point, nonlinear mathematical models must be developed to study the vibration response of the rotor.

This section explores the effects of lubricant temperature and different unbalance masses on disk 1 in the critical speed region and specifically at a rotation speed of 1750 rpm). These parameters are studied in more detail to illustrate their influence on the system's vibration response. To provide a clearer understanding, orbit plots of the shaft at the left bearing position will be presented in comparative diagrams.

Figure 20 presents the orbit plots of the shaft at a rotational speed of 1750 rpm for three different lubricant temperatures: 20 °C, 30 °C, and 40 °C. As expected, an increase in lubricant temperature results in a higher vibration amplitude of the rotor. The need for nonlinear modeling becomes evident, since the vibration amplitude in the critical speed region is very high. To the knowledge of the authors, the main reason for this phenomenon is that the increase in lubricant temperature reduces the damping capability of the bearing, which gives higher vibration amplitudes.

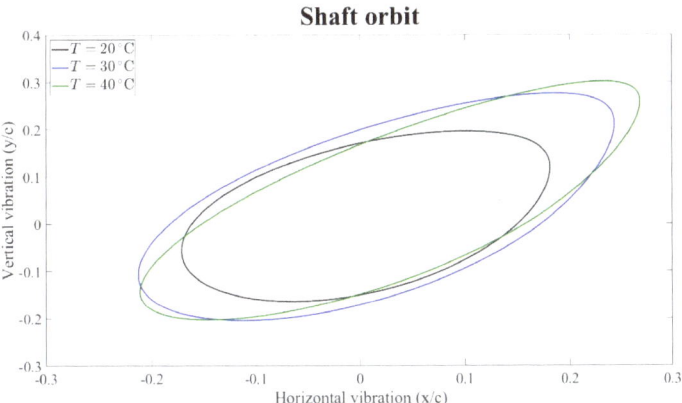

Figure 20. Vibration amplitude at 1750 rpm for three different lubricant temperatures under 1 g unbalance mass on the left disk.

Figure 21 presents the orbit plots of the shaft at a rotational speed of 1750 rpm for three different unbalance masses on the left disk (disk 1). As expected, the increase in unbalance mass leads to a higher vibration amplitude of the rotor. The need for nonlinear modeling becomes evident, since the combination of larger unbalance and operation in the critical speed region led to a very high vibration amplitude with potential failure of the system.

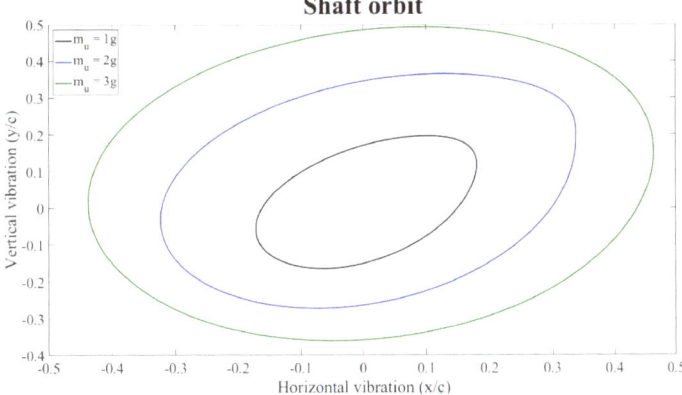

Figure 21. Vibration amplitude at 1750 rpm for three different unbalance masses on the left disk.

4. Development of Artificial Neural Networks for Unbalance Estimation

The primary objective of this section is to employ artificial neural networks (ANNs) for unbalance estimation in a rotating system using nonlinear mathematical models. A comprehensive database of vibration data is necessary in order to develop this data-driven system. The datasets are generated using the mathematical models described earlier, simulating various unbalances in the two disks for different speeds and different lubricant temperatures. The unbalance mass varies from 0.5 to 3 g at a radius r = 30 mm. The model produces two displacement vibration signals in the horizontal and vertical directions at the left bearing position. The flowchart of this research work for the development of the smart system is presented in Figure 22.

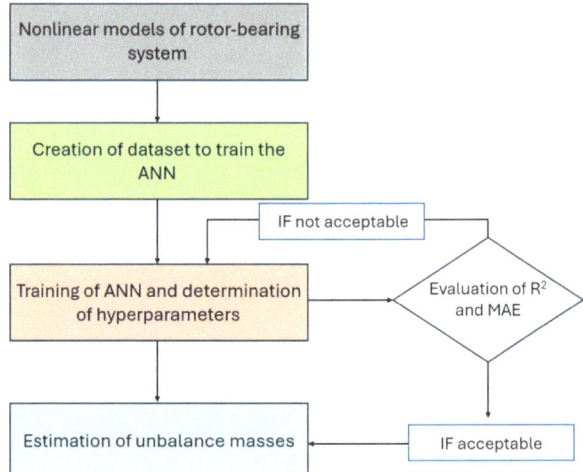

Figure 22. Flowchart for the development of an ANN system to estimate the unbalance masses of the rotating system.

In this research work, a feed-forward multilayer perceptron (MLP) will be used due to its simplicity, short training time, and high accuracy. Figure 23 depicts the feed-forward multilayer perceptron for estimating unbalance according to rotation speed, vibration response, and lubricant temperature.

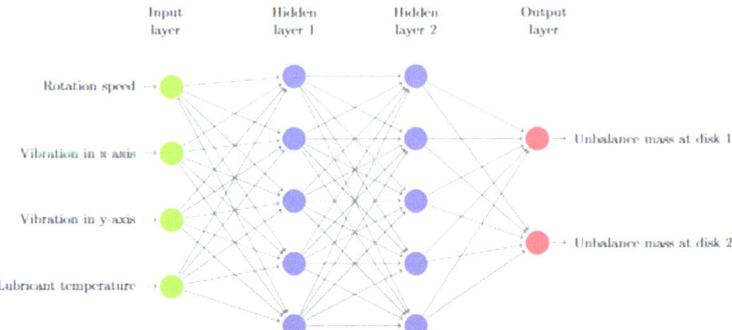

Figure 23. ANN system for unbalance estimation.

Generally, an ANN undergoes supervised training, where it is supplied with a dataset generated from the FEM model to achieve the desired network behavior.

To evaluate the performance of the regression neural network, the coefficient of determination R^2 and Mean Absolute Error (MAE) will be used. A higher coefficient of determination R^2 indicates a better fit, making it a useful metric for assessing the prediction capability of the model [41]. The coefficient values range between zero and one $\left(0 \leq R^2 \leq 1\right)$, with the highest value indicating the best fit.

The R^2 is calculated according to the following equation:

$$R^2 = 1 - \frac{\sum_{i=1}^{m}(X_i - Y_i)^2}{\sum_{i=1}^{m}\left(\overline{Y} - Y_i\right)^2} \quad (26)$$

where X_i is the target values of the dataset, Y_i is the predicted values from the ANN, m is the total number of training samples, and $\overline{Y}$ is the mean of Y_i values.

MAE measures the average magnitude of errors between the predicted and actual values and provides a straightforward and interpretable measure of prediction accuracy. Mean Absolute Error (MAE) is computed as follows:

$$\text{MAE} = \frac{1}{m}\sum_{i=1}^{m}|Y_i - X_i| \qquad (27)$$

where X_i is the actual values of the dataset, Y_i is the predicted values from the ANN, and m is the total number of training samples.

The ANN consists of four inputs and two outputs. The inputs are the synchronous vibration amplitude (1×) in the horizontal and vertical directions in the left bearing, the rotation speed, and the lubricant temperature on the left bearing. The outputs of the network are the unbalance masses at the two disks. The loss function is the 'mean squared error', the optimization algorithm is 'Adam', the activation function for hidden layers is ReLU, and the output layer is 'linear'. The network has two hidden layers with 30 neurons each.

The dataset for the ANN was created for different unbalances, lubricant temperatures, and rotational speeds, and the total number of samples is 15,120. In more detail, six unbalance masses at each disk were simulated (at each disk separately), for rotation speeds from 200 rpm to 3000 rpm (the value of speed is chosen to be on the safe side well below the stability threshold) with a 20 rpm step, and three lubricant temperatures of 20 °C, 30 °C, 40 °C. Of these data, 70% were used for training, 15% for validation, and 15% for testing.

In the Tables 2–4, the performance of the networks is presented using the R^2 and MAE metrics for training, validation, and test segment.

Table 2. Performance metrics for the regression neural network (training segment).

Metrics	Regression Neural Network	
	Output 1 (m_1)	Output 2 (m_2)
MAE	0.0078 g	0.0091 g
R^2	0.957	0.964

Table 3. Performance metrics for the regression neural network (validation segment).

Metrics	Regression Neural Network	
	Output 1 (m_1)	Output 2 (m_2)
MAE	0.0086 g	0.0084 g
R^2	0.955	0.959

Table 4. Performance metrics for the regression neural network (test segment).

Metrics	Regression Neural Network	
	Output 1 (m_1)	Output 2 (m_2)
MAE	0.0092 g	0.0075 g
R^2	0.928	0.932

The most important results are about the test segment, since the neural network is 'tested' with data that have not 'seen' before and it is a good indicator regarding the ability of the ANN to avoid overfitting and predict the desired quantities under 'new' conditions (input values).

Overfitting is a very common problem in machine and deep learning. The models which are over-trained on the training dataset result in weights adjusted to the 'noise' of the training data instead of learning "a general predicting rule" [42]. In order to achieve better performance for the networks and avoid overfitting problems, several methods were applied during the training, such as the early stopping method and feature scaling [43]. In more detail, the early stopping method stops the training after several optimization steps, leading to decreased validation loss. Feature scaling, on the other hand, is performed during the preprocessing stage and involves scaling each feature in the dataset within the range of zero to one.

5. Discussion

The implementation of artificial neural network (ANN) models for unbalance estimation using simulation data has been studied in the last few years due to the importance of unbalance faults in rotor–bearing systems. However, the unbalance estimation based on nonlinear models which also consider the lubricant temperature variation has not been studied in detail.

However, in recent years, some remarkable works integrated mathematical models and AI algorithms for fault detection but to the knowledge of the authors, there are not published works that combine these models (both AI and nonlinear mathematical models) with good accuracy for unbalance estimation. The development of a methodology for unbalance estimation using nonlinear mathematical models is the main novelty of this work since it provides very accurate results (R^2 higher than 0.928 for all the outputs) and makes it feasible to incorporate it in real-time condition monitoring systems.

Future work will deal with the development of mathematical models for bearings with wear and misalignment coupled with rotordynamics models. The creation of these models will allow condition monitoring systems to perform predictions with great accuracy and for various faults in the system, significantly enhancing the performance of the rotating systems while reducing the maintenance costs.

6. Conclusions

The application of ANNs for unbalance estimation under the nonlinear bearing approach is the main novelty of this work. To develop such accurate systems, a proper and detailed nonlinear analysis of the rotating system was performed to gain a better understanding of the nonlinear dynamics. More specifically, the following are observed:

- Nonlinear metrics like Poincare maps, bifurcation diagrams, and waterfall plots are very useful for understanding the behavior of the rotating system under various operating conditions.
- The two different solvers, ODE23S and the Newmark method, provide similar and robust results.
- The vibration response is affected by the lubricant temperature and rotation speed and, therefore, all these phenomena should be studied and taken into account in the simulations.
- The utilization of artificial neural networks provides a very effective tool for online monitoring of unbalance due to fast and accurate calculation. The neural network achieves great performance (R^2 is greater than 0.928 for all the outputs.)

Author Contributions: Conceptualization, P.N. and I.T.; methodology, P.N. and I.T.; writing—original draft preparation, I.T.; writing—review and editing, P.N. All authors have read and agreed to the published version of the manuscript.

Funding: This research received no external funding.

Data Availability Statement: The original contributions presented in the study are included in the article; further inquiries can be directed to the corresponding author.

Conflicts of Interest: The authors declare no conflicts of interest.

References

1. Vázquez, J.A.; Barrett, L.E.; Flack, R.D. A Flexible Rotor on Flexible Bearing Supports: Stability and Unbalance Response. *J. Vib. Acoust.* **2000**, *123*, 137–144. [CrossRef]
2. Chen, J.; Lin, C.; Peng, D.; Ge, H. Fault Diagnosis of Rotating Machinery: A Review and Bibliometric Analysis. *IEEE Access* **2020**, *8*, 224985–225003. [CrossRef]
3. Shao, H.; Xia, M.; Han, G.; Zhang, Y.; Wan, J. Intelligent Fault Diagnosis of Rotor-Bearing System under Varying Working Conditions with Modified Transfer Convolutional Neural Network and Thermal Images. *IEEE Trans. Ind. Inform.* **2021**, *17*, 3488–3496. [CrossRef]
4. Chen, X.; Wang, S.; Qiao, B.; Chen, Q. Basic Research on Machinery Fault Diagnostics: Past, Present, and Future Trends. *Front. Mech. Eng.* **2017**, *13*, 264–291. [CrossRef]
5. Qiao, Z.; Chen, K.; Zhou, C.; Ma, H. An improved fault model of wind turbine gear drive under multi-stage cracks. *Simul. Model. Pract. Theory* **2023**, *122*, 102679. [CrossRef]
6. Polyakov, R.; Majorov, S.; Kudryavcev, I.; Krupenin, N. Predictive Analysis of Rotor Machines Fluid-Film Bearings Operability. *Vibroeng. Procedia* **2020**, *30*, 61–67. [CrossRef]
7. Wang, T.; Han, Q.; Chu, F.; Feng, Z. Vibration Based Condition Monitoring and Fault Diagnosis of Wind Turbine Planetary Gearbox: A Review. *Mech. Syst. Signal Process.* **2019**, *126*, 662–685. [CrossRef]
8. Alsaleh, A.; Sedighi, H.M.; Ouakad, H.M. Experimental and Theoretical Investigations of the Lateral Vibrations of an Unbalanced Jeffcott Rotor. *Front. Struct. Civ. Eng.* **2020**, *14*, 1024–1032. [CrossRef]
9. Wang, J.; Ye, L.; Gao, R.X.; Li, C.; Zhang, L. Digital Twin for Rotating Machinery Fault Diagnosis in Smart Manufacturing. *Int. J. Prod. Res.* **2018**, *57*, 3920–3934. [CrossRef]
10. Mereles, A.; Alves, D.S.; Cavalca, K.L. Bifurcations and limit cycle prediction of rotor systems with fluid-film bearings using center manifold reduction. *Nonlinear Dyn.* **2023**, *111*, 17749–17767. [CrossRef]
11. Visnadi, L.B.; De Castro, H.F. Influence of bearing clearance and oil temperature uncertainties on the stability threshold of cylindrical journal bearings. *Mech. Mach. Theory* **2019**, *134*, 57–73. [CrossRef]
12. Ahmed, O.; El-Sayed, T.; Sayed, H. Finite element analyses of rotor/bearing system using second-order journal bearings stiffness and damping coefficients. *J. Vib. Control* **2023**, *30*, 3961–3984. [CrossRef]
13. Eling, R.; Wierik, M.T.; Van Ostayen, R.; Rixen, D. Towards Accurate Prediction of Unbalance Response, Oil Whirl and Oil Whip of Flexible Rotors Supported by Hydrodynamic Bearings. *Lubricants* **2016**, *4*, 33. [CrossRef]
14. Alves, D.S.; Cavalca, K.L. Investigation into the influence of bearings nonlinear forces in unbalance identification. *J. Sound Vib.* **2021**, *492*, 115807. [CrossRef]
15. De Castro, H.F.; Cavalca, K.L.; Nordmann, R. Whirl and Whip Instabilities in Rotor-Bearing System Considering a Nonlinear Force Model. *J. Sound Vib.* **2008**, *317*, 273–293. [CrossRef]
16. Jing, J.; Meng, G.; Yi, S.; Xia, S.-Y. On the Non-Linear Dynamic Behavior of a Rotor–Bearing System. *J. Sound Vib.* **2004**, *274*, 1031–1044. [CrossRef]
17. Ma, H.; Wang, X.; Niu, H.; Li, H. Effects of Different Disc Locations on Oil-Film Instability in a Rotor System. *J. Vibroeng.* **2014**, *16*, 3248–3259. Available online: https://www.extrica.com/article/15094 (accessed on 15 November 2023).
18. Alves, D.S.; Cavalca, K.L. Numerical Identification of Nonlinear Hydrodynamic Forces. In *Proceedings of the 10th International Conference on Rotor Dynamics—IFToMM. IFToMM 2018*; Cavalca, K., Weber, H., Eds.; Mechanisms and Machine Science; Springer: Cham, Switzerland, 2019; Volume 60. [CrossRef]
19. Ma, H.; Li, H.; Zhao, X.; Niu, H.; Wen, B. Effects of Eccentric Phase Difference between Two Discs on Oil-Film Instability in a Rotor–Bearing System. *Mech. Syst. Signal Process.* **2013**, *41*, 526–545. [CrossRef]
20. Kartheek, A.; Vijayan, K.; Sun, X.; Marburg, S. Stochastic Analysis of Flexible Rotor Supported on Hydrodynamic Bearings. *Mech. Syst. Signal Process.* **2023**, *203*, 110699. [CrossRef]
21. Luo, H.; Bo, L.; Peng, C.; Hou, D. Detection and quantification of oil whirl instability in a rotor-journal bearing system using a novel dynamic recurrence index. *Nonlinear Dyn.* **2022**, *111*, 2229–2261. [CrossRef]
22. Sayed, H.; El-Sayed, T. Nonlinear dynamics and bifurcation analysis of journal bearings based on second order stiffness and damping coefficients. *Int. J. Non-Linear Mech.* **2022**, *142*, 103972. [CrossRef]
23. Reddy, M.C.S.; Sekhar, A.S. Identification of Unbalance and Looseness in Rotor Bearing Systems Using Neural Networks. Available online: http://www.nacomm2011.ammindia.org/files/papers/nacomm2011_attachment_183_1.pdf (accessed on 10 December 2023).
24. Gohari, M.; Kord, A. Unbalance Rotor Parameters Detection Based on Artificial Neural Network. *Int. J. Acoust. Vib.* **2019**, *24*, 113–118. [CrossRef]
25. Katsaros, K.P.; Nikolakopoulos, P.G. Performance Prediction Model for Hydrodynamically Lubricated Tilting Pad Thrust Bearings Operating under Incomplete Oil Film with the Combination of Numerical and Machine-Learning Techniques. *Lubricants* **2023**, *11*, 113. [CrossRef]
26. Miraskari, M.; Hemmati, F.; Gadala, M.S. Nonlinear Dynamics of Flexible Rotors Supported on Journal Bearings—Part I: Analytical Bearing Model. *J. Tribol.* **2017**, *140*, 021704. [CrossRef]
27. Garoli, G.Y.; De Castro, H.F. Analysis of a rotor-bearing nonlinear system model considering fluid-induced instability and uncertainties in bearings. *J. Sound Vib.* **2019**, *448*, 108–129. [CrossRef]

28. Capone, G. Orbital motions of rigid symmetric rotor supported on journal bearings. *La Mecc. Ital.* **1986**, *199*, 37–46.
29. Capone, G. Analytical description of fluid-dynamic force field in cylindrical journal bearing. *Energ. Elettr.* **1991**, *3*, 105–110.
30. DuBois, G.B.; Ocvirk, F.W. The Short Bearing Approximation for Plain Journal Bearings. *Trans. Am. Soc. Mech. Eng.* **1955**, *77*, 1173–1178. [CrossRef]
31. Friswell, M.I.; Penny, J.E.T.; Garvey, S.D.; Lees, A.W. *Dynamics of Rotating Machines*; Cambridge University Press: Cambridge, UK, 2010. [CrossRef]
32. Zorzi, E.S.; Nelson, H.D. Finite element simulation of Rotor-Bearing systems with internal damping. *J. Eng. Power* **1977**, *99*, 71–76. [CrossRef]
33. Kaps, P.; Poon, S.W.H.; Bui, T.D. Rosenbrock methods for Stiff ODEs: A comparison of Richardson extrapolation and embedding technique. *Computing* **1985**, *34*, 17–40. [CrossRef]
34. Bathe, K. Finite Element Procedures. 1995. Available online: http://archives.umc.edu.dz/handle/123456789/116606 (accessed on 1 October 2024).
35. Newmark, N.M. A method of computation for structural dynamics. *J. Eng. Mech. Div.* **1959**, *85*, 67–94. [CrossRef]
36. Qian, D.; Liu, Z.; Yan, J.; Sun, L.; Wang, Y. Numerical and Experimental Research on Periodic Solution Stability of Inclined Rotor Journal Bearing System. In Proceedings of the ASME 2011 Turbo Expo: Turbine Technical Conference and Exposition. Volume 6: Structures and Dynamics, Parts A and B, Vancouver, BC, Canada, 6–10 June 2011; ASME: New York, NY, USA, 2011; pp. 321–330. [CrossRef]
37. Golmohammadi, A.; Safizadeh, M.S. A machine learning-based approach for detection of whirl instability and overheating faults in journal bearings using multi-sensor fusion method. *J. Braz. Soc. Mech. Sci. Eng.* **2023**, *45*, 162. [CrossRef]
38. Chang-Jian, C. Bifurcation analysis of a rotor-bearing system with temperature-dependent viscosity. *J. Low Freq. Noise Vib. Act. Control* **2023**, *43*, 75–88. [CrossRef]
39. Alves, D.S.; Fieux, G.; Machado, T.H.; Keogh, P.S.; Cavalca, K.L. A parametric model to identify hydrodynamic bearing wear at a single rotating speed. *Tribol. Int.* **2021**, *153*, 106640. [CrossRef]
40. Larsson, R.; Larsson, P.O.; Eriksson, E.; Sjöberg, M.; Höglund, E. Lubricant properties for input to hydrodynamic and elastohydrodynamic lubrication analyses. *Proc. Inst. Mech. Eng. Part J J. Eng. Tribol.* **2000**, *214*, 17–27. [CrossRef]
41. Ikumi, T.; Galeote, E.; Pujadas, P.; de la Fuente, A.; López-Carreño, R. Neural Network-Aided Prediction of Post-Cracking Tensile Strength of Fibre-Reinforced Concrete. *Comput. Struct.* **2021**, *256*, 106640. [CrossRef]
42. Dietterich, T. Overfitting and Undercomputing in Machine Learning. *ACM Comput. Surv.* **1995**, *27*, 326–327. [CrossRef]
43. Pang, B.; Nijkamp, E.; Wu, Y.N. Deep Learning with TensorFlow: A Review. *J. Educ. Behav. Stat.* **2019**, *45*, 227–248. [CrossRef]

Disclaimer/Publisher's Note: The statements, opinions and data contained in all publications are solely those of the individual author(s) and contributor(s) and not of MDPI and/or the editor(s). MDPI and/or the editor(s) disclaim responsibility for any injury to people or property resulting from any ideas, methods, instructions or products referred to in the content.

Article

Carrying Capacity of Spherical Hydrostatic Bearings Including Dynamic Pressure

Shengdong Zhang [1,*], Dongjiang Yang [2], Guangming Li [3], Yongchao Cheng [3], Jichao Li [1], Fangqiao Zhao [4] and Wenlong Song [1]

1. School of Mechanical and Electrical Engineering, Jining University, Qufu 273199, China; jichaojnxy@163.com (J.L.); wlsong@jnxy.edu.cn (W.S.)
2. Shandong Xinneng Shipbuilding Co., Ltd., Jining 273500, China; ydj7@163.com
3. Wuhan Secondary Institute of Ships, Wuhan 430064, China; talgming@163.com (G.L.); 15997433862@126.com (Y.C.)
4. Jinan Xiangderui Intelligent Technology Co., Ltd., Jinan 250199, China; pchengzhizao@163.com
* Correspondence: sdzluck@163.com

Abstract: To calculate the carrying capacity of spherical hydrostatic bearings, a numerical calculation model was presented. The influence law of dynamic pressure effect on the carrying characteristics of liquid hydrostatic spherical bearings is revealed. Under general working conditions, the dynamic pressure effect caused by the radial eccentricity of the bearing has little influence on the bearing load characteristics parameters; when the minimum width of the gap between the sealing edge is very small and the bearing rotational speed is high, the dynamic pressure effect is more obvious.

Keywords: spherical hydrostatic bearing; numerical calculation model; carrying capacity; Navier–Stokes equation; dynamic pressure

1. Introduction

Bearings, as the pivotal components of rotating machinery, necessitate a profound understanding of their performance to accurately predict and ensure the functionality of the entire system. In an era that pursues high-speed, high-power output, and stable operation, the precise prediction and control of bearing behavior become paramount. Statistics reveal that over 40% of mechanical failures can be traced back to bearing issues, underscoring the urgency of research aimed at optimizing bearing performance [1–3]. To address this challenge, exhaustive and intensive research and analysis have emerged as crucial avenues for enhancing bearing performance and reducing failure rates. Hydrostatic bearings, distinguished by their low operational friction, minimal viscous dissipation, exceptional load-bearing capacity, and high stiffness, occupy an indispensable position in the realms of heavy-duty and high-speed equipment, encompassing machine tools, precision measurement devices, hydraulic piston pumps, motor systems, telescope structures, gyroscope gimbals, dynamometers, radar tracking systems, and marine propulsion units [4–7].

Notably, spherical bearings exhibit significant advantages over other configurations due to their unique self-aligning capability and the superiority of withstanding both radial and thrust loads simultaneously. This characteristic ensures that spherical bearings maintain stable operational states even in the presence of angular misalignment, remaining unaffected. Consequently, scholars such as Mayer [8] and Elescandarany [9] have adopted a novel approach to theoretically analyze the behavior of this type of bearing and grasp its properties. Unusually, the bearing vibration was studied through fluid dynamics rather than mechanical means, revealing the influence of eccentricity, inertia, restrictor type, valve seat configuration, and supply pressure on performance. New and unique formulas were derived to predict frequency, stiffness, and damping, elucidating how these parameters generate the self-aligning characteristics of the bearing. To effectively reduce power loss and improve stiffness, Kazama [10]

Citation: Zhang, S.; Yang, D.; Li, G.; Cheng, Y.; Li, J.; Zhao, F.; Song, W. Carrying Capacity of Spherical Hydrostatic Bearings Including Dynamic Pressure. *Lubricants* **2024**, *12*, 346. https://doi.org/10.3390/lubricants12100346

Received: 7 September 2024
Revised: 8 October 2024
Accepted: 8 October 2024
Published: 10 October 2024

Copyright: © 2024 by the authors. Licensee MDPI, Basel, Switzerland. This article is an open access article distributed under the terms and conditions of the Creative Commons Attribution (CC BY) license (https://creativecommons.org/licenses/by/4.0/).

designed assembled and clearance thrust spherical bearings using capillary tubes and orifice limiters. Various factors such as surface roughness, centripetal inertia, viscosity, and valve seat configuration of bearings have an impact on the bearing load [11–15]. Elescandarany [16–18] introduced innovative techniques for designing fitted spherical bearings, both with and without restrictors and, thus, expanding design possibilities. Rajashekar [19] explored the lubrication characteristics of squeeze films, focusing on the influence of surface roughness and applied pressure. The study used a revised version of the Reynolds equation that included surface roughness and couple stress to examine their effects on bearing lubrication and load capacity. Results showed that higher couple stress led to increased load capacity. In parallel, the effects of velocity slip and viscosity on the lubrication of the squeeze film in spherical bearings were studied by Raghavendra et al. [20]. The genetic algorithms to fine tune the spherical radius and support angle [21] and to improve the performance of spherical bearings by adjusting parameters such as bearing angle, spherical radius, and oil supply vortex radius [22] has been reported in the literature. Luo [23,24] introduced methods to evaluate the static load rating of spherical plain bearings and conducted extensive life tests on new joint bearings with specialized equipment. Spherical bearings have notable applications in ship engineering [23–32], particularly self-aligning ball bearings used as axial load bearings in hydrofoil ship cylinder flanges, which endured dynamic load tests and achieved a designed carrying capacity of 500 kN. Lastly, Wang, Qian, Li, and colleagues [33–35] employed spherical hydrostatic bearings for supporting large marine propellers during static balance tests. Their thorough analysis of bearing capacity, leakage, and pressure distribution emphasized the precision, stability, and user-friendliness of this method for propellers' static balance testing.

2. Calculation Method for Dynamic Pressure in Spherical Bearings

When two inclined surfaces move relative to each other and the gap between them is filled with fluid, the fluid is dragged into the gradually converging gap due to its viscosity. This results in a localized increase in pressure within the converging gap, a phenomenon known as the dynamic pressure effect. The pressure generated by this effect is referred to as dynamic pressure. Several conditions are necessary for the dynamic pressure effect to occur: there must be an appropriately sized wedge-shaped gap between the moving surfaces, the gap must be filled with a viscous fluid, and the surfaces must have a sufficient relative motion speed.

When a high-speed rotating hydrostatic spherical bearing experiences horizontal eccentricity, all conditions for the dynamic pressure effect are met. Consequently, dynamic pressure is generated within the sealing edge gaps. This dynamic pressure effect disrupts the pressure distribution along the sealing edges, creating unbalanced torques and impacting the stable operation of the spherical bearing.

The spherical hydrostatic bearing comprises a bearing housing and a bearing body, as illustrated in Figure 1.

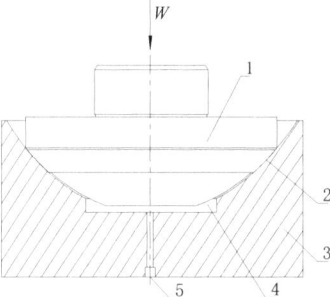

Figure 1. The structure of spherical hydrostatic bearing. 1—bearing body; 2—oil spill gap; 3—bearing housing; 4—high-pressure oil cavity; 5—oil supply hole. Adapted from [35].

The oil film can be depicted as a narrow gap between surfaces of spherical bearings. A cutting plane, considered along the axis, is defined with the sphere's radius labeled as R. The high-pressure oil chamber angle is denoted as $2\theta_1$, while the oil seal angle on one side is marked as θ_2. The bearing is suspended within the high-pressure hydraulic oil. For the sake of simplicity, the analysis focuses exclusively on the vertical eccentricity of the bearing, neglecting horizontal eccentricity. Figure 2 depicts the vertical distance associated with spherical eccentricity and the subsequent oil seal edge, with the values $R = 350$ mm, $\theta_1 = 31°$, $\theta_2 = 8°$.

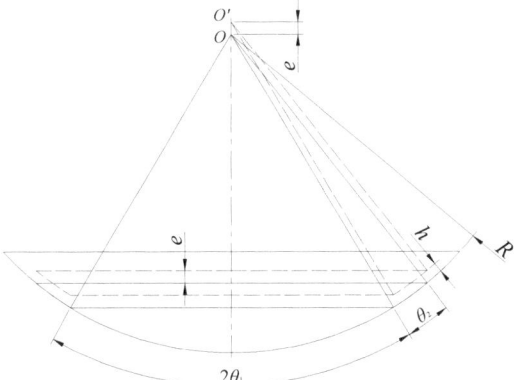

Figure 2. Schematic of the spherical bearing oil sealing edge. Adapted from [35].

A spherical coordinate system is established in the spherical bearing oil seal model, as shown in Figure 3.

The pressure resulting from the dynamic pressure effect can be determined by solving the Reynolds equation. Under isothermal conditions, the Reynolds equation in spherical coordinates can be expressed as follows:

$$\frac{\partial}{\partial \varphi}\left(\frac{\rho h^3}{\mu}\frac{\partial p}{\partial \varphi}\right) + \sin\theta\frac{\partial}{\partial \theta}\left(\sin\theta\frac{\rho h^3}{\mu}\frac{\partial p}{\partial \theta}\right) = 6\omega r^2 \sin^2\theta\frac{\partial(\rho h)}{\partial \varphi} + 12 r^2 \sin^2\theta\frac{\partial(\rho h)}{\partial t} \quad (1)$$

Assuming the hydraulic oil is an incompressible fluid, for incompressible steady flow, Equation (1) can be simplified as follows:

$$\frac{\partial}{\partial \varphi}\left(h^3\frac{\partial p}{\partial \varphi}\right) + \sin\theta\frac{\partial}{\partial \theta}\left(\sin\theta h^3\frac{\partial p}{\partial \theta}\right) = 6\mu\omega r^2 \sin^2\theta\frac{\partial h}{\partial \varphi} \quad (2)$$

Equations (1) and (2) can be further decomposed into:

$$\frac{\partial}{\partial \varphi}\left(h^3\frac{\partial p}{\partial \varphi}\right) + \sin^2\theta\frac{\partial}{\partial \theta}\left(h^3\frac{\partial p}{\partial \theta}\right) + \sin\theta\cos\theta h^3\frac{\partial p}{\partial \theta} = 6\mu\omega r^2 \sin^2\theta\frac{\partial h}{\partial \varphi} \quad (3)$$

When the bearing body becomes eccentric, let the radial eccentricity be denoted as e_1 and the axial (vertical) eccentricity be denoted as e_2. The gap width at the sealing edge can then be expressed as:

$$h = e_1 \sin\theta\cos\varphi + e_2 \cos\theta \quad (4)$$

The gap, denoted as h, is determined by the angle, θ, and the spherical radius, R; radial eccentricity, $e_1 = e$, is shown in Figure 2, and axial (vertical) eccentricity, e_2, is illustrated in Figure 4.

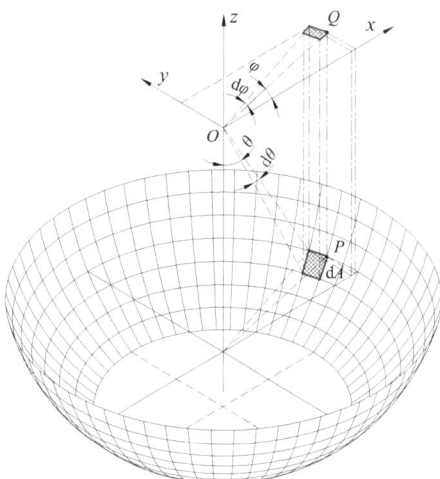

Figure 3. Discrete model of spherical bearing and parameter definition. Adapted from [35].

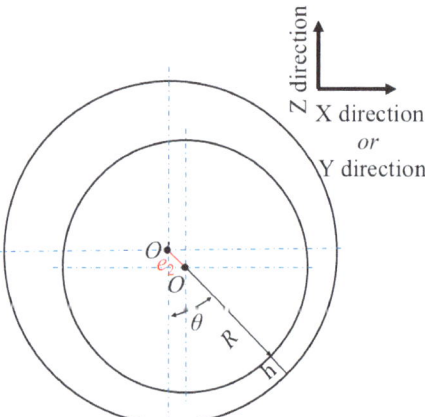

Figure 4. Slit schematic diagram of spherical eccentricity.

The above Reynolds equation is a partial differential equation involving two variables: the gap width and the pressure. Since the gap width is already expressed in Equation (4), solving the equation essentially means solving for the pressure. Researchers have proposed various effective methods for solving the Reynolds equation, such as the finite difference method, the finite volume method, and the finite element method. Among these, the finite difference method is widely acknowledged for its early application, simplicity, effectiveness, and the flexibility in controlling computational errors.

The basic concept of the finite difference method involves dividing the solution domain into a finite number of computational nodes, where the unknown quantities at each node are expressed in terms of the values at neighboring nodes. Typically, this entails replacing the derivatives of the unknown quantities with difference approximations derived from adjacent nodes, thereby converting the differential equation into a system of linear algebraic equations at the nodes. Through iterative computations, the values of the unknown quantities at all nodes are determined. In the finite difference method, boundary conditions are set as known values or specified relationships at certain boundary nodes.

The pressure value at node (i,j), denoted as $p(i,j)$, can be expressed as [36,37]:

$$p_{i,j} = A_{i,j}p_{i-1,j} + B_{i,j}p_{i+1,j} + C_{i,j}p_{i,j-1} + D_{i,j}p_{i,j+1} + E_{i,j} \tag{5}$$

In Equation (5), the coefficients $A_{i,j}$, $B_{i,j}$, $C_{i,j}$, $D_{i,j}$, and $E_{i,j}$, can be derived from Equations (3) and (4).

To solve the spherical bearing oil-sealing edge model using numerical calculation methods, the physical model needs to be discretized into a finite number of elements first. The circumferential angle of the oil-sealing edge, ranging from 0 to 2π, is discretized into m elements, with a circumferential step size of $d\varphi = 2\pi/m$, and the node numbers ranging from 1 to m. The axial angle, from θ_1 to $(\theta_1 + \theta_2)$, is discretized into n elements, with an axial step size of $d\theta = \theta_2/n$, and the node numbers ranging from 1 to $n+1$. Additionally, the thickness of the oil-sealing edge gap, h, is discretized into k elements, with a thickness step size of $dh = h/k$, and the node numbers ranging from 1 to $k+1$, as illustrated in Figure 5.

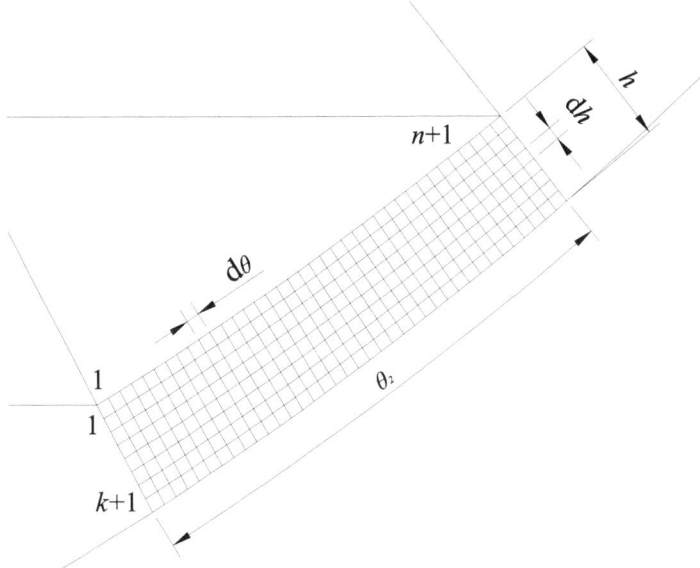

Figure 5. Discrete model of the oil-sealing edge gap.

There are various methods to solve the above system of algebraic equations. Among them, the iterative method stands out for its simplicity, stability, and relatively low computational cost. The iterative method involves assigning an initial estimated value to the unknown quantity at each node and then using Equation (5) to calculate an updated $p_{i,j}$ value. This updated value replaces the original estimate, and the process is repeated for all nodes. Through multiple iterations, increasingly accurate p values are obtained, eventually achieving the desired precision.

After completing the calculations for all nodes in each iteration, it is necessary to determine whether the results have achieved the required accuracy. The convergence criterion used in this paper can be expressed as:

$$\frac{\sum_{i=2}^{m}\sum_{j=2}^{n}\left|p_{i,j}^{(k)} - p_{i,j}^{(k-1)}\right|}{\sum_{i=2}^{m}\sum_{j=2}^{n}\left|p_{i,j}^{(k)}\right|} \leq [\Delta] \tag{6}$$

In the equation, [Δ] represents the allowable convergence error, set at 0.001.

After calculating the dynamic pressure in the sealing edge gap, the forces exerted by this dynamic pressure on the spherical bearing body can be determined using numerical integration methods.

From the derivation of the Reynolds equation, it is assumed that there is no axial flow. In hydrostatic spherical bearings, axial flow is primarily caused by hydrostatic pressure differences. On the other hand, dynamic pressure values are generally small, and the influence of hydrostatic pressure on dynamic pressure calculations is significant. Therefore, when calculating the dynamic pressure effect, the hydrostatic pressure difference should be set to zero. This precludes solving for both circumferential and axial flows simultaneously in the dynamic pressure effect calculation.

3. Results and Discussion

In this paper, using the hydrostatic spherical bearing model described in [35] as an example, we examine the impact of dynamic pressure effects on the characteristic parameters of hydrostatic spherical bearings. Based on the analysis of the dynamic pressure calculation method, the following assumptions are made for the computational model:

To avoid the influence of surface elastic deformation on the results, the surface elastic deformation of bearing components is not considered in the calculations.

As seen from Equation (3), both the viscosity of the hydraulic oil and the bearing rotational speed are important factors affecting the dynamic pressure effect, appearing as coefficients on the right-hand side of the equation. Since the viscosity's variation range is small and relatively stable, we assume a constant hydraulic oil viscosity and primarily discuss the dynamic pressure effect relative to changes in the bearing rotational speed.

For the calculation of the dynamic pressure effect, we assume that the pressure in the high-pressure oil chamber is zero, thereby eliminating any axial flow driven by hydrostatic pressure differences.

Given a radial eccentricity, e_1 = 0.25 mm, and an axial eccentricity, e_2 = 0.25 mm, for the bearing body, along with a rotational speed of 100 r/min, we input these values along with other parameters listed in Tables 1 and 2 from reference [35] into the computational program. The solutions yielded the distribution of the oil film gap width and the dynamic pressure after accounting for the bearing body's eccentricity. The results are illustrated in Figures 6 and 7.

Based on the above computational results, it is evident that after the bearing body experiences eccentricity, the gap width at the oil-sealing edge is significantly reduced in the direction of radial eccentricity. Along the same axial cross-section, the gap width at the oil-sealing edge's outlet is smaller compared to the inlet. The minimum gap width occurs at the oil-sealing edge's outlet in the direction of radial eccentricity.

From the pressure distribution diagram, it can be observed that most of the oil film gap within the oil-sealing edge maintains a relatively low pressure. On the side where the gap width decreases along the direction of the bearing body's rotation, local pressure increases are observed near the region of the minimum gap width. This high-pressure zone, which tends to be near the oil-sealing edge's outlet, reaches a maximum pressure of approximately 0.14 MPa. The elevated pressure in this area is attributed to the dynamic pressure effect. Analyzing the dynamic pressure distribution for different radial eccentricities reveals that as the radial eccentricity increases, the high-pressure zone becomes progressively smaller, the pressure gradient increases, and the maximum dynamic pressure rises accordingly.

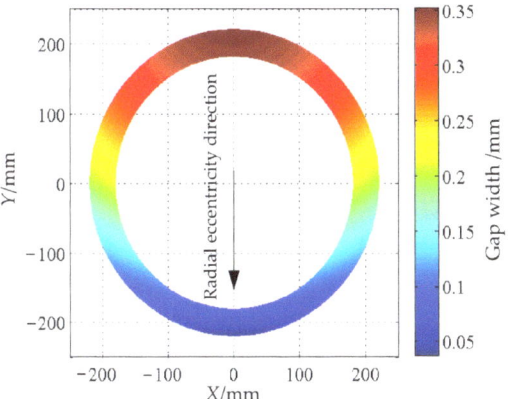

Figure 6. Top view of the width distribution of the sealing edge gap.

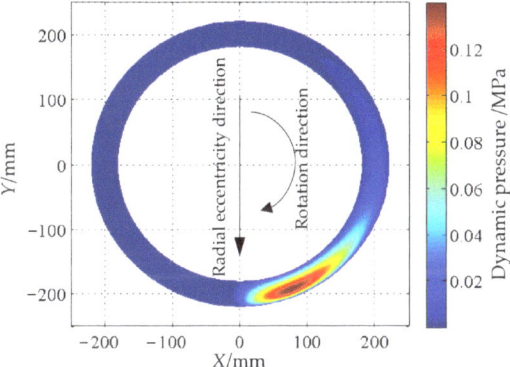

Figure 7. Top view of dynamic pressure distribution at the sealing edge.

The maximum dynamic pressure reflects the intensity of the dynamic pressure effect and, thus, serves as a measure of its strength. The minimum gap width at the oil-sealing edge is a crucial factor in determining the dynamic pressure effect. When the bearing body undergoes axial and radial eccentricity, it essentially alters the dynamic pressure effect by changing this minimum gap width.

The transversal distribution of the pressure of the spherical hydrostatic bearings is shown in Figure 8.

The primary force in a spherical hydrostatic bearing, namely the axial carrying capacity, emanates from the Z-direction. The transversal distribution of pressure in the X and Y directions plays a secondary role in determining the bearing's carrying capacity. Therefore, only the axial carrying capacity is researched in this paper.

With a constant axial eccentricity, $e_2 = 0.25$ mm, we incrementally increase the radial eccentricity, which results in a corresponding reduction in the minimum gap width at the oil-sealing edge. We then investigate how the maximum dynamic pressure varies with the minimum gap width at different rotational speeds. The resulting variation curves are presented in Figure 9.

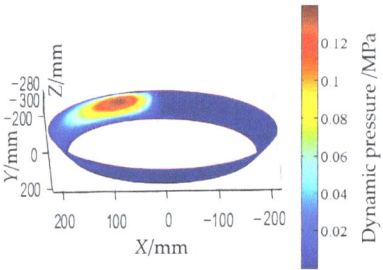

Figure 8. The transversal distribution of the pressure of the spherical hydrostatic bearings.

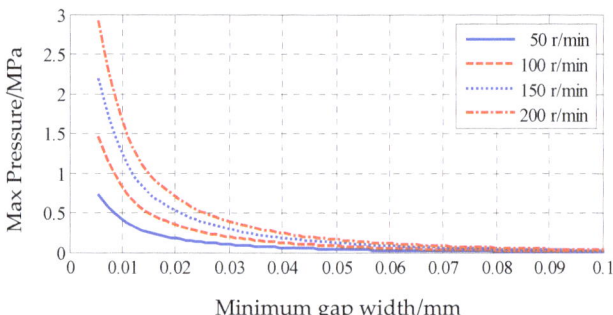

Figure 9. The variation of maximum pressure with the minimum width of the sealing edge gap at different speeds.

The curves indicate that the minimum gap width at the oil-sealing edge and the maximum dynamic pressure exhibit an approximately inverse relationship. At a given rotational speed, as the minimum gap width decreases, the maximum dynamic pressure increases, with the rate of change also accelerating. When comparing the maximum dynamic pressure at different rotational speeds relative to the minimum gap width, it is evident that, for the same gap width, higher bearing rotational speeds result in higher maximum dynamic pressures. As the gap width narrows, the influence of rotational speed on the maximum dynamic pressure becomes more pronounced. When the gap width is very small, the maximum dynamic pressure is significantly high and highly sensitive to changes in rotational speed. At higher rotational speeds, the maximum dynamic pressure can become substantial, which may adversely affect the stable operation of the bearing.

For hydrostatic spherical bearings, the dynamic pressure resulting from the dynamic pressure effect acts on the bearing surface and contributes to an increased load-carrying capacity to some extent. By integrating the computed dynamic pressures, we can determine the axial load-carrying capacity generated by the dynamic pressure effect. The curves illustrating how the axial dynamic load-carrying capacity varies with the minimum gap width at the oil-sealing edge under different rotational speeds are shown in Figure 10.

The variation of the axial dynamic load-carrying capacity due to the dynamic pressure effect follows a similar pattern to that of the maximum dynamic pressure with respect to bearing rotational speed and the minimum gap width at the oil-sealing edge. When the minimum gap width is small, the axial dynamic load-carrying capacity is significantly higher and changes rapidly with the width. This effect is particularly pronounced at higher rotational speeds where the axial dynamic load-carrying capacity is relatively large. However, compared to the static load-carrying capacity that a hydrostatic spherical bearing typically withstands during normal operation, the load-carrying capacity generated by the dynamic pressure effect is relatively minimal and can be considered negligible.

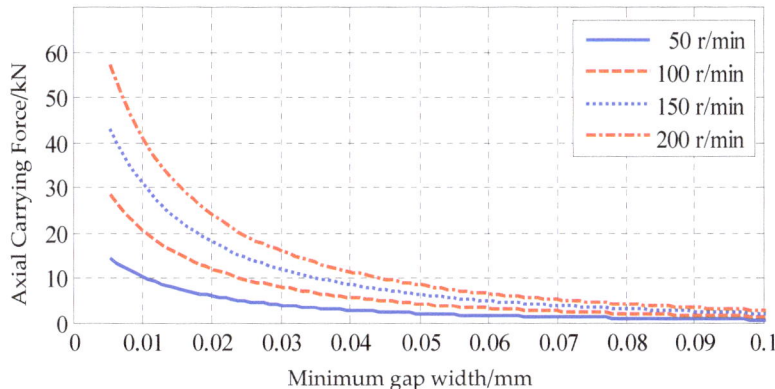

Figure 10. Variation of axial dynamic pressure bearing capacity with minimum width of sealing edge gap at different speeds.

4. Conclusions

The impact of dynamic pressure effects on the load-carrying performance of liquid hydrostatic spherical bearings can be significant. Analyzing the dynamic pressure distribution reveals that local high-pressure areas emerge near the minimum gap on the side where the gap width decreases in the direction of the bearing body's rotation. These localized high-pressure zones generate unbalanced torque, potentially affecting the stable operation of the bearing.

Furthermore, the high pressure near the minimum gap width can prevent further reduction in the gap, counteracting the external forces causing bearing eccentricity and ultimately stabilizing the bearing body at a certain eccentric position. In terms of the forces generated by the dynamic pressure effect, the axial component acting on the bearing body is relatively small and can be considered negligible.

From the observed trends, it is evident that the smaller the minimum gap width and the higher the rotational speed, the stronger the dynamic pressure effect and the greater its influence. Therefore, it is advisable to minimize large eccentricities in the bearing body to avoid significant dynamic pressure effects.

Author Contributions: Conceptualization, S.Z. and G.L.; methodology, F.Z.; data curation, S.Z. and J.L.; writing—original draft preparation, S.Z. and G.L.; writing—review and editing, S.Z., D.Y., Y.C., G.L., F.Z. and W.S. All authors have read and agreed to the published version of the manuscript.

Funding: This research and APC were funded by the Shandong Province Higher Education Undergraduate Teaching Reform Research Project (No. Z2023085), the "Hundred Outstanding Talents" of Jining University (No. 2023ZYRC56), Science and Technology Innovation Team Foundation of Jining University (23KCTD07), the Key Research and Development Program of Jining of China (2022HHCG014), and the Scientific Research Foundation of Jining University (2022HHKJ02, 2023HHKJ04).

Data Availability Statement: Data are contained within the article.

Conflicts of Interest: Author D.Y. was employed by the company Shandong Xinneng Shipbuilding Co., Ltd. Author F.Z. was employed by the company Jinan Xiangderui Intelligent Technology Co., Ltd. The other authors conducted the research in the absence of any commercial or financial relationships that could be construed as a potential conflict of interest.

References

1. Deng, H.; Hu, C.; Wang, Q.; Wang, L.; Wang, C. Friction and wear analysis of the external return spherical bearing pair of axial piston pump/motor. *Mech. Ind.* **2020**, *21*, 104. [CrossRef]
2. Lin, S.C. Friction and Lubrication of sliding bearings. *Lubricants* **2023**, *11*, 226. [CrossRef]
3. Li, J.; Yamada, S.; Kishiki, S.; Yamazaki, S.; Watanabe, A.; Terashima, M. Experimental and numerical study of spherical sliding bearing (SSB)-part 2: Friction models. In Proceedings of the 17th World Conference on Earthquake Engineering, 17WCEE, Sendai, Japan, 28 September–2 October 2021.
4. Ryan, K.L.; Mojidra, R. Analysis of effect of vertical ground shaking in bridges isolated with spherical sliding bearings. *Earthq. Eng. Struct. Dyn.* **2023**, *52*, 5010–5032. [CrossRef]
5. Deng, H.; Zhu, P.; Hu, C.; He, T. Study on dynamic lubrication characteristics of the external return spherical bearing pair under full working conditions. *Machines* **2022**, *10*, 107. [CrossRef]
6. Tomar, A.K.; Sharma, S.C. Non-Newtonian lubrication of hybrid multi-recess spherical journal bearings with different geometric shapes of recess. *Tribol. Int.* **2022**, *171*, 107579. [CrossRef]
7. Yang, Y.; Zhang, Y.; Ju, J. Study on the mechanical properties of a type of spherical bearing. *J. Theor. Appl. Mech.* **2021**, *59*, 539–550. [CrossRef]
8. Meyer, D. Reynolds equation for spherical bearings. *J. Tribol.* **2003**, *125*, 203–206. [CrossRef]
9. Elescandarany, A.W. Analytical study of the spherical hydrostatic bearing dynamics through a unique technique. *Sci. Rep.* **2023**, *13*, 19364. [CrossRef]
10. Kazama, T. Design of hydrostatic spherical bearings in fluid film lubrication. *J. Tribol.* **2000**, *1228*, 66–69. [CrossRef]
11. Yacout, A.W.; Ismail, A.S.; Kassab, S.Z. The combined effects of the centripetal inertia and the surface roughness on the hydrostatic thrust spherical bearings performance. *Tribol. Int.* **2007**, *40*, 522–532. [CrossRef]
12. Elescandarany, A.W.Y. The effect of the fluid film variable viscosity on the hydrostatic thrust spherical bearing performance in the presence of centripetal inertia and surface roughness. *Int. J. Mech. Eng. Appl.* **2018**, *6*, 1–12. [CrossRef]
13. Yacout, A.W.; Ismail, A.S.; Kassab, S.Z. The surface roughness effect on the hydrostatic thrust spherical bearings performance (part 2 un-recessed clearance type). In Proceedings of the ASME, International Mechanical Engineering Congress and Exposition, IMECE2006-13004, Chicago, IL, USA, 5–10 November 2006. [CrossRef]
14. Yacout, A.W.; Ismail, A.S.; Kassab, S.Z. The surface roughness effect on the hydrostatic thrust spherical bearings performance (part 3 recessed clearance type of bearings). In Proceedings of the ASME International Mechanical Engineering Congress and Exposition, IMECE 2007-41013, Seattle, WA, USA, 11–15 November 2007. [CrossRef]
15. Elescandarany, A.W. Externally pressurized thrust spherical bearing performance under variable lubricant viscosity (un-recessed clearance type). *Int. J. Mech. Eng. Appl.* **2020**, *8*, 45–64. [CrossRef]
16. Elescandarany, A.W.Y. Design of the hydrostatic thrust spherical bearing with restrictors (fitted type). *Int. J. Mech. Eng. Appl.* **2019**, *7*, 34–45. [CrossRef]
17. Elescandarany, A.W.Y. Design of self-restriction hydrostatic thrust spherical bearing (fitted type). *Int. J. Mech. Eng. Appl.* **2019**, *7*, 111–122. [CrossRef]
18. Elescandarany, A.W. Kugel ball as an interesting application of designing the hydrosphere. *Int. J. Mech. Eng. Appl.* **2021**, *9*, 25–32. [CrossRef]
19. Rajashekar, M.; Kashinath, B. Effect of surface roughness on MHD couple stress squeeze-film characteristics between a sphere and a porous plane surface. *Adv. Tribol.* **2012**, *2012*, 935690. [CrossRef]
20. Rao, R.R.; Gowthami, K.; Kumar, J.V. Effects of velocity-slip and viscosity variation in squeeze film lubrication of spherical bearings. In Proceedings of the International Conference on Advances in Tribology and Engineering Systems: ICATES, Ahmedabad, India, 15–17 October 2013; Springer: Delhi, India, 2013; pp. 35–47. [CrossRef]
21. Tao, J.Z.; Yin, G.F.; Wang, F.Y. An optimism method of air static pressure ball bear parameters. *China Mech. Eng.* **2004**, *15*, 48–50. [CrossRef]
22. Guo, X.P.; He, H.T.; Zhao, L. Optimization design of non-liquid friction spherical sliding bearing based on ANSYS. *Bearing* **2007**, *2*, 4–5. [CrossRef]
23. Luo, L.; Wang, X. Finite element analysis of self-Lubricating joint bearing liner wear. In Proceedings of the 2017 International Conference on Computer Technology, Electronics and Communication (ICCTEC), Dalian, China, 19–21 December 2017; pp. 245–249. [CrossRef]
24. Luo, L.; Wang, X.; Liu, H.; Zhu, L. Number simulation analysis of self-lubricating joint bearing liner wear. *Int. J. Interact. Des. Manuf.* **2019**, *13*, 23–34. [CrossRef]
25. Royston, T.J.; Basdogan, I. Vibration transmission through self-aligning (spherical) rolling element bearings: Theory and experiment. *J. Sound Vib.* **1998**, *215*, 997–1014. [CrossRef]
26. Geng, K.; Lin, S. Effect of angular misalignment on the stiffness of the double-row self-aligning ball bearing. *Proc. Inst. Mech. Eng. Part C J. Mech. Eng. Sci.* **2020**, *234*, 946–962. [CrossRef]
27. Ambrożkiewicz, B.; Syta, A.; Gassner, A.; Georgiadis, A.; Litak, G.; Meier, N. The influence of the radial internal clearance on the dynamic response of self-aligning ball bearings. *Mech. Syst. Signal Process.* **2022**, *171*, 108954. [CrossRef]
28. Zhuo, Y.; Zhou, X.; Yang, C. Dynamic analysis of double-row self-aligning ball bearings due to applied loads, internal clearance, surface waviness and number of balls. *J. Sound Vib.* **2014**, *333*, 6170–6189. [CrossRef]

29. Parmar, V.; Saran, V.H.; Harsha, S.P. Effect of an unbalanced rotor on dynamic characteristics of double-row self-aligning ball bearing. *Eur. J. Mech. A/Solids* **2020**, *82*, 104006. [CrossRef]
30. Parmar, V.; Saran, V.H.; Harsha, S.P. Nonlinear vibration response analysis of a double-row self-aligning ball bearing due to surface imperfections. *Proc. Inst. Mech. Eng. Part K J. Multi-Body Dyn.* **2020**, *234*, 514–535. [CrossRef]
31. Liu, Z.L.; Lu, J.; Zhou, J.N. The research on the design of largescale static pressure spherical bearing used in propeller balancing installation. *J. Wuhan Univ. Technol.* **2003**, *27*, 429–432.
32. Liu, Z.L.; Xiao, H.L.; Zhou, J.N. Design of large-sized static pressure ball bearing. *Ship Eng.* **1999**, *5*, 25–27.
33. Wang, C.T.; Wu, U.J.; Chen, L. Algorithm improvement and experimental research for propeller static balancing detection. *Sci. Technol. Eng.* **2015**, *15*, 294–297.
34. Qian, L.B.; Jiang, Y.H. Design of mechanical prop based on space and tracks movement. *Equip. Manuf. Technol.* **2012**, *3*, 8–9.
35. Zhang, S.; Yang, D.; Li, G.; Cheng, Y.; Chen, G.; Zhang, Z.; Li, J. Carrying capacity of spherical hydrostatic bearings including elastic deformation. *Lubricants* **2024**, *12*, 97. [CrossRef]
36. Huang, W.X.; Li, J.M.; Xiao, Z.Y. *Engineering Fluid Mechanics*; Chemical Industry Press: Beijing, China, 2009.
37. Xu, X.L.; Deng, H.S.; Wang, C.L. Theory study of spherical interstitial flow. *China Hydraul. Pneum.* **2004**, *10*, 3–5.

Disclaimer/Publisher's Note: The statements, opinions and data contained in all publications are solely those of the individual author(s) and contributor(s) and not of MDPI and/or the editor(s). MDPI and/or the editor(s) disclaim responsibility for any injury to people or property resulting from any ideas, methods, instructions or products referred to in the content.

Article

Power Losses of Oil-Bath-Lubricated Ball Bearings—A Focus on Churning Losses

Florian de Cadier de Veauce [1], Yann Marchesse [1,*], Thomas Touret [1], Christophe Changenet [1], Fabrice Ville [2], Luc Amar [3] and Charlotte Fossier [4]

1. LabECAM, ECAM LaSalle Campus de Lyon, University of Lyon, 69321 Lyon, France; florian.de-cadier-de-veauce@ntn-snr.fr (F.d.C.d.V.); thomas.touret@ecam.fr (T.T.); christophe.changenet@ecam.fr (C.C.)
2. INSA Lyon, CNRS, LaMCoS, UMR5259, 69621 Villeurbanne, France; fabrice.ville@insa-lyon.fr
3. CETIM, 60300 Senlis, France; luc.amar@cetim.fr
4. NTN Europe, 74010 Annecy, France; charlotte.fossier@ntn-snr.fr
* Correspondence: yann.marchesse@ecam.fr

Abstract: This study investigates the power losses of rolling element bearings (REBs) lubricated using an oil bath. Experimental tests conducted on two different deep-groove ball bearings (DGBBs) provide valuable insights into the behaviour of DGBBs under different oil levels, generating essential data for developing accurate models of power losses. Observations of the oil bath dynamics reveal the formation of an oil ring at high oil levels, as observed for planetary gear trains, leading to modifications in the oil flow behaviour. The experiments demonstrate that oil bath lubrication generates power losses comparable to injection lubrication when the oil level is low. However, as the oil level increases, so do the power losses due to increased drag within the bearing. This study presents a comprehensive model for calculating drag losses. The proposed drag power loss model accounts for variations in oil level and significantly improves loss predictions. A comparison of existing models with the experimental results shows good agreement for both bearings, demonstrating the effectiveness of the developed model in accounting for oil bath height in loss calculations.

Keywords: rolling element bearings; churning power losses; experiments; model; drag; oil bath lubrication

Citation: de Cadier de Veauce, F.; Marchesse, Y.; Touret, T.; Changenet, C.; Ville, F.; Amar, L.; Fossier, C. Power Losses of Oil-Bath-Lubricated Ball Bearings—A Focus on Churning Losses. *Lubricants* **2024**, *12*, 362. https://doi.org/10.3390/lubricants12110362

Received: 20 September 2024
Revised: 18 October 2024
Accepted: 19 October 2024
Published: 23 October 2024

Copyright: © 2024 by the authors. Licensee MDPI, Basel, Switzerland. This article is an open access article distributed under the terms and conditions of the Creative Commons Attribution (CC BY) license (https:// creativecommons.org/licenses/by/ 4.0/).

1. Introduction

Energy consumption is particularly high in the transportation sector, accounting for about 28% of all energy produced globally, with 75% of this energy used by road vehicles. Many moving parts still lose about a third of their energy due to friction and wear [1,2]. Reducing friction and optimising lubrication significantly contributes to the energy performance of mechanical components. In this context, driven by the increasing electrification of vehicles, high rotational speeds are achieved. This implies greater power losses in mechanical transmissions. Rolling element bearings (REBs) have become key elements in high-performance electric motor systems [3], but they are also significant sources of losses in mechanical transmissions, especially at high speeds [4]. While the sources of losses related to the load applied to the bearings are well known, the origin of load-independent losses is less obvious, but could be explained by hydrodynamic phenomena [5] at moderate speeds, whereas drag effects are predominant at higher speeds [4].

Several studies on REB power losses have been performed and led to the development of global models such as the Harris [6] and SKF [7] models. But some works on oil-jet-lubricated REB power losses demonstrated that these global models need some correction [4,5,8,9]. Power losses can also be calculated by a local model for each local source: sliding [10,11], hydrodynamic rolling [12–16], and drag [17–19]. The sliding contribution is important at high radial load, but is quite low under limited load. Hydrodynamic rolling has two components:

the first is load-dependent and occurs in the loaded contacts in the REB, while the second is independent of the load and occurs in the unloaded contacts [5,12–14,20]. Power losses due to the shearing of lubricant in the cage is frequently neglected for normal operating conditions [21–23]. For injection lubrication, drag power loss is calculated for aligned spheres moving in oil–air mixtures [24,25]. The oil–air volume fraction calculation is based on Parker's work [25] and has been modified by Niel [4,26]. This drag power loss model has been validated through experiments [4,24]. This contribution has also been widely studied through numerical tests as computational fluid dynamics [26–29].

However, most of these studies considered oil-jet-lubricated REBs. Only a few studies on power losses in oil-bath-lubricated DGBBs exist. Since lubrication does not impact sliding or hydrodynamic rolling as long as the contacts are fully flooded, only the drag power loss is reconsidered to adapt the model from injection to oil bath lubrication. SKF already takes into account lubrication in its drag loss model. Oil bath lubrication is primarily considered with the height of the bath, and a demonstration was presented by Morales [30]. When the REB is lubricated by injection, the height must be fixed at half the diameter of the lowest rolling element, and the drag torque obtained must be multiplied by two [7]. The Harris model originally proposed one value of f_0 according to the REB and lubrication types [6]. For example, the value of parameter f_0 for a DGBB is 4.0 and 2.0, respectively, for injection and oil bath [6]. According to global models, passing from injection to oil bath lubrication should halve the load-independent power losses of REBs. The Schaeffler company suggested modifying the load-independent component M_0 from the Harris model according to the lubrication used [31]. Two different f_0 values are proposed between injection and oil bath lubrication for each bearing series. In the case of oil bath lubrication, this parameter is multiplied by a coefficient proportional to the height of the bath. In the study by Peterson et al. [32], REB power losses were measured under different operating conditions and a method to isolate drag losses was proposed. By subtracting the results from two tests at different oil bath levels, with one very low, it is possible to obtain a drag loss difference, and then study that power loss source experimentally.

In the context of oil bath lubrication, Peterson et al. demonstrated that when the oil level is high, an oil ring is formed [32]. Hannon et al. presented results related to the quantity of oil in an oil-bath-lubricated bearing, showing a phenomenon of oil aspiration into the bearing and the beginning of ring formation, although these tests were conducted at low speeds [33]. This phenomenon of oil ring formation has also been observed in splash-lubricated epicyclic gear trains [34]. This particular flow pattern within the bearing when an oil bath is employed must have an impact on the churning power losses, and this raises the question of the suitability of existing models for correctly predicting these losses.

The aim of this study was to provide a better understanding of churning power losses in oil-bath-lubricated deep-groove ball bearings through experimental results and to propose a model to predict this source of dissipation by considering the oil level. First, the experimental study of oil-bath-lubricated REBs with a limited radial load applied is presented. The test rig used, the procedure, and the two DGBBs tested are described, followed by power loss measurements. The influence of lubrication type and oil bath level on the REB power losses are depicted. Then, a new drag loss model is developed. Finally, this drag power loss model is added to the Harris model and compared to measurements.

2. Experiments on Power Loss in Rolling Element Bearings

2.1. Test Rig and Test Procedure

Niel et al. [4] developed a test bench for studying both the power losses and the thermal behaviour of rolling element bearings under different operating conditions. A scheme of the test bench is given in Figure 1 and a photo of the measurement region is given in Figure 2. A more detailed presentation is available in references [5,20], but the test apparatus and the measurement protocol are briefly introduced here.

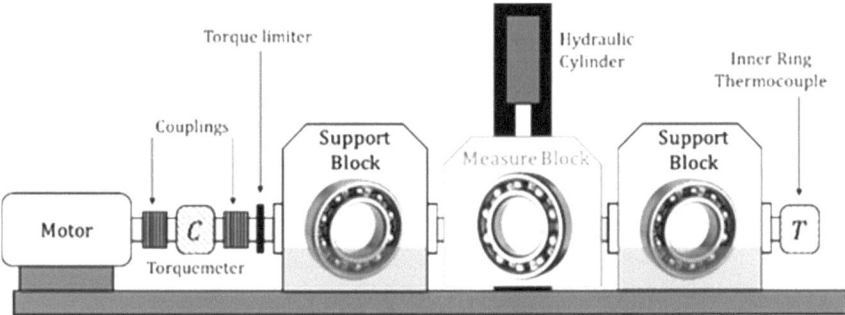

Figure 1. Scheme of the test bench.

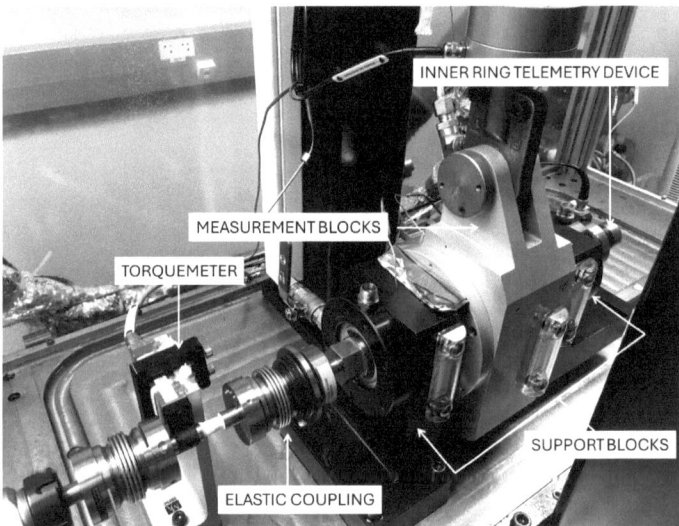

Figure 2. Photograph of the test bench and the blocks.

The test bench comprises a motor rotating a shaft equipped with a torque meter positioned between two bellows couplings. The signal is measured with strain gauges, and the information is transmitted by induction. The torque meter is followed by two support blocks (in black in Figure 2) surrounding the measuring block containing the REB to be studied. Identical bearings in the two support blocks are lubricated by an oil bath with the same oil level. As they generate resistive torque, one must be able to estimate it whatever the operating condition. A specific calibration campaign was therefore conducted, including three radial loads (1, 3, and 5 kN) at rotational speeds from 3200 rpm to 9700 rpm, during which the outer ring temperature of each bearing in the blocks and the total power were measured. This allowed the formulation of the four-support block torques based on both the load and the rotational speed to be proposed. During the experiment, once the tested bearing has been placed in the measurement block and the rotational speed and the load specified, (i) the torque meter measures the total torque on the shaft line, (ii) the torque generated by the bearings located in the two blocks is evaluated using the calibration formulation, and (iii) the torque generated by the tested bearing is estimated from the subtraction of the two previous torque values. A flowchart illustrates this procedure in Figure 3. A radial load can be applied to the measurement block using a hydraulic cylinder. No axial load was applied in this study. Labyrinth seals or lip seals can be installed in the blocks, with the latter mainly used in oil bath lubrication tests where the oil level is significant.

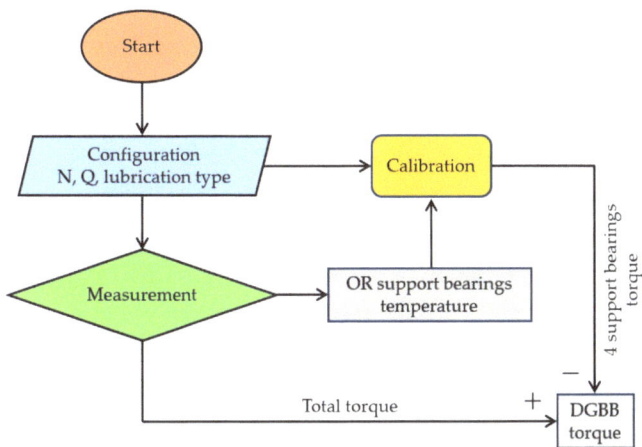

Figure 3. Flowchart of the measurement protocol (N, rotational speed; Q, radial load; OR: outer ring).

In this study, rotational speed evolves from 3200 rpm to 12,100 rpm, and a radial load, corresponding to 7% of the static capacity (C_o) of the tested bearings, is applied to avoid cage slipping and vibrations. The temperature of the outer ring (OR) of each DGBB, the external surface of the measurement block, and the ambient air are measured with type K or type T thermocouples. The temperature of the oil bath is measured when the tested DGBB is lubricated by oil bath lubrication. The temperature of the inner ring (IR) of the DGBB in the measurement block is measured by telemetry. The range and precision of the sensors are summarised in Table 1.

Table 1. Characteristics of sensors on the test bench.

Sensors	Range	Accuracy
Torque meter	0 Nm to 10 Nm	0.02 Nm
Thermocouple	−40 °C to +125 °C	0.5 °C
Load	0 kN to 20 kN	0.4%

In the case of oil bath lubrication, oil evacuations of the block can be closed to have an oil sump at the bottom of the block. The amount of oil in the bath is adjusted using a gauge on the block indicating the oil sump height.

Test campaigns were conducted on two bearing references (6311 and 6208) to study the losses of DGBBs lubricated through oil bath lubrication. For each tested bearing, four oil levels were applied. The oil level was defined in relation to the number of immersed balls under non-rotating conditions. To achieve this, it was decided that the static bearing be placed in such a position that a single ball was at its lowest point (Figure 4). This position is in accordance with the recommendation given by bearing manufacturers when they suggest the oil level (i.e., half the diameter of the lowest rolling element). The lowest level corresponds to a half-immersed ball (0.5 B), and then increases with one fully immersed ball (1 B), three immersed balls (3 B), and half of the immersed DGBB (0.5 R) (Figure 4).

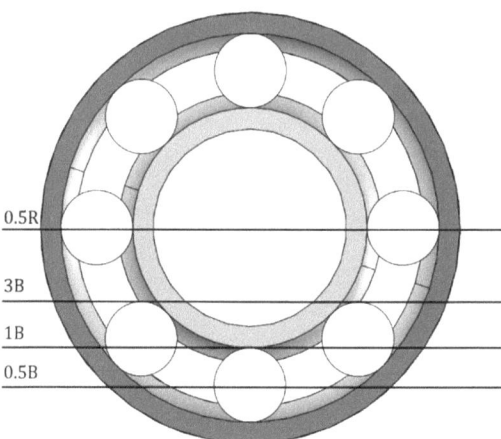

Figure 4. Representation of the four oil levels applied to bearing 6311.

2.2. Bearing Geometry

The two DGBBs depicted in Figure 5, with references 6311 and 6208, were tested. The dimensions of the tested bearings are provided in Table 2. The internal geometry was known for both tested bearings. Some information provided by the manufacturers is given in their catalogues such as the outer diameter d_o, the inner diameter d_i, the mean diameter d_m (equal to $(d_o + d_i)/2$), the width B, and the static load rating C_o. Additional information about the internal geometry, given by the bearing manufacturer, is provided in Table 2 such as the ball diameter D and the number of balls Z. The relative space L/D between the balls is given as $L = \pi d_m / Z$.

Figure 5. Bearing 6311 (**left**) and 6208 (**right**).

Table 2. Characteristic and dimensions of the studied bearings.

Characteristic	DGBB 6311	DGBB 6208
d_o	120 mm	80 mm
d_i	55 mm	40 mm
d_m	87.5 mm	60 mm
D	20.6 mm	11.9 mm
Z	8	9
B	29 mm	18 mm
C_o	45 kN	17.8 kN
L/D	1.66	1.76

Bearing 6311 was chosen for its large dimensions, especially the ball diameter, which highlights load-independent power losses such as drag. Bearing 6208, on the other hand, is smaller and closer to the target industrial application. For example, the ratio of the volumes of the bearings is 3.8 with the volume expressed as $V = \pi \cdot (d_o^2 - d_i^2) \cdot B/4$, and the ball diameter ratio is 1.73.

The cages have identical designs for both bearings, with only their dimensions being different. They are asymmetrical plastic cages adapted to high speed.

As previously explained, both bearings were oil-bath-lubricated and the oil level varied. The oil properties are defined in Table 3. Several experimental campaigns were carried out to test these two bearings. The influence of various parameters was observed and measured.

Table 3. Oil properties.

Kinematic Viscosity at 40 °C (cSt)	Kinematic Viscosity at 100 °C (cSt)	Density at 15 °C (kg/m³)
36.6	7.8	864.6

2.3. Power Loss Measurement Results

The lubrication mode can influence bearing power losses, as the phenomena related to oil flow are different. Therefore, several models for calculating bearing losses predict differences depending on lubrication modes. As mentioned in the introductory section, some models suggest an increase in power losses when the bearing is lubricated by injection rather than oil bath, whereas a decrease is pointed out by others. To verify these hypotheses, tests were conducted with the oil-bath-lubricated DGBBs 6311 and 6208, with an oil level corresponding to a half-submerged ball. Indeed, the SKF model suggests that the drag power loss will be half of that obtained for an oil-jet-lubricated REB. The results of the tests were then compared with measurements obtained in injection tests from [5], carried out using the same test rig and procedure and under equivalent operating conditions. Power losses from different tests were compared at the same DGBB temperature, defined by the mean of OR and IR temperatures. The losses between the two lubrication modes were the same for each bearing and rotational speed, contrary to the predictions of the existing global models (Figure 6).

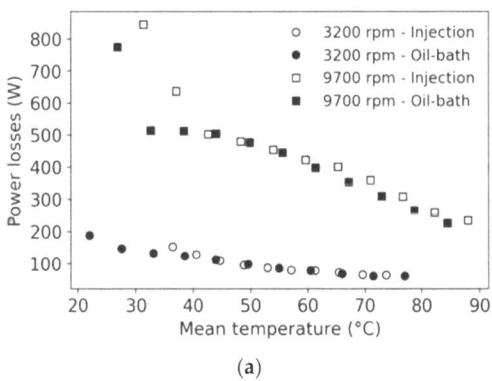
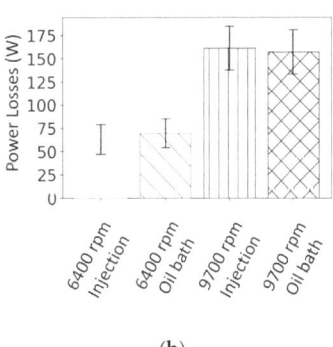

Figure 6. Comparison of bearing losses between injection and oil bath lubrication with half-submerged ball ((a) 6311 DGBB at several mean temperatures; (b) 6208 DGBB at 50 °C).

The mean relative difference between injection losses and oil bath losses was less than 10.2% for both bearings. The errors were less than the measurement uncertainty. Some observations made in injection tests [5] are therefore still valid when DGBBs are oil-bath-lubricated with a half-submerged ball, such as speed or load influences.

Only a low oil bath level was previously considered. Then, tests with different oil levels were conducted. As mentioned previously, the oil bath level is relative to the number of submerged rolling elements, ranging from a half-submerged ball (industry recommendation) to a half-submerged DGBB. DGBBs 6311 and 6208 were tested at 6400 rpm and for each immersion level. One mainly notices at first that the larger bearing (i.e., 6311) leads to higher power losses considering identical mean temperature, which was probably due to the bigger size of the rolling elements. The losses increase with the oil level in the DGBB (Figures 7 and 8). Moving from a half-submerged ball to a fully submerged ball and then to three submerged balls corresponds to an increase in power losses of 7.5% and between 11% and 20%, respectively, for both DGBBs. When half of the bearing is submerged, the losses increase between 24% and 36%.

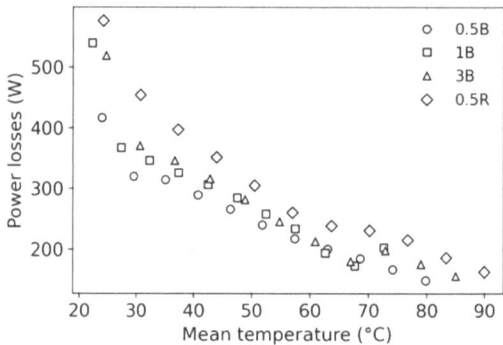

Figure 7. Influence of oil level on losses of bearing 6311 at 6400 rpm.

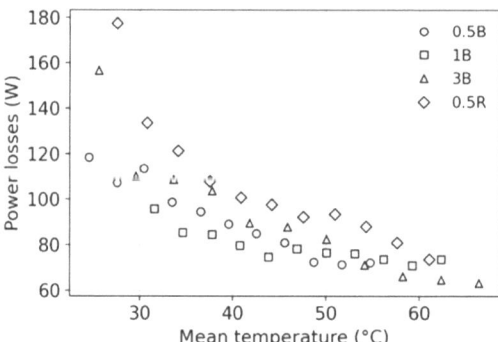

Figure 8. Influence of oil level on losses of bearing 6208 at 6400 rpm.

This increase in power losses is assumed to have no influence on the friction coefficient and therefore on sliding power losses, nor on hydrodynamic rolling power losses, as the lubrication remains fully flooded. These power loss differences are only due to the increase in the amount of oil in the block and in the DGBB cavity. However, the current models do not consider the oil level in their formulations, and this could lead to significant differences between measurements and predictions from these models. The purpose of the present investigation was to add a term considering the oil level to the classical model.

Oil bath lubrication generates losses close to those obtained by injection when the bath height is relatively low, i.e., 0.5 B. The results and models of an injection-lubricated DGBB [5] can therefore be reused in oil bath lubrication. However, these models must be modified to work in the case of higher oil levels. Based on these experimental results, the study of power losses in an oil-bath-lubricated REB can, thus, be separated into two cases: the losses of a bearing with low submersion, as recommended by manufacturers and industrials; and the case with higher oil levels. The next section presents a model developed to consider drag losses according to the oil level.

3. Drag Loss Model

Fluid dynamic drag power losses are due to the displacement of rolling elements within a fluid. In the case of an isolated sphere moving in an infinite medium, drag power losses are expressed by [17]:

$$P_{\text{drag}} = \frac{1}{2} \rho \, A \, C_D \, u^3 \tag{1}$$

with the density ρ, the sphere velocity u, the reference area A, and the drag coefficient C_D depending on the Reynolds number Re defined by the following equation:

$$Re = \frac{u \, D}{\nu} \tag{2}$$

where D is the sphere diameter, u is the tangential speed of the sphere, and ν is the kinematic viscosity of the fluid. Drag losses also occur in ball bearings [6,24,35,36].

3.1. Low Submersion Level

When the oil bath level is low, and when the DGBB rotates, the bath in the block can remain static, but the oil in the DGBB is assumed to be mixed with air due to the regular passage of the balls, thus forming an oil mist. This mixture of oil and air is such that it can be considered homogeneous. The rolling elements are then moving in a mixture whose physical and rheological properties are mainly defined by its oil volume fraction X [18,24,25,35]. The properties of the mist, such as the effective density ϱ_{eff} and the dynamic viscosity η_{eff}, read [37]:

$$\varrho_{\text{eff}} = \varrho_{\text{oil}} X + \varrho_{\text{air}}(1 - X) \tag{3}$$

$$\eta_{\text{eff}} = \left(\frac{(1 - X_m)}{\eta_{\text{air}}} + \frac{X_m}{\eta_{\text{oil}}} \right)^{-1} \tag{4}$$

with X_m the mass fraction of oil in the mixture such that $X_m = X \cdot \varrho_{\text{oil}} / \varrho_{\text{eff}}$. The calculation of drag power losses in a lubricated ball bearing can be written as:

$$P_{\text{drag}} = \frac{1}{2} Z \, \rho_{\text{eff}} \, C_D \, A \left(\omega_c \frac{d_m}{2} \right)^3 \tag{5}$$

where Z is the number of balls, ω_c is the cage rotation speed that drives the rolling elements, d_m is the average diameter of the bearing, and A is the cross-sectional area perpendicular to the flow modified by the cage thickness.

The drag coefficient C_D must be estimated for bearing applications. A numerical model integrating three rolling elements with periodicity conditions and the different relative movements was developed, considering the displacement and rotation of the balls in the fluid [27]. This coefficient was determined numerically using the CFD (Computational Fluid Dynamics) method for the two bearings studied, at different speeds and for several oil fractions within the bearing [27] (Figure 9). The drag coefficient is expressed as a function of the Reynolds number. A drag coefficient law was defined using simulation results and is valid for the two bearings tested over a speed range from 3200 rpm to 12,100 rpm (Figure 9):

$$C_D = K \cdot Re^{3.52} + 0.23 \tag{6}$$

with the constant $K = -4.89 \times 10^{-17}$. This expression is only valid for $4.0 \times 10^3 < Re < 2.2 \times 10^4$. If Re is greater than 2.2×10^4, $C_D = 0.13$. Otherwise, this formula has not been verified. One observes that the relative difference equals 10% at maximum.

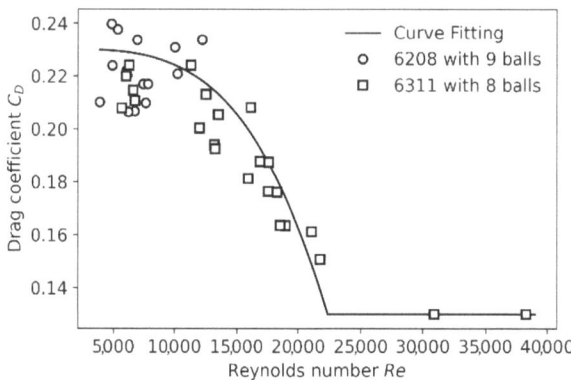

Figure 9. Drag coefficient versus Reynolds number.

This formula will be used for the calculation of the drag coefficient, itself used in the estimation of drag losses.

To estimate the physical properties of a mixture (Equations (3) and (4)), one needs to know its oil volume fraction within the DGBB. This fraction is defined by both the oil and DGBB volumes, V_{bath} and V_{total}, respectively. Thus, a simplified analytical model was developed to calculate the oil volume in the bearing, the number of submerged balls, and the available volume in each bearing. This model is based on the bearing geometry and the static oil level. When one increases the oil level value in the DGBB cavity, the ratio between the volume occupied by the fluid and the total volume corresponding to the oil volume fraction is estimated. The results were verified with CAD models (Figure 10), and an error on the bath volumes below 5% was obtained.

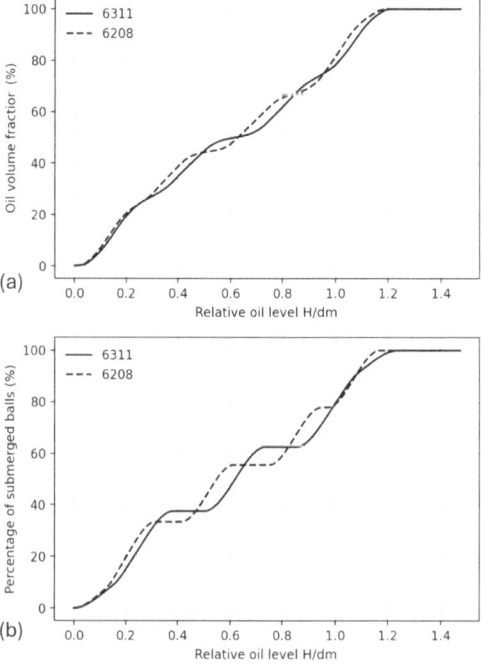

Figure 10. Evolution of (**a**) bath volume in the bearing and (**b**) number of submerged balls depending on the relative oil level.

The amount of oil in the DGBB is estimated from the static bath level, and the oil fraction of the air–oil mixture X is thus defined by:

$$X = V_{bath}/V_{total} \tag{7}$$

with V_{bath} and V_{total} representing the volume of the bath at rest in the bearing and the total volume available in the bearing, respectively. The estimation of the oil volume fraction using the static bath level is debatable, since when the DGBB rotates, the oil located in the cavity is expelled by centrifugal effects and is replaced by the oil from the sump. Therefore, the volume occupied by the oil in the cavity must be different from the one considering static bath. The air–oil fraction might be higher for a rotating bearing due to oil replenishment from the surrounding oil bath. This replenishment for low oil levels has been neglected. This fraction is assumed constant for a fixed oil bath level. The drag power losses can, therefore, be evaluated using Equations (3)–(7).

3.2. High Submersion Level

When the oil level is high, the oil bath behaviour can be different. Peterson et al. [32] and Hannon et al. [33] observed, for example, an initiation or the formation of an oil ring in the DGBB cavity. This was also the case in the present study when the DGBB was half-submerged (0.5 R). This phenomenon was also observed by Boni et al. [34] in splash-lubricated epicyclic gear trains [34]. A bearing [32] and an epicyclic gear train [34], both oil-bath-lubricated, exhibit similar phenomena and flows (Figure 11). This allows for an analogy to be drawn between these two systems, which have a similar architecture, with the ring and sun gears corresponding to the rings of a DGBB, the planets to the rolling elements, and the planet-carrier to the cage (Figure 12).

This analogy allows us to draw inspiration from studies conducted on epicyclic gear trains lubricated by an oil bath. An oil ring is not necessarily observed, and a condition for its formation has been developed by Boni et al. [34] based upon both (i) a relation between Froude and Reynolds numbers and (ii) a critical bath volume ($V_{crit.}$) above which an oil ring can form. The former criterion is always met within the speed and temperature ranges of the tests with DGBB s 6311 and 6208. The critical volume corresponds to the volume of oil such that if the oil ring existed, it would at least be in contact with the periphery of the satellite carrier.

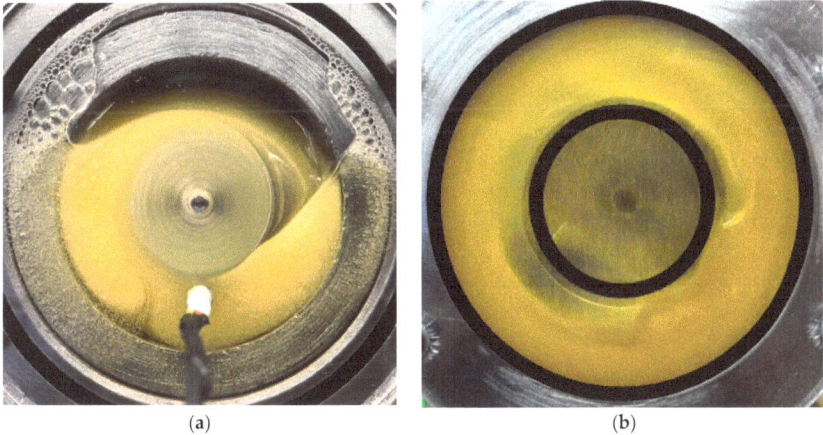

(a)   (b)

Figure 11. Photos of (**a**) a bearing [32] (reprinted from [32], copyright (2024), with permission from Elsevier) and (**b**) an oil-bath-lubricated epicyclic gear train [34].

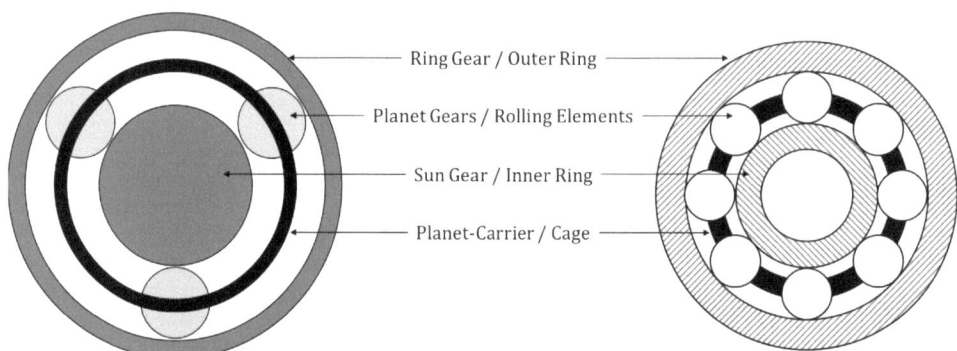

Figure 12. Analogy between an epicyclic gear train (**left**) and a bearing (**right**).

The drag power loss calculation must be modified when an oil ring is present in the DGBB cavity so that only the contribution of oil in the ring is considered: the properties of pure oil are used, $\rho_{\text{eff}} = \rho_{\text{oil}}$, in Equation (5), and the cross-sectional area is calculated based on the oil ring thickness h relative to the rolling elements. The air drag is neglected, and the drag coefficient is always determined from relationship (6).

3.3. Drag Power Loss Model Formulations

Drag power losses are then expressed by two distinct equations depending on the presence of a ring:

$$V < V_{\text{crit}} : P_{\text{drag}} = \frac{1}{2} Z \rho_{\text{eff}}(X) C_D A \left(\omega_c \frac{d_m}{2} \right)^3 \qquad (8)$$

$$V \geq V_{\text{crit}} : P_{\text{drag}} = \frac{1}{2} Z \rho_{\text{oil}} C_D A(h) \left(\omega_c \frac{d_m}{2} \right)^3 \qquad (9)$$

An example of drag loss calculation for the 6311 and 6208 DGBBs (at 6400 rpm with oil at 50 °C) depending on the relative oil level is given in Figure 13. The marks on the curves correspond to immersions 0.5 B, 1 B, 3 B, and 0.5 R. The transition from an air–oil mixture to an oil ring is visible, since a sudden increase in power losses occurs for a relative level equal to 0.491 and 0.422, respectively, for the 6311 and 6208 bearings. This value corresponds to an immersion greater than three submerged balls for the two tested bearings. The oil quantity can, however, vary without modifying the number of balls in the bath over a small level range; the plateau is visible in Figure 10b. The model result coincides with the experimental observations: an oil ring forms when the bearing is half-immersed.

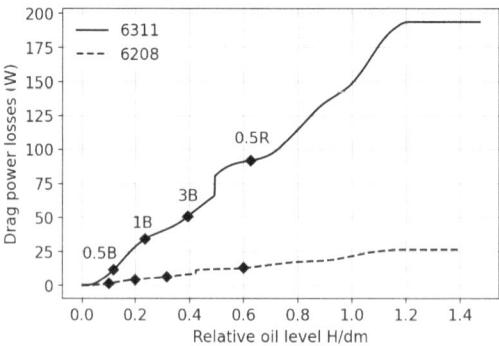

Figure 13. Calculation of drag losses of bearings at 50 °C and 6400 rpm.

It is possible to experimentally isolate the drag power losses due to the increase in oil level by subtracting the power losses obtained for large heights from the power losses of the tests with a half-submerged ball [32]. Thus, the difference in drag losses is measurable. Figure 14 shows the difference in the measured drag losses at 6400 rpm compared to those calculated by the developed drag model. The drag power losses predicted by the model are consistent with the experimentally estimated values. Negative values were found for the 6208 bearing, which can be explained by the very low orders of magnitude of the power losses and the high uncertainty. The absolute error of the power losses calculated by the model compared to the experimental power losses is less than the measurement uncertainty (26.8 W) for each oil level.

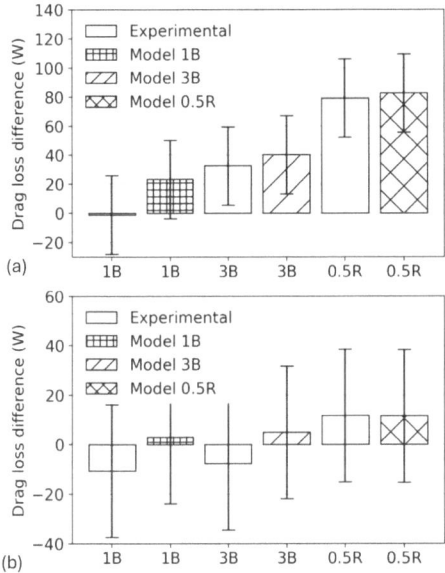

Figure 14. Calculation of drag losses of bearing (**a**) 6311 and (**b**) 6208 at 50 °C and 6400 rpm.

The existing power loss models can be used with the proposed drag power loss model considering the oil quantity in the bearing. This model provides good results compared to measurements at both low and high immersions. In the case of very high oil levels, an analogy with epicyclic gear trains was made and showed the formation of an oil ring on the outer ring. This phenomenon was considered in the drag modelling. This drag loss model can, thus, be added to the global Harris model or a local model to better take into account the oil bath height in the calculation.

4. Comparison of Harris Model with Experiments

Since the local model and the Harris model give very close results once the f_0 parameter is experimentally adjusted [5], only the Harris model is presented here, to which the previously presented drag power loss model is added. The f_0 values for DGBBs 6311 and 6208 are, respectively, 1.65 and 1.40. The results of this model were compared to the measurements made during tests on the 6311 and 6208 bearings for several oil levels. The addition of drag in the Harris model allows the oil level to be considered in the power loss calculation (Figure 15). Moreover, one clearly observes that the curves obtained at different oil levels remain parallel when the mean temperature increases, highlighting the poor influence of the viscosity on the added drag loss. The relative error is less than 7.5% for the 6311 DGBB and less than 11.5% for the 6208 bearing for less than three submerged balls, and 16.2% for half of the submerged 6208 DGBB. The results from the model were also compared with the measured power losses of the 6311 DGBB at two different rotating speeds (8000 rpm and 9700 rpm) and various oil levels (Figure 16). These results are still in good agreement with the experiments, with the mean relative error equal to 6.9%. When the rotation speed and the oil level equal 9700 rpm and 0.5 B, respectively (Figure 16), one notices a considerable decrease in the power losses with the mean temperature followed by 500 W constant value from 20 °C to 45 °C, which was not observed for the other curves. Since the oil level is low, it is not due to the drag power. Therefore, this behaviour might be connected to hydrodynamic effects in the contact. However, it has been demonstrated in previous investigations [20] that the contact temperature has a significant role in the lubrication regime. Therefore, the mean temperature cannot be used here to consider this fact.

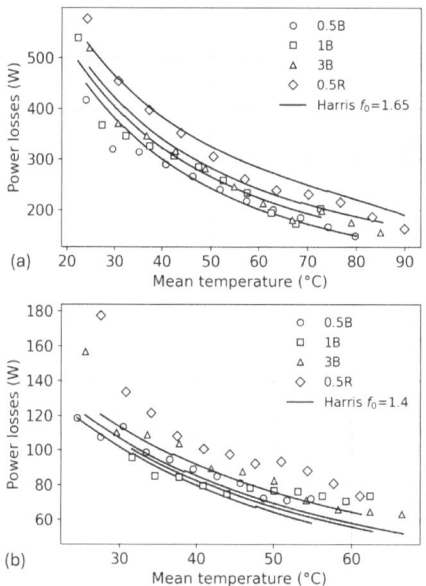

Figure 15. Comparison of measurements and Harris model with drag for bearings (**a**) 6311 and (**b**) 6208, with various immersion levels at 6400 rpm (for curves, the higher they are, the higher the level of immersion).

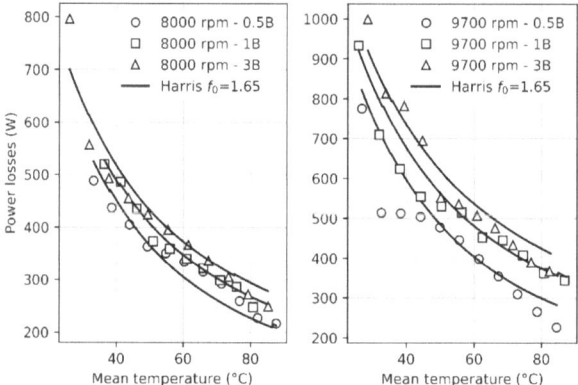

Figure 16. Comparison of measurements and Harris model with drag for bearing 6311 with various immersion levels at 8000 rpm and 9700 rpm (for curves, the higher they are, the higher the level of immersion).

5. Conclusions

This study investigated the power losses of lightly loaded DGBBs, focusing on the influence of lubrication mode, particularly oil bath lubrication (i.e., churning power losses). The experiments conducted on bearings 6311 and 6208 revealed insights into the behaviour of bearings under different oil levels, providing valuable data for modelling power losses. The measured power losses for DGBBs lubricated by an oil jet or an oil bath with a low oil level were equivalent under the same operating conditions. For higher oil levels, the power losses increased. The observations of the oil bath and the development of a drag loss model have enhanced our understanding of the factors affecting bearing performance. The oil bath had two different behaviours depending on the oil quantity. An oil–air mixture was assumed in the DGBBs for low levels, while an oil ring was observed for important oil volumes, as already found for epicyclic gear trains [34]. Hence, a drag loss model that is based on the oil level was proposed that significantly improves the accuracy of loss predictions compared to existing models. Overall, this research contributes to the optimisation of bearing design and lubrication strategies, essential for improving energy efficiency in transportation systems. This study only focused on DGBB power losses, but thermal aspects can also be included. Ring temperatures between the different lubrication processes could be compared and predicted with a thermal network, for example [4,24,38–40].

Author Contributions: Conceptualisation, F.d.C.d.V., Y.M., C.C. and F.V.; methodology, F.d.C.d.V., Y.M., T.T., C.C. and F.V.; software, F.d.C.d.V.; validation, F.d.C.d.V., Y.M., T.T., C.C. and F.V.; formal analysis, F.d.C.d.V., Y.M., C.C., F.V., L.A. and C.F.; investigation, F.d.C.d.V., Y.M., C.C., F.V., L.A. and C.F.; writing—original draft preparation, F.d.C.d.V.; writing—review and editing, F.d.C.d.V., Y.M., C.C. and F.V.; All authors have read and agreed to the published version of the manuscript.

Funding: This research received no external funding.

Data Availability Statement: The original contributions presented in the study are included in the article, further inquiries can be directed to the first author.

Acknowledgments: The authors are grateful to Association Nationale de la Recherche et de la Technologie (ANRT) (CIFRE No. 2020/1522). The authors would like also to thank both CETIM and NTN Europe for their financial support.

Conflicts of Interest: Luc Amar was employed by the company CETIM. Charlotte Fossier was employed by the company NTN Europe. The remaining authors declare that the research was conducted in the absence of any commercial or financial relationships that could be construed as a potential conflict of interest.

References

1. Holmberg, K.; Erdemir, A. Influence of tribology on global energy consumption, costs and emissions. *Friction* **2017**, *5*, 263–284. [CrossRef]
2. Holmberg, K.; Erdemir, A. The impact of tribology on energy use and CO_2 emission globally and in combustion engine and electric cars. *Tribol. Int.* **2019**, *135*, 389–396. [CrossRef]
3. Levillain, A.; Ameye, S. Splash Lubrication Simulation of a High-Speed EV Reducer Using a SPH Tool. In Proceedings of the SIA SImulation Numérique "La Simulation Numérique au Cœur de L'innovation Automobile", Digital Edition, 7–8 April 2021; Available online: https://www.sia.fr/evenements/231-sia-simulation-numerique (accessed on 18 October 2024).
4. Niel, D.; Changenet, C.; Ville, F.; Octrue, M. Thermomecanical study of high speed rolling element bearing: A simplified approach. *Proc. Inst. Mech. Eng. Part J J. Eng. Tribol.* **2019**, *233*, 541–552. [CrossRef]
5. de Cadier de Veauce, F.; Darul, L.; Marchesse, Y.; Touret, T.; Changenet, C.; Ville, F.; Amar, L.; Fossier, C. Power Losses of Oil-Jet Lubricated Ball Bearings with Limited Applied Load: Part 2—Experiments and Model Validation. *Tribol. Trans.* **2023**, *66*, 822–831. [CrossRef]
6. Harris, T.A. *Rolling Bearing Analysis*, 3rd ed.; John Wiley & Sons Inc.: New York, NY, USA, 1991; ISBN 0-471-51349-0.
7. SKF Group. *Rolling Bearings*; SKF Group: Göteborg, Sweden, 2013; 1375p.
8. Fernandes, C.M.; Marques, P.M.; Martins, R.C.; Seabra, J.H. Gearbox power loss. Part I: Losses in rolling bearings. *Tribol. Int.* **2015**, *88*, 298–308. [CrossRef]
9. Dindar, A.; Hong, I.; Garg, A.; Kahraman, A. A Methodology to Measure Power Losses of Rolling Element Bearings under Combined Radial and Axial Loading Conditions. *Tribol. Trans.* **2022**, *65*, 137–152. [CrossRef]
10. Jones, A.B. Ball Motion and Sliding Friction in Ball Bearings. *J. Basic Eng.* **1959**, *81*, 1–12. [CrossRef]

11. Coulomb, C.A. *Théorie Des Machines Simples*; Hachette Livre BNF: Paris, France, 1821.
12. Biboulet, N.; Houpert, L. Hydrodynamic force and moment in pure rolling lubricated contacts. part 2: Point contacts. *Proc. Inst. Mech. Eng. Part J J. Eng. Tribol.* **2010**, *224*, 777–787. [CrossRef]
13. Houpert, L. Piezoviscous-rigid rolling and sliding traction forces, application: The rolling element-cage pocket contact. *J. Tribol.* **1987**, *109*, 363–370. [CrossRef]
14. Brewe, D.E.; Hamrock, B.J. Analysis of Starvation Effects on Hydrodynamic Lubrication in Nonconforming Contacts. *Trans. ASME J. Lubr. Technol.* **1982**, *104*, 410–417. [CrossRef]
15. Tevaarwerk, J.; Johnson, K.L. The influence of fluid rheology on the performance of traction drives. *ASLE Trans. J. Lubr. Technol.* **1979**, *101*, 266–274. [CrossRef]
16. Johnson, K.L. Regimes of Elastohydrodynamic Lubrication. *J. Mech. Eng. Sci.* **1970**, *12*, 9–16. [CrossRef]
17. Schlichting, H.; Gersten, K. *Boundary-Layer Theory*, 8th ed.; Springer: Berlin/Heidelberg, Germany, 2016.
18. Rumbarger, J.; Filetti, E.; Gubernick, D. Gas Turbine Engine Mainshaft Roller Bearing-System Analysis. *J. Lubr. Technol.* **1973**, *95*, 401–416. [CrossRef]
19. Gupta, P.K. *Advanced Dynamics of Rolling Elements*; Springer: New York, NY, USA, 1984; ISBN 978-1-4612-5276-4.
20. Darul, L.; Touret, T.; Changenet, C.; Ville, F. Power Loss Analysis of an Oil-Jet Lubricated Angular Contact Ball Bearing: Theoretical and Experimental Investigations. *Lubricants* **2024**, *12*, 14. [CrossRef]
21. Russell, T.; Sadeghi, F. The effects of lubricant starvation on ball bearing cage pocket friction. *Tribol. Int.* **2022**, *173*, 107630. [CrossRef]
22. Aamer, S.; Sadeghi, F.; Russell, T.; Peterson, W.; Meinel, A.; Grillenberger, H. Lubrication, Flow Visualization, and Multiphase CFD Modeling of Ball Bearing Cage. *Tribol. Trans.* **2022**, *65*, 1088–1098. [CrossRef]
23. Arya, U.; Sadeghi, F.; Aamer, S.; Meinel, A.; Grillenberger, H. In Situ Visualization and Analysis of Oil Starvation in Ball Bearing Cages. *Tribol. Trans.* **2023**, *66*, 965–978. [CrossRef]
24. Pouly, F.; Changenet, C.; Ville, F.; Velex, P.; Damiens, B. Power Loss Predictions in High-Speed Rolling Element Bearings Using Thermal Networks. *Tribol. Trans.* **2010**, *53*, 957–967. [CrossRef]
25. Parker, R.J. Comparison of Predicted Performance of Angular and Experimental Thermal Contact Ball Bearings. *NASA Tech. Pap.* **1984**, *2275*, 1–16.
26. Marchesse, Y.; Changenet, C.; Ville, F. Numerical Investigations on Drag Coefficient of Balls in Rolling Element Bearing. *Tribol. Trans.* **2014**, *57*, 778–785. [CrossRef]
27. Marchesse, Y.; Changenet, C.; Ville, F. Computational Fluid Dynamics Methodology to Estimate the Drag Coefficient of Balls in Rolling Element Bearings. *Dynamics* **2024**, *4*, 303–321. [CrossRef]
28. Peterson, W.; Russell, T.; Sadeghi, F.; Berhan, M.T.; Stacke, L.E.; Ståhl, J. A CFD investigation of lubricant flow in deep groove ball bearings. *Tribol. Int.* **2021**, *154*, 106735. [CrossRef]
29. Cao, W.; Nelias, D.; Lyu, Y.; Boisson, N. Numerical investigations on drag coefficient of circular cylinder with two free ends in roller bearings. *Tribol. Int.* **2018**, *123*, 43–49. [CrossRef]
30. Morales-Espejel, G.; Wemekamp, A. An engineering drag losses model for rolling bearings. *Proc. Inst. Mech. Eng. Part J J. Eng. Tribol.* **2022**, *237*, 1–16. [CrossRef]
31. Schaeffler Technologies. *Lubrication of Rolling Bearings*; Schaeffler Technologies: Fort Mill, SC, USA, 2013.
32. Peterson, W.; Russel, T.; Sadeghi, F.; Tekletsion Berhan, M. Experimental and analytical investigation of fluid drag losses in rolling element bearings. *Tribol. Int.* **2021**, *161*, 107106. [CrossRef]
33. Hannon, W.M.; Barr, T.A.; Froelich, S.T. Rolling-element bearing heat transfer—Part III: Experimental validation. *J. Tribol.* **2015**, *137*, 13. [CrossRef]
34. Boni, J.B. Modélisation Thermique d'un Train Epicycloïdal Lubrifié par Barbotage. Ph.D. Thesis, University of Lyon, Lyon, France, 2020.
35. Nelias, D.; Seabra, J.; Flamand, L.; Dalmaz, G. Power loss prediction in high-speed roller bearings. *Tribol. Ser.* **1994**, *27*, 465–478.
36. Paleu, V.; Nelias, D. On Kerosene Lubrication of Hybrid Ball Bearings. In Proceedings of the International Conference on Diagnosis and Prediction in Mechanical Engineering Systems, Galati, Romania, 26–27 October 2007; pp. 50–56.
37. Isbin, H.S.; Sher, M.; Eddy, K.C. Void Fractions in Two-phase Steam-water Flow. *Aiche J.* **1957**, *3*, 136–142. [CrossRef]
38. Takabi, J.; Khonsari, M.M. Experimental testing and thermal analysis of ball bearings. *Tribol. Int.* **2013**, *60*, 93–103. [CrossRef]
39. Giannetti, G.; Meli, E.; Rindi, A.; Ridol, A.; Shi, Z.; Tangredi, A.; Facchini, B.; Fondelli, T.; Massini, D. Modeling and experimental study of power losses in a rolling bearing. *Proc. Inst. Mech. Eng. Part J J. Eng. Tribol.* **2020**, *234*, 1332–1351. [CrossRef]
40. Kerrouche, R.; Dadouche, A.; Mamou, M.; Boukraa, S. Power Loss Estimation and Thermal Analysis of an Aero-Engine Cylindrical Roller Bearing. *Tribol. Trans.* **2021**, *64*, 1079–1094. [CrossRef]

Disclaimer/Publisher's Note: The statements, opinions and data contained in all publications are solely those of the individual author(s) and contributor(s) and not of MDPI and/or the editor(s). MDPI and/or the editor(s) disclaim responsibility for any injury to people or property resulting from any ideas, methods, instructions or products referred to in the content.

 lubricants

Article

End-to-End Intelligent Fault Diagnosis of Transmission Bearings in Electric Vehicles Based on CNN

Yong Chen [1,*], Guangxin Li [2], Anhe Li [2] and Bolin He [2]

1. State Key Laboratory of Featured Metal Materials and Life-Cycle Safety for Composite Structures, Guangxi University, Nanning 530004, China
2. Tianjin Key Laboratory of Power Transmission and Safety Technology for New Energy Vehicles, School of Mechanical Engineering, Hebei University of Technology, Tianjin 300130, China; ligx1229@163.com (G.L.); 13064047382@163.com (A.L.); h664191433@163.com (B.H.)
* Correspondence: chenyong1585811@163.com

Abstract: Environmental noise and transmission components can cause significant interference in vibration signals, rendering the extraction of bearing fault features challenging in service scenarios. Traditional fault diagnosis methods rely heavily on professional domain knowledge, prior models, and signal preprocessing methods. The accuracy of fault diagnosis depends on the quality of the fault-sensitive features extracted by vibration signal preprocessing methods. An improved convolutional neural network (CNN) end-to-end intelligent fault diagnosis model based on raw vibration data (RVDCNN) is proposed. The time-domain vibration signal of the transmission bearing is converted into a continuous two-dimensional numerical matrix, and a two-dimensional CNN model is constructed through network structure optimization. The original time-domain vibration signal numerical matrix of the bearing is trained and tested to extract and learn abstract fault features of different fault types, and then the fault classification of the bearing is achieved. To verify the generalizability of the RVDCNN intelligent fault diagnosis model, it is applied to the recognition of rolling bearings in the two-speed mechanical automatic transmission of electric vehicles, achieving recognition accuracy of 99.11% for seven types of bearings.

Keywords: end to end; raw vibration data; fault diagnosis; transmission bearings

1. Introduction

In recent years, prognostic and health management (PHM) has become an important field in the intelligent manufacturing industry [1]. As an additional external excitation source for electric vehicle transmission, the drive motor adds tangential electromagnetic force, axial electromagnetic force, and electromagnetic torque fluctuations and excitations during transmission. These excitations act on the bearings and gears through the rotor and have an undeniable impact on the working conditions and vibration characteristics of the bearings [2]. The working speed of the motor is significantly higher than that in traditional internal combustion engines. In addition, the rapid power response of the motor increases the impact excitation of the gear, and the recovery of the braking energy also causes frequent changes in the direction of the load borne by the gear. The multi-level factors brought by driving motors accelerate the fatigue damage and fault evolution of the gears and bearings, which creates higher requirements for the early fault identification and classification diagnosis of bearings.

The vibration characteristics of transmission bearings exhibit complex features, such as nonstationarity and nonlinearity, and their vibration signals are often disturbed by the vibration or noise of other components, making it difficult to accurately identify and extract fault features directly from the original vibration signals.

Guo et al. [3] proposed a modulation signal bispectrum analysis method based on non-Gaussian noise suppression, using autoregressive filters as preprocessing units to effectively process non-Gaussian noise while retaining the advantage of suppressing Gaussian noise, achieving efficient and accurate performance in extracting fault features. Ziani et al. [4] achieved the detection of bevel gears under variable load conditions by successfully integrating EMD, the Teager–Kaiser energy operator, and an impact detector. This method effectively extracts the fault features from complex vibration signals and has important reference value for the fault feature extraction of rolling bearings. In their research on the fault diagnosis of rolling bearings, Zhang et al. [5] adopted the ensemble empirical mode decomposition technique and selected the singular value entropy as the key criterion; this enabled the accurate identification and classification of different fault characteristics of rolling bearings through the in-depth analysis of vibration signals. Zhen et al. [6] introduced a fault detection technique for the analysis of nonstationary vibration signals. This method leverages weighted average ensemble empirical mode decomposition and modulation signal bispectrum analysis to effectively reduce Gaussian noise and decompose the intrinsic modulation components within the vibration signal. Consequently, it enables the detection of faults in both the inner and outer races of rolling bearings.

The feature extraction diagnostic method performed through signal processing heavily relies on the professional experience and knowledge of engineers or advanced manual signal processing methods. Due to the rapid advancements in machine learning within the vision and speech recognition domains, intelligent fault diagnosis techniques for PHM have gained widespread adoption in mechanical fault diagnosis. These methods are favored for their adaptive learning capabilities, automated feature extraction processes, and robust nonlinear regression abilities.

The existing fault diagnosis approaches based on traditional machine learning usually combine signal processing methods, and the diagnosis process is as follows: first, the fault features are extracted and enhanced; then, traditional machine learning algorithms are used to identify bearing faults.

In their research on the fault diagnosis of rolling bearings, Amar et al. [7] converted the vibration signals of rolling bearings into spectral images. In order to enhance the key features in the images, a two-dimensional average filter was applied to process the spectral images. Finally, an ANN was used to successfully classify the spectral images of bearings with different fault types. Lei et al. [8] proposed an improved distance evaluation technique and selected six sensitive features from the temporal and spectral characteristics of bearing signals as the input dataset for an adaptive neural fuzzy inference system, achieving bearing fault classification. Liu et al. [9] used the multi-scale entropy feature index of rolling bearings and implemented fault detection using BPNN. Muruganatham et al. [10] used singular values of the bearing condition as characteristic indicators and implemented the fault diagnosis of bearings using BPNN. Khazaee et al. [11] collected the vibration and sound signals of planetary gearboxes and then used wavelet analysis to extract features from the temporal to time–frequency domains. After signal processing, the data from each sensor were used as input to the ANN classifier for primary fault diagnosis. The output of the classification was used as input for the Dempster–Shafer rule, which was used for the fusion of classifiers, thus achieving high accuracy in the final classification. Li et al. [12] extracted 10 temporal features from the vibration signals of bearings and then conducted fault diagnosis research on the bearings using an ANN optimized with the firefly algorithm. In their study of the early fault diagnosis of rotors, Bin et al. [13] first used a wavelet packet transform (WPT) to process the original vibration signal; they then reconstructed the wavelet coefficients obtained through WPT processing using EMD to obtain the energy characteristics of each component; finally, they used BPNN to achieve the efficient and accurate diagnosis of early rotor faults.

With the significant achievements in bearing fault diagnosis research utilizing signal processing and traditional machine learning techniques, it has been possible, to some extent, to reduce the dependence on factors such as fault mechanisms and knowledge experience models and improve the accuracy of fault diagnosis. However, there are also certain limitations, which mainly include the following.

(1) The precision of fault diagnosis largely depends on the choice of feature indicators and the design of the feature components. However, in the case of strong background noise and weak fault features, selecting sensitive feature indicators is a challenging research task [14].

(2) Due to the presence of weak early fault signals, low signal-to-noise ratios, and the varying operational conditions of transmission bearings, there is a complex mapping relationship between the sensitive features that characterize the degrees of faults. However, the nonlinear feature learning ability of traditional machine learning is limited, making it difficult to fully explore the fault information contained in the vibration signals of transmission bearings [15].

Cheng et al. [16] proposed a data-driven intelligent fault diagnosis method for rotating machinery based on a new continuous wavelet transform local binary CNN, and they established an end-to-end diagnostic mechanism. Li et al. [17] proposed a feature fusion algorithm for bearing fault diagnosis based on an integrated deep CNN and the improved Dempster–Shafer theory. Their algorithm used the root mean square of the spectral features of two sensors as input data and achieved good results on the open-source CWRU bearing dataset. Miao et al. [18] converted the vibration signal into an angle domain. Then, the corner domain signal was converted into corresponding envelope and squared envelope spectral features and fused into a red–green–blue color image to enhance the sample features and expand the differences between various health states. Finally, a CNN was constructed to complete fault identification. Pang et al. [19] proposed an intelligent diagnosis method for planetary gear faults based on a deep CNN and vibration bispectrum. Raouf et al. [20] introduced a feature aggregation network into a two-dimensional CNN and used scale map images to detect faults in the servo motor bearings of industrial robots.

Transfer learning has shown great potential in dealing with data scarcity problems and has become a new research hotspot. Pan et al. [21] proposed a residual service life prediction method combining a multi-head attention network and adaptive meta-transfer learning, which achieved the accurate residual service life prediction of low-temperature bearings in rocket engines during the steady-state stage. Chen et al. [22] proposed an online unsupervised anomaly detection framework that did not rely on professional knowledge or labeled historical data. To address the issue of data scarcity, they proposed an adaptive self-transfer learning algorithm based on Gaussian processes, which modeled monitoring data using uncertainty information and achieved the fault diagnosis and monitoring of steam turbines. Fang et al. [23] proposed a method based on transfer learning and deep transfer clustering, which achieved the high-precision diagnosis of unknown faults. The application of transfer learning in various fields provides new ideas in solving the problem of rocket engine fault diagnosis. Li et al. [24] proposed an extreme learning machine based on transfer learning to align the distribution differences in data from turbofan engines, and they verified the effectiveness and feasibility of the method through fault diagnosis experiments on turbofan engines. Jamil et al. [25] proposed an instance-based weight deep transfer learning method that could update source and target machine training samples separately, thereby achieving the high-precision fault detection of wind turbine gearboxes.

However, in service scenarios, environmental noise, electromagnetic excitation, and other transmission components of the gearbox can cause significant interference in vibration signals, making it difficult to extract clear fault features. Traditional fault diagnosis methods rely heavily on professional domain knowledge, prior models, and signal preprocessing methods. The accuracy of fault diagnosis relies on the quality of fault-sensitive feature extraction by vibration signal preprocessing methods. Here, given sufficient sample data, a novel intelligent diagnosis approach is proposed, which leverages the original time-domain vibration signal for end-to-end fault diagnosis, using a convolutional neural network (CNN) as the underlying model framework. This study constructs a two-dimensional CNN network structure with strong feature extraction abilities and optimizes the hyperparameters to achieve the high-precision fault diagnosis and classification of transmission bearings.

This study proposes the RVDCNN intelligent fault diagnosis model based on raw vibration data. The time-domain vibration signals of transmission bearings are converted into continuous two-dimensional numerical matrices. A two-dimensional CNN model is constructed through network structure optimization to train and test the original time-domain vibration signal numerical matrices of bearings, extract and learn abstract fault features of different fault types, and then achieve the fault classification of bearings. To verify the generalization capacity of the RVDCNN intelligent fault diagnosis model, it is utilized for the diagnosis and identification of faults in rolling bearings within a two-speed mechanical automatic transmission system in an electric vehicle, achieving multi-type and high-precision diagnosis and recognition and overcoming the difficulties associated with advanced signal preprocessing technology and professional diagnostic experience.

2. Methodology

2.1. Data Reconstruction

Moving away from signal processing methods based on professional knowledge or expert experience, and using the original time-domain vibration signals of bearings as training data, the original time-domain signals are reorganized into a two-dimensional numerical matrix, and abstract numerical feature information is extracted and mined from the time-domain signals through a two-dimensional deep convolutional kernel. In order to reduce the number of training samples and improve the diagnostic efficiency, the size of the input matrix is defined by periodically sampled data points.

The calculation of the data collection volume within one cycle is as follows:

$$M = f_s \times \frac{60}{N} \quad (1)$$

where M is the number of signal data in one cycle. f_s is the sampling frequency of the vibration signals. N is the rotation speed of the rotating machinery.

The process of segmenting and reassembling the original vibration signal data is shown in Figure 1. Divide one-dimensional raw vibration data into several signal segments based on the size of the model input samples as shown in the red dot box. After segmentation, independent one-dimensional signal segments are obtained, as shown in the red line box. Assuming that the size of the input numerical matrix is n and the number of samples is y, n^2 data points are extracted by sliding in sequence from a signal of length $y \times n^2$. The n^2 original time-domain vibration data are flattened from left to right and top to bottom into a matrix in sequence, and the one-dimensional original vibration signal is reset to obtain a y two-dimensional numerical matrix of size n.

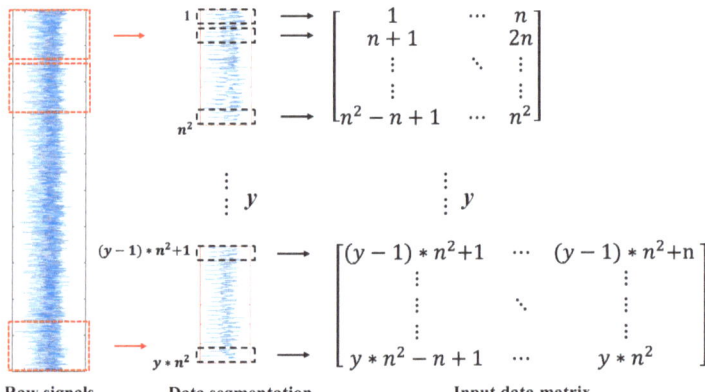

Figure 1. Vibration signal segmentation and recombination process.

2.2. Model Theory

The convolutional layer contains multiple convolutional kernels, which constitute a matrix, also known as a convolutional filter. In the convolutional layers, a convolution operation is performed on the local area of the input signal to obtain a corresponding two-dimensional feature map. Different features are extracted using different kernels for each convolutional layer.

The convolution is defined as follows:

$$Y_j^l = f\left(\sum_{i=1}^{j} Y_j^{l-1} \times W_{ij}^l\right) \quad (2)$$

where Y_j^{l-1} is the $(l-1)$th layer's jth element. W_{ij}^l is the convolutional kernel weight matrix. $f(*)$ is the activation function, which is defined as the ReLu function.

$$f(x) = \begin{cases} 0, x \leq 0 \\ x, x > 0 \end{cases} \quad (3)$$

After the operation of the convolutional layer, the number of feature maps increases rapidly. It is imperative to establish a pooling layer to reduce the dimensions of the feature maps and the parameters of the network after the convolutional layer. There are two methods commonly used in the pooling layer, namely average pooling and max pooling. In this study, average pooling is applied to process each feature map.

The pooling layer is calculated as follows:

$$Y_j^l = avgdown\left(Y_j^{l-1}\right) \quad (4)$$

There is a fully connected layer after the combination of two convolutional layers and two pooling layers. The fully connected layer is similar to the convolutional layer and applies different classification steps. The neuron nodes of the fully connected layer are connected to all of the neuron nodes of the feature maps from the former pooling layer. If the number of output labels is k, the output of softmax regression can be represented as follows:

$$output = \begin{bmatrix} p(1|x; W_1, b_1) \\ p(2|x; W_2, b_2) \\ \cdots \\ p(K|x; W_K, b_K) \end{bmatrix} = \frac{1}{\sum_{j=1}^{K} exp(W_j x + b_j)} \begin{bmatrix} exp(W_1 x + b_1) \\ exp(W_2 x + b_2) \\ \cdots \\ exp(W_K x + b_K) \end{bmatrix} \quad (5)$$

where W_j is the weight matrix and b_j is the bias.

2.3. Intelligent Fault Diagnosis Method

A CNN is a type of deep feedforward neural network that has been widely used in fields such as visual image recognition, natural language processing, and fault diagnosis due to its unique network structure, unique computing principles, and powerful nonlinear feature extraction capabilities. It is currently one of the most widely used deep learning models. A CNN has the characteristics of local connections, weight sharing, pooling operations, and a multi-level structure. Different hierarchical structures can be used to mine numerical features from different dimensions. This study focuses on the LeNet-5 model, further simplifying the network structure by removing one convolutional layer and a fully connected layer. In addition to the characteristics of bearing vibration signals, appropriate convolutional kernel sizes and quantities are selected to construct a high-precision fault diagnosis model that is suitable for electric vehicle transmission bearings.

A two-dimensional CNN is a deep learning structure composed of convolutional layers, pooling layers, and fully connected layers. The convolutional kernel of the convolutional layer is a two-dimensional structure that has excellent feature extraction abilities due to its ability to capture the shift characteristics of the input data [26]. CNNs have great potential and abilities in complex nonlinear numerical feature extraction and high-precision recognition [27], providing new ideas for the end-to-end fault diagnosis of transmission bearings.

The intelligent fault diagnosis process is shown in Figure 2. Firstly, bearings with different fault types are embedded into the transmission, and acceleration vibration signal data are collected under the same operating conditions. Then, the time-domain vibration data are sequentially segmented and reassembled into a fixed-size two-dimensional numerical matrix. By constructing a CNN network structure and optimizing the network's hyperparameters, a model with strong adaptive feature extraction capabilities is obtained. Finally, the original two-dimensional numerical matrix of vibration signals is used as the sample set to achieve the high-precision diagnosis and recognition of rolling bearings in electric vehicle transmission.

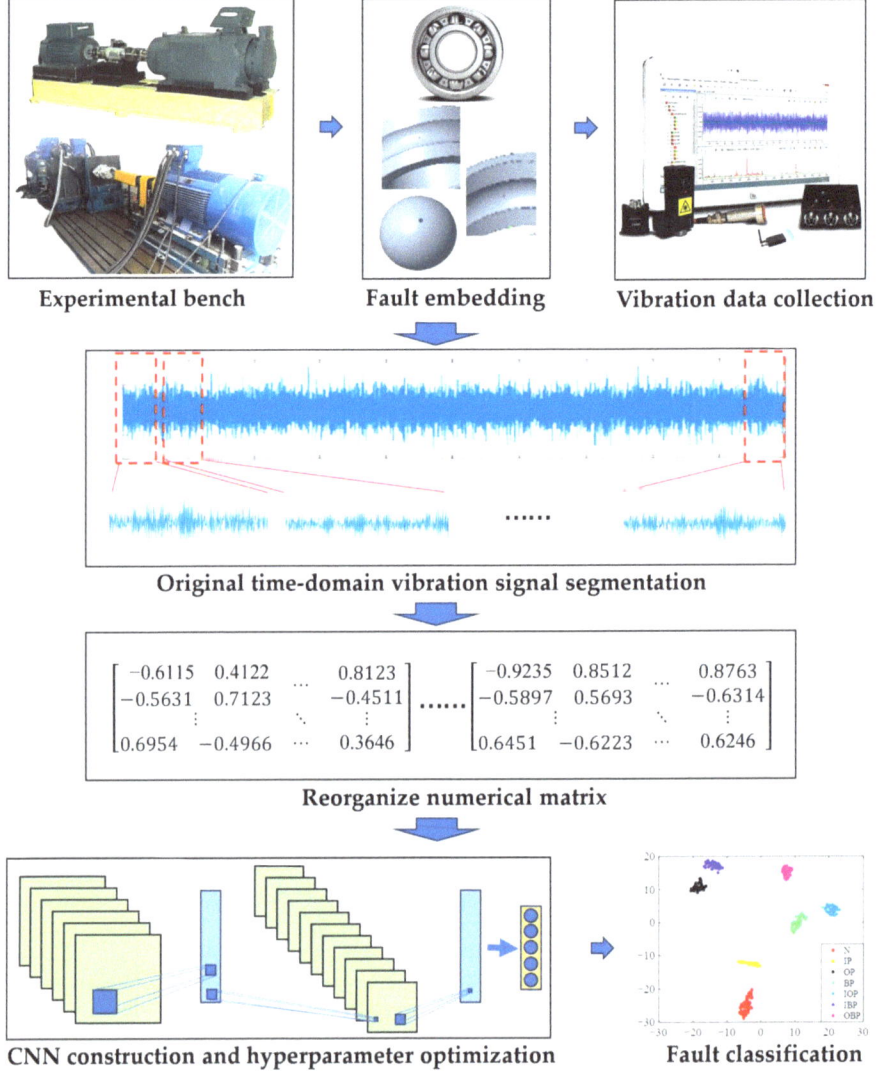

Figure 2. Intelligent diagnostic process.

3. Analysis and Selection of CNN Model Parameters

Based on the advantages of CNNs in adaptive feature learning, a two-dimensional deep CNN intelligent diagnostic model is constructed using the framework of "convolutional layer–pooling layer–convolutional layer–pooling layer–fully connected layer".

We optimize the CNN's hyperparameters using fault bearing data from the open-source Case Western Reserve University (CWRU) Bearing Data Center. The CWRU fault bearing sampling frequency is f_s = 12 kHz. The minimum speed in the experiment is 1730 r/min and the maximum speed is 1797 r/min, so the number of data acquired in one rotation is $M_{bearing}$ = 400~416. The input speed of the electric vehicle transmission is significantly increased, leading to a substantial decrease in the quantity of data points collected after one rotation of the gears and bearings and a decrease in the data volume required for a single training sample in the end-to-end intelligent diagnostic model. Each sample's data size should include at least one vibration signal of the bearing rotation cycle to ensure that the sample contains the characteristics of the bearing cycle sampling point data. Therefore, the input numerical matrix size of the model is designated as 24 × 24, which means that each sample contains 576 data points.

In order to exploit the potential of deep learning, improve the diagnostic accuracy, and obtain better CNN models, the selection of the model's structural parameters is crucial. Research in the literature has shown that the network layer structures and hyperparameters of CNN models have a complex impact on the diagnostic accuracy [28]. Shallow network layers can lead to insufficient feature information extracted by the convolutional layers and low classification accuracy. When the depth of the network layers is too great, the weight of the convolutional kernel increases, which not only increases the time cost but may also cause overfitting. If the convolutional kernel size is too small, it can lead to fragmented features after pooling and reduce the sensitivity of feature recognition. If the convolutional kernel size is too large, it will also increase the number of weights, the computational time, and the probability of overfitting. Therefore, in order to obtain a better deep CNN structure, it is necessary to define a reasonable range of hyperparameters.

Using the CWRU rolling bearing fault dataset as the training and test set, we explore the influence of the convolutional kernel number and size and the pooling function on the diagnostic results.

The datasets used for training and testing are presented in Table 1. Considering the deep feature extraction abilities of the CNN model, the data type is selected as a weak fault diameter of 0.18 mm located on the outer race, and it is divided into four labels under different working conditions. The original time-domain signal of each label is taken for 9.6 s, with a total of 115,200 data points, forming 200 samples, with 576 data points per sample. The fault forms are the same, but the working conditions are different, and the feature recognition of the model is refined and sensitive to small differences.

Table 1. CWRU rolling bearing outer race failure dataset.

Fault Location	Fault Size (mm)	Working Condition (r/min, hp)	Number of Samples	Sample Size
outer race	0.18	1730, 12	625	24 × 24
outer race	0.18	1750, 8	625	24 × 24
outer race	0.18	1772, 4	625	24 × 24
outer race	0.18	1797, 0	625	24 × 24

3.1. Determining the Quantity of Convolutional Kernels

Adopting a structure consisting of a convolutional layer, pooling layer, another convolutional layer, another pooling layer, and a fully connected layer, a two-dimensional deep CNN model is constructed, where the activation function is ReLU, the pooling function is average pooling, the quantity of neurons in the fully connected layer is set to 4, the step size is set to 1, and the learning rate is set to 0.01, with weight decay of 0.005, momentum value of 0.9, and a dropout rate of 0.8. The batch processing volume is 10, and 50 iterations are performed. The first convolutional layer has a kernel size of 5×5, with numbers of 6, 8, and 10, respectively. The size and quantity of the second convolutional layer are both $5 \times 5 \times 24$, with 80% of the sample size used as training data and 20% as testing data. After conducting five diagnoses, the average value is computed. The outcomes of these diagnoses are displayed in Table 2.

Table 2. Effect of the number of convolutional kernels in convolutional layer 1.

Type	1	2	3
Convolutional Layer	$5 \times 5 \times 6$	$5 \times 5 \times 8$	$5 \times 5 \times 10$
Activation Function		ReLU	
Pooling Layer		average pooling 2×2	
Convolutional Layer		$5 \times 5 \times 24$	
Activation Function		ReLU	
Pooling Layer		average pooling 2×2	
Fully Connected Layer		10	
Result	87.00%	94.20%	94.20%

With six convolutional kernels, the diagnostic accuracy is 87.00%, which is lower than the 94.20% achieved with eight convolutional kernels. However, further increasing the number to 10 does not enhance the diagnostic results. Therefore, eight convolutional kernels in the first layer is the optimal choice.

After determining the optimal number of convolutional kernels for the first layer, five experiments are carried out to assess the effect of varying the number of convolutional kernels in the second layer on the diagnostic outcomes. In the second layer, the number of convolutional kernels is set to 10, 16, 20, 24, and 30, respectively, while all other parameters remain constant. The diagnostic accuracy results are summarized in Table 3, revealing a similar trend to that observed with the first layer's convolutional kernels.

Table 3. Impact of varying the number of convolutional kernels in the second convolutional layer.

Type	1	2	3	4	5
Convolutional Layer			$5 \times 5 \times 8$		
Activation Function			ReLU		
Pooling Layer			average pooling 2×2		
Convolutional Layer	$5 \times 5 \times 10$	$5 \times 5 \times 16$	$5 \times 5 \times 20$	$5 \times 5 \times 24$	$5 \times 5 \times 30$
Activation Function			ReLU		
Pooling Layer			average pooling 2×2		
Fully Connected Layer			10		
Result	83.80%	88.40%	90.40%	94.20%	89.60%

As the number of convolutional kernels increases, the fault recognition rate improves to a certain point. Specifically, the diagnostic accuracy is gradually enhanced as the number of convolutional kernels increases from 10 to 24, suggesting that an increased number of convolutional kernels aids in the better extraction of fault features. However, when the number reaches 30, the recognition accuracy declines to 89.60%. Consequently, for the second layer, a more suitable choice for the number of convolutional kernels is 24.

3.2. Selection of Convolutional Kernel Size

To investigate the influence of the convolutional kernel size on the model's diagnostic results, experiments were conducted based on a configuration with eight convolutional kernels in the first layer and 24 in the second layer. The size of the first convolutional kernel was varied as follows: $5 \times 5, 7 \times 7, 9 \times 9, 11 \times 11$, and 13×13. Corresponding to each of the first convolutional kernel sizes, the second convolutional kernel size was also varied.

When the size of the first convolutional kernel was 5 × 5, the tested sizes for the second convolutional kernel were 3 × 3, 5 × 5, 7 × 7, and 9 × 9. When the size of the first convolutional kernel was 7 × 7, the tested sizes for the second convolutional kernel were 2 × 2, 4 × 4, 6 × 6, and 8 × 8. When the size of the first convolutional kernel was 9 × 9, the tested sizes for the second convolutional kernel were 3 × 3, 5 × 5, and 7 × 7. When the size of the first convolutional kernel was 11 × 11, the tested sizes for the second convolutional kernel were 2 × 2, 4 × 4, and 6 × 6. When the size of the first convolutional kernel was 13 × 13, the tested sizes for the second convolutional kernel were 3 × 3 and 5 × 5.

A total of 16 deep convolutional neural network models were evaluated, and their fault diagnosis results are presented in Table 4. When the second layer's convolutional kernel size was 3 × 3, and when comparing models 1, 9, and 15, which had first-layer convolutional kernel sizes of 5 × 5, 9 × 9, and 13 × 13, respectively, the diagnostic results were 90.30%, 93.80%, and 95.80%. These results indicate a positive trend as the convolutional kernel size increases. Additionally, when comparing models 3, 11, 5, and 12, as well as models 7 and 14, it can be observed that when the second convolutional kernel's size remains constant, an increase in the first convolutional kernel's size leads to an improvement in the model's diagnostic accuracy.

Table 4. The influence of different convolutional kernel sizes on the diagnostic results.

Type	1	2	3	4	5	6	7	8
Convolutional kernel	5 × 5					7 × 7		
Pooling layer	average pooling 2 × 2							
Convolutional kernel	3 × 3	5 × 5	7 × 7	9 × 9	2 × 2	4 × 4	6 × 6	8 × 8
Pooling layer	average pooling 2 × 2							
Output size	4 × 4	3 × 3	2 × 2	1 × 1	4 × 4	3 × 3	2 × 2	1 × 1
Result	90.30%	94.20%	92.80%	92.40%	91.80%	95.40%	92.50%	95.60%
Type	9	10	11	12	13	14	15	16
Convolutional kernel	9 × 9			11 × 11			13 × 13	
Pooling layer	average pooling 2 × 2							
Convolutional kernel	3 × 3	5 × 5	7 × 7	2 × 2	4 × 4	6 × 6	3 × 3	5 × 5
Pooling layer	average pooling 2 × 2							
Output size	3 × 3	2 × 2	1 × 1	3 × 3	2 × 2	1 × 1	2 × 2	1 × 1
Result	93.80%	95.80%	95.20%	93.80%	94.20%	96.40%	95.80%	95.40%

When the second convolutional kernel size is 5 × 5, and when comparing models 2, 10, and 16, it is observed that as the first convolutional kernel size increases from 5 × 5 to 9 × 9, the diagnostic results improve. However, when the first convolutional kernel size is increased to 13 × 13, the diagnostic accuracy decreases by 0.40%. Similarly, when the second convolutional kernel size is 4 × 4, upon comparing models 6 and 13, it is found that as the first convolutional kernel size increases from 7 × 7 to 11 × 11, the diagnostic results decrease by 1.20%.

When the size of the first convolutional kernel remains constant, and when comparing models 1 and 4, it is observed that as the size of the second convolutional kernel increases, the diagnostic result improves from 90.30% to 94.20%. However, further increases to 7 × 7 and 9 × 9 result in a decrease in the diagnostic accuracy. Similarly, when the first convolutional kernel size is 7 × 7, upon comparing models 5 and 8, it is found that as the second convolutional kernel size increases from 2 × 2 to 8 × 8, the diagnostic results show an initial improvement, followed by a decrease and then another improvement. On the other hand, comparing models 12, 13, and 14, when the first convolutional kernel size is 11 × 11, the accuracy increases with the size of the second convolutional kernel, reaching the maximum fault recognition accuracy of 96.40%. Lastly, when the first convolutional kernel size is 13 × 13, changes in the size of the second convolutional kernel do not significantly improve the diagnostic results.

The results of this comprehensive comparison of the models indicate that, while increasing the size of the convolutional kernel can expand the convolutional receptive field, which aids in feature learning and extraction, this relationship is not linear. In fact, blindly increasing the size of the convolutional kernel may result in a decrease in diagnostic accuracy, producing the opposite effect. Therefore, it is crucial to choose the network structure parameters carefully. Upon comparing the models, it is found that the recognition accuracy is optimal when the first convolutional kernel size is 11 × 11 and the second convolutional kernel size is 6 × 6.

3.3. RVDCNN Model Structure

The proposed two-dimensional deep convolutional neural network's structure is shown in Figure 3. The network consists of two convolutional layers, two pooling layers, and one fully connected layer. The input layer is a 24 × 24 numerical matrix of the original vibration signals, and the first convolutional layer consists of eight large-sized 11 × 11 convolutional kernels. The pooling layer uses average pooling to maintain feature homogenization. The second convolutional layer consists of 24 small-sized 6 × 6 convolutional kernels. The first convolutional layer uses large-sized convolutional kernels, while the second convolutional layer has three times the number of kernels as the first layer. These configurations are beneficial in extracting features that reflect the different health conditions of rotating machinery. After the pooling layer, the number of neurons in the fully connected layer depends on the fault label of the diagnostic object. As an adaptive parameter, in this section, the number of neurons in the fully connected layer is 4~10, and the results of the fault types are classified.

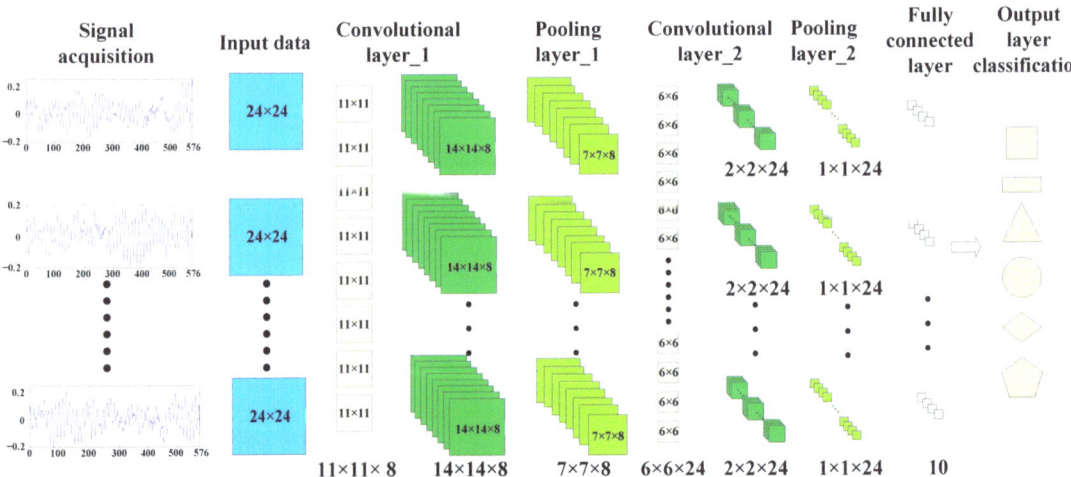

Figure 3. RVDCNN model structure.

The model adopts the cross-entropy loss function and mini-batch gradient descent optimization algorithm. The signal data used as input for the model are time-domain vibration acceleration data collected on the surface of the gearbox or gearbox housing. Each sample undergoes an initial layer of large-scale convolution and average pooling to generate eight 7 × 7 abstract numerical feature matrices. Then, after the second layer of deep convolution and average pooling, 24 abstract numerical features are generated. Finally, a one-dimensional feature matrix consisting of 4~10 values is generated through a fully connected layer.

4. Experimental Verification of Bearing Fault Diagnosis

4.1. Experimental Verification of CWRU Bearing Dataset

In this set of experiments, the ability to perform feature extraction from raw time-domain numerical matrices is tested. Ten datasets are applied to diagnose and analyze the fault bearings using the proposed model. A comprehensive description of the dataset is given in Table 5. There are ten fault types in each working condition, which are labeled as NO, IR7, IR14, IR21, B7, B14, B21, OR7, OR14, and OR21. Each dataset comprises a total of 2000 samples, with 80% allocated for model training and the remaining 20% for testing. The model undergoes 80 iterations, and the batch size is configured to 10.

Table 5. Description of dataset.

Fault Location	None	Inner Race			Ball			Outer Race		
Fault diameter (mm)	0	0.18	0.36	0.53	0.18	0.36	0.53	0.18	0.36	0.53
Label	NO	IR7	IR14	IR21	B7	B14	B21	OR7	OR14	OR21
1797 r/min, 0 hp Dataset AA	200	200	200	200	200	200	200	200	200	200
1772 r/min, 1 hp Dataset BB	200	200	200	200	200	200	200	200	200	200
1750 r/min, 2 hp Dataset CC	200	200	200	200	200	200	200	200	200	200
1730 r/min, 3 hp Dataset DD	200	200	200	200	200	200	200	200	200	200

The diagnosis results obtained with the four datasets for ten repeated trials are shown in Figure 4. Table 6 presents the average accuracy and standard deviation for the four datasets. The best average classification result among the four datasets appears in dataset DD. The average accuracy for dataset DD is 99.78%, with a standard deviation of 0.30. As shown in Table 6, datasets AA, BB, and CC perform well with raw time-domain signals too, with the average accuracy being around 99.05%, 98.83%, and 99.03%, respectively.

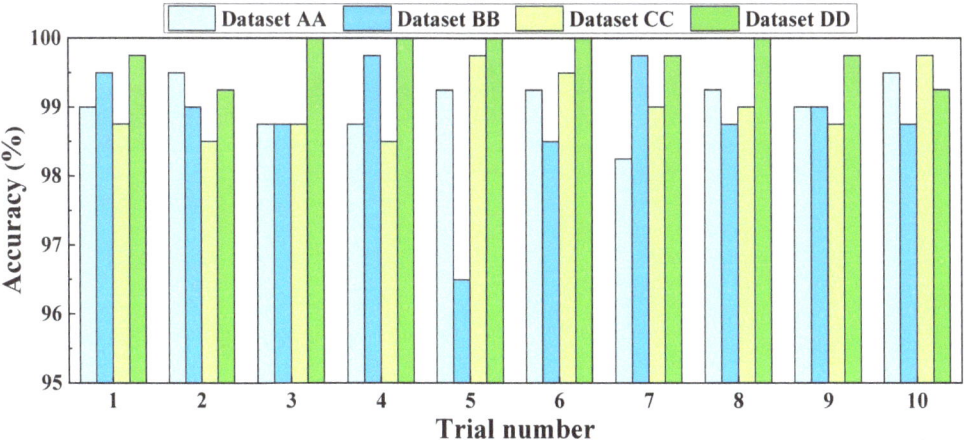

Figure 4. Diagnosis results of four datasets for ten trials.

Table 6. Average accuracy and standard deviation of four datasets.

Dataset	Average Accuracy (%)	Standard Deviation
AA	99.05	0.39
BB	98.83	0.93
CC	99.03	0.48
DD	99.78	0.21

To illustrate the diagnostic ability of the RVDCNN model proposed in this paper with raw signal numerical matrices, the results are compared with those of an image CNN [29], frequency-domain DNN [30], and time–frequency-domain CNN [31], as presented in Table 7. The average accuracies and the standard deviations for ten trials and the number of input data points for each dataset are shown in Table 7.

Table 7. Diagnostic results for CWRU bearing datasets.

Dataset	RVDCNN	Image CNN [29]	Frequency DNN [30]	CSCoh-CNN [31]
DD	99.78 ± 0.21	99.79 ± 0.08	99.74 ± 0.16	97.68 ± 0.98
Data points (million)	1.152	39.3216	4.8	14.4

A fault diagnosis method based on a data-driven approach was proposed by Wen et al. [29], which converts signals into images with a size of 64 × 64. Dataset DD is selected as the CNN input data and contains 2400 samples. There are 39.3216 million data points used for training and testing, which is 34.13 times greater than in the proposed method. The method of Wen et al. [29] performed well, but the amount of input data was large, leading to a time-consuming process.

Jia et al. [30] put forward an intelligent fault diagnosis DNN model based on frequency-domain signals. There were 200 samples for each health type, where 1200 Fourier coefficients were included in each sample. Meanwhile, in the present research, each sample included 576 data points, which reduced the amount of data by 76.02%, and it did not require the Fourier transform method. The diagnostic results were 0.04% higher than those of the DNN, and it had a greater advantage regarding the size of the dataset.

Chen et al. [31] used a two-dimensional map to represent cyclic spectral coherence (CSCoh) features. The 2D CSCoh maps of 10 bearing health conditions were obtained using the cyclic spectral analysis method, and discriminative patterns for specific types of bearing faults were provided. Then, they used a CNN model to learn the features and achieve fault classification. The results of the four datasets are shown in Table 6. The accuracy of the CSCoh-CNN was 2.10% lower than that of the proposed RVDCNN for dataset DD. Regarding the data points, the requirement of RVDCNN is 8% of that of the CSCoh-CNN method, providing it with a great advantage in terms of the computing time.

Table 7 shows that RVDCNN, with the capacity for intelligent feature learning, achieves similar diagnostic accuracies with the smallest number of data points compared with other feature extraction methods for bearing data. The proposed method does not require advanced artificial feature extraction methods and can mitigate the influence of the feature extraction method on fault diagnosis. It achieves satisfactory performance when directly using raw time-domain signals, with the advantage of a simple structure and fewer calculations.

4.2. Experimental Verification on Two-Speed Mechanical Automatic Transmission Bearing Fault Diagnosis

4.2.1. Two-Speed Mechanical Automatic Transmission

The faulty bearing is intended for use in a two-speed automatic transmission system for purely electric vehicles, as illustrated in Figure 5. The two-speed mechanical automatic transmission system consists of an input shaft system, an intermediate shaft system, a differential, front housing, rear housing, a shifting mechanism, and a parking mechanism. The input shaft system comprises an input gear shaft and supporting bearings, incorporating the first and second drive gears directly into the shaft. The intermediate shaft system, on the other hand, consists of the first and second driven gear, the main reduction gear, a synchronizer system, and supporting bearings. The red bearing on the right side of the middle shaft system is the faulty bearing in this experiment, and the outer race is installed on the rear housing for the easy arrangement of vibration acceleration sensors.

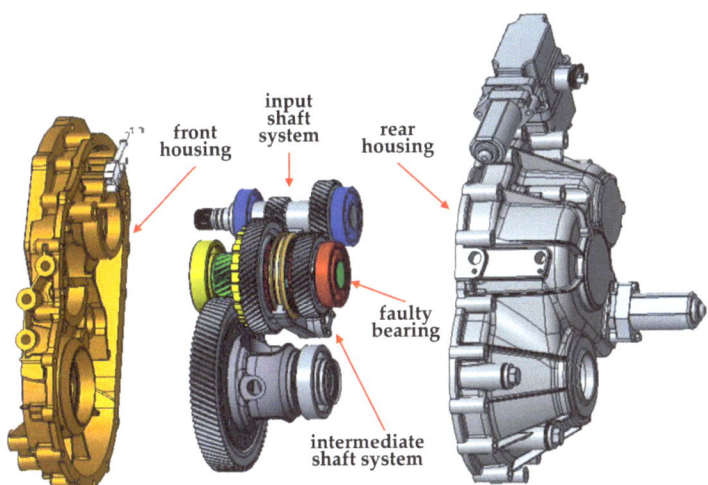

Figure 5. Two-speed mechanical automatic transmission system.

The maximum working conditions and gear ratio parameters of the transmission system are shown in Table 8. The maximum input speed is 12,000 r/min and the maximum input torque is 250 Nm when implemented in a purely electric passenger car. The bearings inside the transmission mainly include two types, deep groove ball bearings and needle roller bearings, with a total of eight bearing supports. The faulty bearing model is 6307.

Table 8. Structural parameters of two-speed mechanical automatic transmission system.

Parameter Name	Value
Maximum speed (r/min)	12,000
Maximum torque (Nm)	250
1st gear ratio	3.000
2nd gear ratio	1.194
Final drive ratio	3.905

4.2.2. Three-Motor Powertrain Comprehensive Performance Test Bench

To collect bearing vibration acceleration data under the working conditions of two-speed mechanical automatic transmission in electric vehicles, a comprehensive performance test bench for a three-motor powertrain is built. The test bench is shown in Figure 6 and consists of a drive motor, a loading motor, a gearbox, a torque sensor, a cooling system, a temperature control system, a drive unit, and a loading unit. The two-speed mechanical automatic transmission system is fixed to the flange fixture plate of the experimental platform by bolts, and two half-shafts are connected to two loading motors to simulate the working conditions of the transmission system and conduct bearing vibration acceleration experiments under different health states.

Figure 6. Comprehensive performance test bench for three-motor powertrain.

4.2.3. Rolling Bearing Dataset

The experimental bearing is a deep groove ball bearing located on the right side of the intermediate transmission shaft of the automatic transmission, with a brand and model number of SKF 6307, and the parameters are shown in Table 9. For convenience in disassembling and assembling the bearings, pitting damage faults were created on the inner race, outer race, and roller. The bearing model was replaced with the 6307-2RZ, and the cage was an integrated structure composed of nylon material.

Table 9. Deep groove ball bearing data.

Parameter Name	Value
Inside diameter (mm)	35
Outside diameter (mm)	80
Width (mm)	21
Number of rollers	8

To construct a database of multiple types of bearing faults, including single and composite faults, six types of faulty bearings were produced using laser processing technology, as shown in Table 10. Circular pitting faults were machined on the inner and outer raceway and roller surfaces. This included single faults in the inner race, outer race, and roller positions, as well as composite faults in the inner and outer race, inner roller, and outer roller. The pitting diameter of a single fault was 0.53 mm, and the pitting diameter of a composite fault was 0.18 mm. The damage diameter of the fault was derived from the CWRU dataset, with values of 0.18 mm and 0.53 mm, respectively. The damage diameter of the composite fault was 0.18 mm, and the fault depth was 0.15 mm. Figure 7 shows the disassembly and labeling of the six types of faulty bearings in Table 10. The location and diameter of the fault were marked on the outer race surface through laser engraving.

Table 10. Fault types of deep groove ball bearings in two-speed automatic mechanical transmission.

Location of Pitting Fault	Fault Size (mm)
Inner race	0.53
Roller	0.53
Outer race	0.53
Inner and outer race	0.18
Inner race and rolling	0.18
Outer race and rolling	0.18

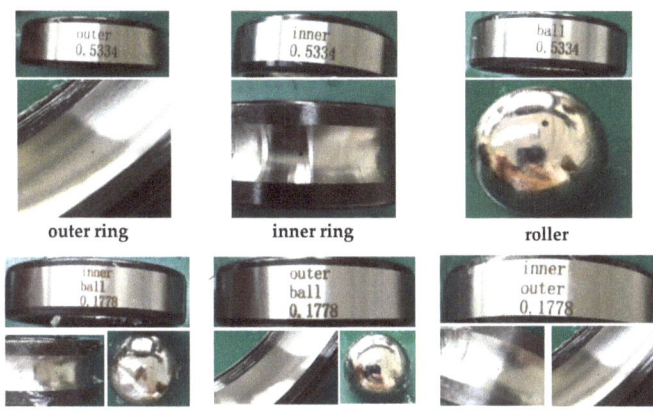

Figure 7. Physical diagram of single and composite fault bearings.

Based on the Chinese automotive driving condition CLTC-P, and according to the comprehensive shifting rules of two-speed automatic transmission, the working range for the second speed was 5000~9000 r/min [32]. Then, the experimental conditions for the faulty bearing of the transmission were formulated, with input speeds of 5000 r/min, 6000 r/min, and 7000 r/min and torque of 32 Nm. Three directional vibration acceleration sensors were arranged at the bearing supports of the input shaft, intermediate shaft, and differential shaft ends of the transmission housing, as shown in Figure 8.

Figure 8. Layout of the vibration acceleration sensors.

The data acquisition system was a 24-channel LMS Test Lab, and the experimental bench control system and data acquisition system worked together to conduct transmission vibration signal acquisition experiments on 7 types of bearings, including single-point corrosion with a diameter of 0.53 mm on the inner race, outer race, and roller, normal bearings, and three composite fault types, under three working conditions.

The vibration signal acquisition time for each type of bearing was 30 s, and the sampling frequency was 16,384 Hz. We selected the Z channel of the sensor located at the end of the intermediate shaft when analyzing and processing the vibration signals. Taking the working conditions of 5000 r/min and 32 Nm as an example, 7 types of bearing vibration acceleration signals were extracted, as shown in Figure 9.

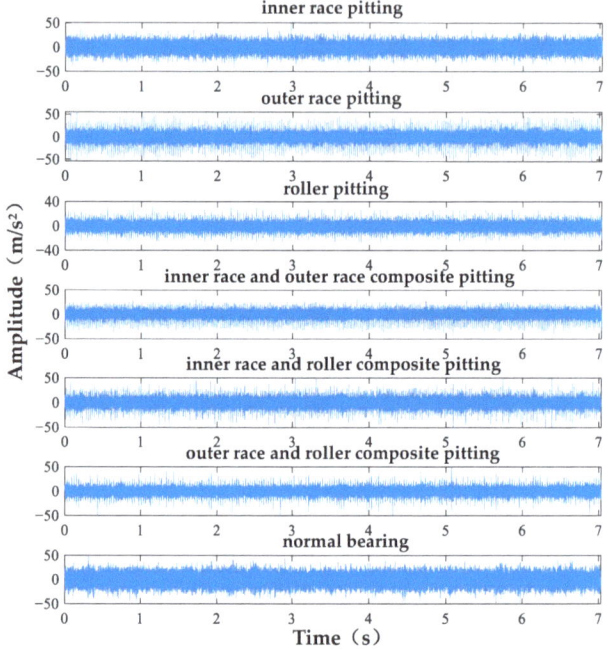

Figure 9. Original vibration signals of 7 health states.

The dataset for the end-to-end intelligent fault diagnosis of rolling bearings in a two-speed mechanical automatic transmission system for purely electric vehicles is shown in Table 11. We use the labels N, IP, BP, OP, IOP, IBP, and OBP to represent normal, inner race pitting, roller pitting, outer race pitting, inner race and outer race composite pitting, inner race and roller composite pitting, and outer race and roller composite pitting fault samples. Under the three different speed conditions, the total sample for each label was composed of data with ratios of 30%, 30%, and 40%, respectively. The original time-domain vibration signal data length for each label was 115,200 data points, which were segmented and reorganized into 200 sample matrices with a size of 24 × 24.

Table 11. Transmission bearing raw signal dataset.

Label	Location of Pitting Fault	Fault Size (mm)	Working Conditions (r/min, Nm)
N	None		
IP	Inner race	0.53	
BP	Roller	0.53	① 5000, 32
OP	Outer race	0.53	② 6000, 32
IOP	Inner and outer race	0.18	③ 7000, 32
IBP	Inner race and rolling	0.18	
OBP	Outer race and rolling	0.18	

4.2.4. Experimental Results

Based on the proposed RVDCNN model, feature adaptive learning and diagnostic recognition were performed on the vibration signals of seven types of deep groove ball bearings applied in the two-speed mechanical automatic transmission of an electric vehicle. Under the conditions of a batch processing volume of 10 and an iteration count of 80, fault diagnosis was performed 10 times, and the results are shown in Figure 10.

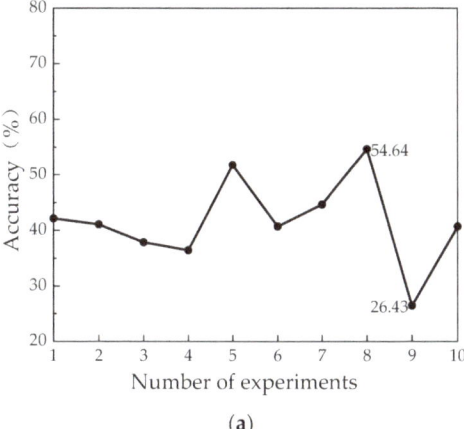
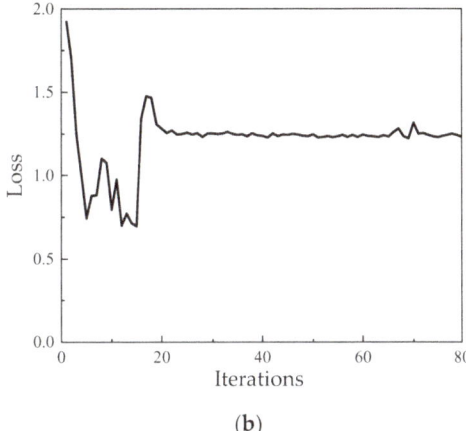

(a)   (b)

Figure 10. Diagnosis results of the model with a batch size of 10: (**a**) 10 diagnostic results; (**b**) variation in loss value with number of iterations.

As shown in Figure 10a, the highest diagnostic accuracy was 54.64%, and the lowest was 26.43%. The diagnostic results were not ideal. Considering the small batch size and large number of training samples, the method required 112 training iterations to complete one iteration in the forward and backward propagation processes, which could easily lead to overfitting. Therefore, the loss function value during the first training process was extracted, as shown in Figure 10b. In the first 20 iterations, the loss value oscillated and then stabilized at around 1.25, without reaching convergence.

We increased the batch sample size to 20, 28, and 40 for the comparison of the bearing fault diagnoses. The relationship between the loss value and iterations is shown in Figure 11. When the batch training sample size was 20, the loss value decreased the fastest in the first 20 iterations, indicating an advantage in terms of the convergence speed and computational time. However, as the number of iterations increased to 30, the loss values of the three tended to be relatively stable, and the difference was not significant. After zooming in on the local image, it could be observed that when the batch training sample size was 40, the loss value tended to converge smoothly. When the batch processing sample sizes for training were 20 and 28, there was slight oscillation and convergence.

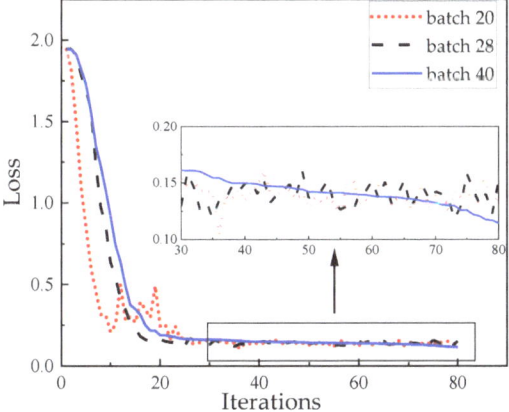

Figure 11. The loss variation with the number of iterations.

Therefore, for the fault diagnosis experiment on deep groove ball bearings in transmission, a batch training sample size of 40 was adopted and 10 diagnostic experiments were conducted on the faulty bearings of the two-speed mechanical automatic transmission, as shown in Table 12. The average accuracy in the 10 experiments reached 99.11%, with a standard deviation of 0.20. The model showed excellent performance in terms of feature extraction, feature recognition, classification, and diagnosis and also had good stability.

Table 12. Results of 10 fault diagnosis tests.

Serial Number	1	2	3	4	5	6	7	8	9	10
Accuracy (%)	99.64	99.29	100	99.29	98.93	99.29	99.29	99.29	98.21	97.86

To reveal the training, recognition, and diagnostic processes of the RVDCNN model for seven types of bearings, taking the second diagnostic experiment as an example, we analyze the visualization results of the classification of bearings with different health types during the iteration process and obtain the confusion matrix of the final diagnostic results. The classification results for different types of bearings under different iteration times are visualized using the t-Stochastic Neighbor Embedding (t-SNE) dimensionality reduction method, as shown in Figure 12. Each color represents a corresponding bearing type, and the horizontal and vertical axes denote dimensionless values.

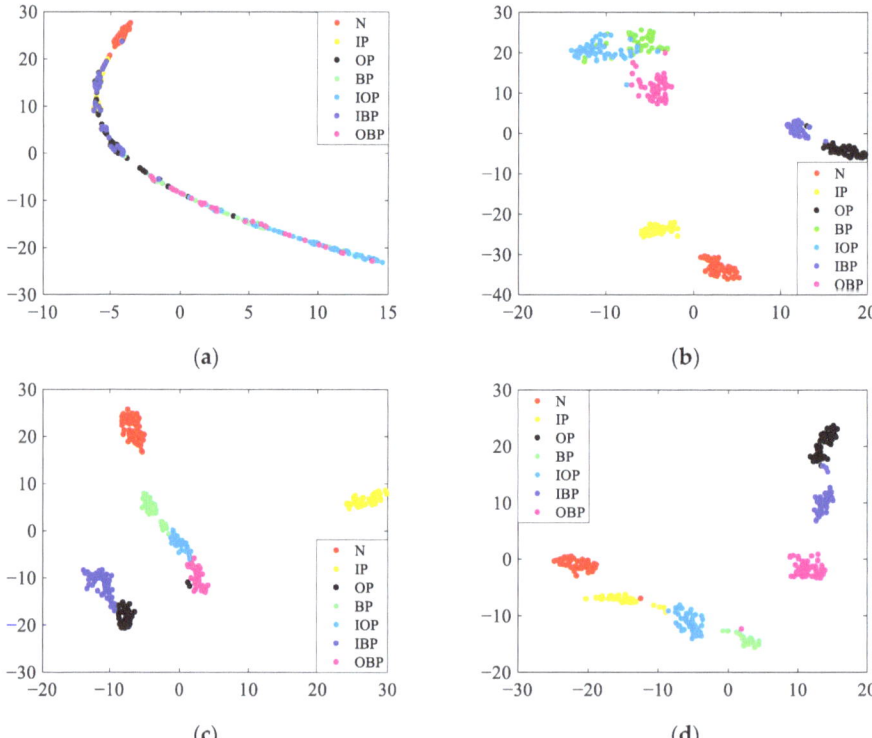

Figure 12. *Cont.*

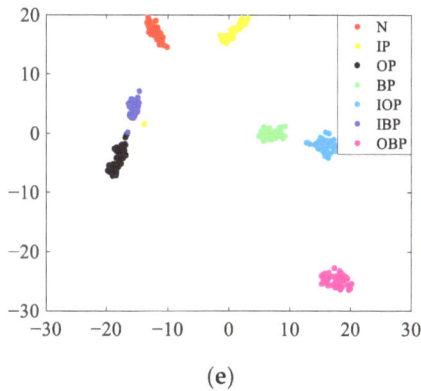
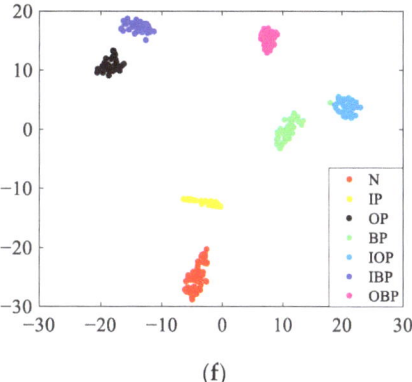

Figure 12. Visualization of classification under different numbers of iterations: (**a**) 5 iterations; (**b**) 10 iterations; (**c**) 20 iterations; (**d**) 30 iterations; (**e**) 40 iterations; (**f**) 50 iterations.

When the number of iterations is five, the seven types of bearings in different states are mixed, which creates confusion. When the number of iterations is 20, as shown in Figure 12, the model's loss value drops rapidly, resulting in a notable improvement in its fault recognition capabilities. Although a preliminary classification of the seven bearing types emerges, there are still numerous classification errors. However, as the number of iterations reaches 30, the overall diagnosis and classification of most fault types is essentially achieved, and a small number of bearings with outer race and roller composite faults are incorrectly identified as single roller fault types. There is a slight conflict and an unclear boundary in the classification of a single fault in the inner race and a composite fault in the inner and outer race. As the number of iterations continues to increase, the recognition and classification boundaries of the seven types of bearings become clearer, and the recognition accuracy is further improved.

Regarding the second diagnostic experiment, the specific sample sizes of seven healthy bearings among 280 test samples are shown in Table 13. There are feature recognition errors for the labels OP and IBP, i.e., there is a classification diagnosis error between the outer race pitting fault sample and the inner race and roller composite fault sample.

Table 13. Second diagnostic test samples.

Label	N	IP	OP	BP	IOP	IBP	OBP
Number of actual samples	46	34	36	35	52	36	41
Number of predicted samples	46	34	38	35	52	34	41

The diagnostic results are represented by a confusion matrix in Figure 13. While the composite fault diagnosis accuracy reaches 94.00% for the inner race and roller, the recognition accuracy for the other six types of bearing faults reaches 100.00%. Moreover, 6% of the samples with inner race and roller composite faults were mistakenly identified as outer race faults, as shown in Table 13. Two samples with the label IBP were assigned the label OP.

Figure 13. Confusion matrix of the second bearing fault diagnosis.

The t-SNE visualization and representation of the confusion matrix is shown in Figure 14. Because the diagnosis result was 99.29%, the overall clustering of the seven different types of bearings is obvious and the boundaries are clear. Purple circles represent IBP samples, and black plus signs represent OP samples. In the magnified images of the OP and IBP samples, it can be observed that there are two sample points with the label OP in the purple group of samples labeled IBP.

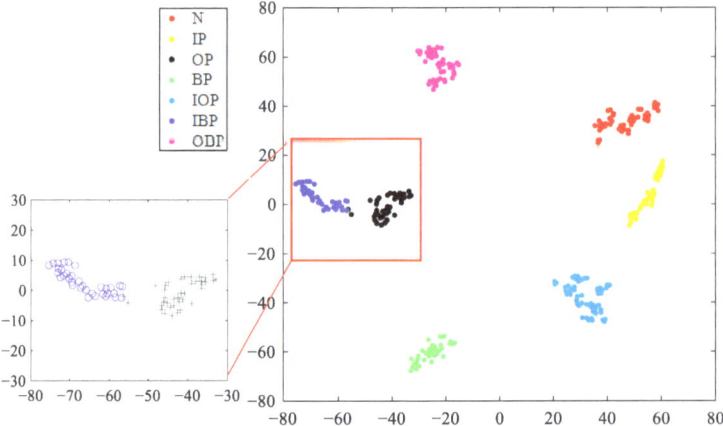

Figure 14. Visualization of the second fault diagnosis results.

5. Conclusions

A two-dimensional deep convolutional neural network intelligent fault diagnosis model (RVDCNN) is proposed for the original time-domain vibration signals of bearings. It addresses the difficulties in establishing personalized diagnosis models for automotive transmission systems, the strong dependence on expert experience, and the insensitivity to the fault characteristics. Using the two-dimensional numerical matrix of the original time-domain vibration signals of the bearings in electric vehicle transmission as input data, the end-to-end intelligent fault identification and diagnosis of rotating components in automatic transmission was achieved. The conclusions are as follows.

(1) An experimental comparison using the CWRU rolling bearing dataset was conducted to analyze how the number and size of the convolutional kernels in each layer of the CNN affect the accuracy of bearing fault diagnosis. The findings indicate that the model achieves the highest fault diagnosis accuracy when the first convolutional layer has eight kernels with a size of 11 × 11 and the second convolutional layer has 24 kernels with a size of 6 × 6.

(2) Compared with feature extraction methods, the proposed end-to-end fault diagnosis model uses raw vibration signals as CNN training samples, requires the smallest amount of data, and achieves higher accuracy than other models.

(3) A comprehensive performance test bench for a three-motor powertrain was built, and vibration experiments were conducted on deep groove ball bearings in a two-speed mechanical automatic transmission system for purely electric vehicles under service conditions. Accuracy of 99.11% was achieved in the 2AMT deep groove ball bearing dataset.

In the future, high-precision intelligent fault diagnosis research will be carried out to address the scarcity and imbalance of fault samples from transmission bearings or gears.

Author Contributions: Conceptualization, Y.C.; methodology, Y.C.; software, G.L.; validation, G.L. and B.H.; formal analysis, G.L.; investigation, G.L. and A.L.; data curation, Y.C. and G.L.; writing—original draft preparation, G.L.; writing—review and editing, Y.C. and G.L.; visualization, G.L. and B.H.; funding acquisition, Y.C. All authors have read and agreed to the published version of the manuscript.

Funding: This research was funded by the Guangxi Key Research and Development Plan Special Project through Grant No. 2023AB07038.

Data Availability Statement: The CWRU bearing data presented in the study are openly available at https://engineering.case.edu/bearingdatacenter/download-data-file (accessed on 11 August 2021). The transmission bearing data from this study will be made available by the authors on request.

Conflicts of Interest: The authors declare no conflicts of interest.

References

1. Raouf, I.; Lee, H.; Kim, H.S. Mechanical fault detection based on machine learning for robotic RV reducer using electrical current signature analysis: A data-driven approach. *J. Comput. Des. Eng.* **2022**, *9*, 417–433. [CrossRef]
2. Cao, Z.; Chen, Y.; Li, G.X.; Zang, L.B.; Wang, D.; Qiu, Z.Z.; Wei, G.Y. Dynamic simulation and experimental study of electric vehicle motor-gear system based on state space method. *Machines* **2022**, *10*, 589. [CrossRef]
3. Guo, J.C.; Zhang, H.; Zhen, D.; Shi, Z.Q.; Gu, F.S.; Ball, A.D. An enhanced modulation signal bispectrum analysis for bearing fault detection based on non-Gaussian noise suppression. *Measurement* **2020**, *151*, 107240. [CrossRef]
4. Ziani, R.; Hammami, A.; Chaari, F.; Felkaoui, A.; Haddar, M. Gear fault diagnosis under non-stationary working mode based on EMD, TKEO, and Shock Detector. *Comptes Rendus Méc.* **2019**, *347*, 663–675. [CrossRef]
5. Zhang, C.; Zhao, R.Z.; Deng, L.F. Rolling bearing fault diagnosis method based on EEMD singular value entropy. *J. Vib. Meas. Diagn.* **2019**, *39*, 353–358.
6. Zhen, D.; Guo, J.C.; Xu, Y.D.; Zhang, H.; Gu, F.S. A novel fault detection method for rolling bearings based on non-stationary vibration signature analysis. *Sensors* **2019**, *19*, 3994. [CrossRef]
7. Amar, M.; Gondal, I.; Wilson, C. Vibration spectrum imaging: A novel bearing fault classification approach. *IEEE Trans. Ind. Electron.* **2014**, *62*, 494–502. [CrossRef]
8. Lei, Y.G.; He, Z.J.; Zi, Y.Y.; Hu, Q. Fault diagnosis of rotating machinery based on multiple ANFIS combination with GAs. *Mech. Syst. Signal Process.* **2007**, *21*, 2280–2294. [CrossRef]
9. Liu, H.H.; Han, M.H. A fault diagnosis method based on local mean decomposition and multi-scale entropy for roller bearings. *Mech. Mach. Theory* **2014**, *75*, 67–78. [CrossRef]
10. Muruganatham, B.; Sanjith, M.A.; Krishnakumar, B.; Murty, S.A.V.S. Roller element bearing fault diagnosis using singular spectrum analysis. *Mech. Syst. Signal Process.* **2013**, *35*, 150–166. [CrossRef]
11. Khazaee, M.; Ahmadi, H.; Omid, M.; Moosavian, A.; Khazaee, M. Classifier fusion of vibration and acoustic signals for fault diagnosis and classification of planetary gears based on Dempster-Shafer evidence theory. *Proc. Inst. Mech. Eng. Part E J. Process Mech. Eng.* **2014**, *228*, 21–32. [CrossRef]
12. Li, W.H.; Weng, S.L.; Zhang, S.H. A firefly neural network and its application in bearing fault diagnosis. *J. Mech. Eng.* **2015**, *51*, 99–106. [CrossRef]

13. Bin, G.F.; Gao, J.J.; Li, X.J.; Dhillon, B.S. Early fault diagnosis of rotating machinery based on wavelet packets—Empirical mode decomposition feature extraction and neural network. *Mech. Syst. Signal Process.* **2012**, *27*, 696–711. [CrossRef]
14. Lei, Y.G.; Jia, F.; Lin, J.; Xing, S.B.; Ding, S.X. An intelligent fault diagnosis method using unsupervised feature learning towards mechanical big data. *IEEE Trans. Ind. Electron.* **2016**, *63*, 3137–3147. [CrossRef]
15. Zhang, K. A Weakly Supervised Deep Learning Method for Incipient Fault Diagnosis of Wind Turbine Gearboxes. Ph.D. Thesis, Chongqing University, Chongqing, China, 2021.
16. Cheng, Y.W.; Lin, M.X.; Wu, J.; Zhu, H.P.; Shao, X.Y. Intelligent fault diagnosis of rotating machinery based on continuous wavelet transform-local binary convolutional neural network. *Knowl.-Based Syst.* **2021**, *216*, 106796. [CrossRef]
17. Li, S.B.; Liu, G.K.; Tang, X.H.; Lu, J.G.; Hu, J.J. An ensemble deep convolutional neural network model with improved D-S evidence fusion for bearing fault diagnosis. *Sensors* **2017**, *17*, 1729–1748. [CrossRef]
18. Miao, J.G.; Wang, J.Y.; Miao, Q. An enhanced multifeature fusion method for rotating component fault diagnosis in different working conditions. *IEEE Trans. Reliab.* **2021**, *70*, 1611–1620. [CrossRef]
19. Pang, X.Y.; Xue, X.Y.; Jiang, W.W.; Lu, K.B. An investigation into fault diagnosis of planetary gearboxes using a bispectrum convolutional neural network. *IEEE/ASME Trans. Mechatron.* **2020**, *26*, 2027–2037. [CrossRef]
20. Raouf, I.; Kumar, P.; Kim, H.S. Deep learning-based fault diagnosis of servo motor bearing using the attention-guided feature aggregation network. *Expert Syst. Appl.* **2024**, *258*, 125137. [CrossRef]
21. Pan, T.Y.; Chen, J.L.; Ye, Z.S.; Li, A.M. A multi-head attention network with adaptive meta-transfer learning for RUL prediction of rocket engines. *Reliab. Eng. Syst. Saf.* **2022**, *225*, 108610. [CrossRef]
22. Chen, Z.; Zhou, D.; Zio, E.; Xia, T.B.; Pan, E.S. Adaptive transfer learning for multimode process monitoring and unsupervised anomaly detection in steam turbines. *Reliab. Eng. Syst. Saf.* **2023**, *234*, 109162. [CrossRef]
23. Fang, H.R.; Liu, H.; Wang, X.; Deng, J.; An, J.L. The method based on clustering for unknown failure diagnosis of rolling bearings. *IEEE Trans. Instrum. Meas.* **2023**, *72*, 3509508. [CrossRef]
24. Li, B.; Zhao, Y.P.; Chen, Y.B. Learning transfer feature representations for gas path fault diagnosis across gas turbine fleet. *Eng. Appl. Artif. Intell.* **2022**, *111*, 104733. [CrossRef]
25. Jamil, F.; Verstraeten, T.; Nowé, A.; Peeters, C.; Helsen, J. A deep boosted transfer learning method for wind turbine gearbox fault detection. *Renew. Energy* **2022**, *197*, 331–341. [CrossRef]
26. Chmidhuber, J. Deep learning in neural networks: An overview. *Neural Netw.* **2015**, *61*, 85–117. [CrossRef]
27. Guo, Y.M.; Zhou, Y.F.; Zhang, Z.S. Fault diagnosis of multi-channel data by the CNN with the multilinear principal component analysis. *Measurement* **2021**, *171*, 108513. [CrossRef]
28. Hu, Y.; Luo, D.Y.; Hua, K.; Lu, H.M.; Zhang, X.G. Overview on deep learning. *CAAI Trans. Intell. Syst.* **2019**, *14*, 1–19.
29. Wen, L.; Li, X.; Gao, L.; Zhang, Y.Y. A new convolutional neural network-based data-driven fault diagnosis method. *IEEE Trans. Ind. Electron.* **2017**, *65*, 5990–5998. [CrossRef]
30. Jia, F.; Lei, Y.G.; Lin, J.; Zhou, X.; Lu, N. Deep neural networks: A promising tool for fault characteristic mining and intelligent diagnosis of rotating machinery with massive data. *Mech. Syst. Signal Process.* **2016**, *72*, 303–315. [CrossRef]
31. Chen, Z.; Mauricio, A.M.R.; Li, W.; Gryllias, K. A Deep Learning method for bearing fault diagnosis based on Cyclic Spectral Coherence and Convolutional Neural Networks. *Mech. Syst. Signal Process.* **2020**, *140*, 106683. [CrossRef]
32. He, B.; Chen, Y.; Wei, Q.; Wang, C.; Wei, C.Y.; Li, X. Performance comparison of pure electric vehicles with two-speed transmission and adaptive gear shifting strategy design. *Energies* **2023**, *16*, 3007. [CrossRef]

Disclaimer/Publisher's Note: The statements, opinions and data contained in all publications are solely those of the individual author(s) and contributor(s) and not of MDPI and/or the editor(s). MDPI and/or the editor(s) disclaim responsibility for any injury to people or property resulting from any ideas, methods, instructions or products referred to in the content.

Article

A Computational Fluid Dynamics-Based Study on the Effect of Bionic-Compound Recess Structures in Aerostatic Thrust Bearings

Fangjian Yuan [1,2], Hang Xiu [1,2,*], Guohua Cao [1,2], Jingran Zhang [1], Bingshu Chen [1], Yutang Wang [3] and Xu Zhou [1]

1. College of Mechanical and Electric Engineering, Changchun University of Science and Technology, Changchun 130022, China; 18334943540@163.com (F.Y.); caogh@cust.edu.cn (G.C.); zhangjingran@cust.edu.cn (J.Z.); bingshuc2003@163.com (B.C.); 18804460095@163.com (X.Z.)
2. Chongqing Research Institute, Changchun University of Science and Technology, Chongqing 404100, China
3. Changchun Institute of Optics, Fine Mechanics and Physics, Chinese Academy of Sciences, Changchun 130033, China; ytwang@ciomp.ac.cn
* Correspondence: xiuhang@cust.edu.cn

Abstract: To investigate the effect of recess structures on the static and dynamic performance of aerostatic thrust bearings and to explore superior designs, this study analyzes the load-capacity theoretical model, identifying that the throttling effect and pressure-holding effect of the recess are the key factors determining the bearings' static performance. Computational fluid dynamics (CFD) was used to evaluate three types of recess structures: a simple-orifice recess (SOR), a rectangular-compound recess (RCR), and a bionic-compound recess (BCR). The results indicate that the BCR structure demonstrates efficient transmission performance by reducing flow resistance and diverting air, while ensuring a reasonable pressure drop as the radial ratio α_i changes. Additionally, the smaller air capacity of the BCR structure contributes to enhanced bearing stability, showing clear advantages in both static and dynamic performance. This research illustrates the practical application of bionics in mechanical design and provides new theoretical foundations and design strategies for improving aerostatic bearing performance.

Keywords: aerostatic thrust bearing; bionics recess; CFD simulation; static and dynamic stability

1. Introduction

During the operation of aerostatic thrust bearings, air is continuously supplied from an external source, serving as a lubricant and replacing traditional rollers. This enables the bearings to be characterized by low noise, high precision, low friction, high speed and long service life, and they are widely used in aerospace, semiconductor processing and ultra-precision testing, among other applications [1,2]. Given the compressibility of air and the stringent requirements for precision, stability, and operational speed, the static performance of high-performance aerostatic thrust bearings—including load capacity, stiffness, and dynamic characteristics—has attracted significant attention from researchers and industry professionals [3]. Among various methods to enhance performance, recesses have emerged as particularly effective, leading to extensive investigation.

Hiromu Hashimoto et al. [4] optimized the design of grooves for thrust air bearings using a sequential quadratic programming method based on objective functions, including air-film thickness, bearing torque, air-film dynamic stiffness, and their combinations, with cubic spline functions representing the groove geometries. They found that the groove geometry optimizing air-film thickness or frictional moment was a helical groove, while the geometry optimizing dynamic stiffness was an improved helical groove. Tomotaka Yoshimura et al. [5] conducted a computational fluid analysis of airflow around the bearings, comparing it with experimental photographs to investigate the cause of tiny fluctuations, termed nano-fluctuations, in the aerostatic thrust bearing confined by the T-shaped groove surface. Yinan Chen et al. [6] modeled four aerostatic bearings with grooves of different geometries by combining grooves with annular aerostatic thrust bearings featuring porous throttling. They investigated the pressure distribution, load-carrying capacity, stiffness, and flow characteristics of the flow field in the bearing gap through computational fluid dynamics simulations. Colombo, F et al. [7]. performed experimental and numerical simulations to analyze the effects of orifice diameter, position, and air supply pressure on the load capacity and air consumption of rectangular chambers. Puliang Yu et al. [8] designed a square microporous array restrictor for aerostatic bearings, revealing that the design effectively reduces turbulent vortices and micro-vibrations through numerical simulation and experimentation. Zhuang [9] examined bearing capacity and pressure distribution at varying eccentricities and design parameters using FDM modeling and CFD simulations. Gao et al. [10] explored the impact of six different chamber shapes on pressure distribution and load capacity at ultra-high speeds (200,000 rpm) through experimental and CFD analyses. Puliang Yu et al. [11] designed a novel primary and secondary orifice restrictor, and the static–dynamic characteristics based on this restrictor were investigated by using large eddy simulations and experiments. The results show that this type of restrictor can effectively suppress the generation of turbulent vortices. Qiao, Y.J. et al. [12] investigated the static performance of aerostatic bearings with trapezoidal groove restrictors and increased the bearing capacity by 18%. Mohamed E. Eleshaky [13] analyzed two structural forms of circular bearings using CFD simulations to investigate the mechanisms behind the pressure-drop phenomenon in aerostatic thrust bearings. Wenjun Li et al. [14] introduced a novel aerostatic bearing design featuring back-flow channels, which was validated through CFD and experimental studies. This new structure demonstrated improved load capacity, stiffness, and stability compared to traditional pocketed orifice-restricted aerostatic bearings. Siyu Gao et al. [15] examined the effects of various orifice length/diameter ratios (OLDR) on pressure, velocity, turbulence intensity, eddy currents, load capacity, stiffness, volumetric flow rate, and orifice flow resistance in aerostatic thrust bearings using computational fluid dynamics (CFD). The results indicated that the performance of aerostatic thrust bearings varied under different OLDRs. Ruzhong Yan [16] proposed six types of groove restrictor designs—linear, extended, S-shaped, elliptical, X-shaped, and mesh—for two common throttle arrangements (linear and rectangular) on the thrust surface. Numerical simulations of different bearing types revealed that the use of groove restrictors significantly enhances bearing capacity, static stiffness, operational accuracy, and longevity. Kai Feng [17] introduced an innovative restrictor structure for aerostatic bearings aimed at suppressing eddy currents by altering airflow characteristics within the groove. Both theoretical and experimental results indicated that the Arc Hole Bearing (AHB) achieved superior stability and the smallest vibration amplitude while maintaining consistent load capacity. In summary, numerous scholars have studied restrictors, focusing on various adjustments to structure parameters. These include optimizing the shapes of dynamic pressure grooves, refining groove designs through adjustments in dimensions and parameters, and altering the size specifications of the air supply orifice. Despite these advancements, there remains relatively limited research on directing fluid flow within bearings to optimize flow patterns and enhance both static and dynamic performance.

Biological structures, particularly vascular branching structures in plant leaves, have evolved to exhibit exceptional mass transfer capabilities in substance transport [18]. These branched networks have garnered attention across various fields, significantly enhancing humanity's ability to adapt to and transform nature while generating substantial economic benefits. For example, Ruofei Zhu [19] developed an artificial tree evaporator based on the Murray network, highlighting the value of branching structures in desalination applications. Guohui Zhou et al. [20] created a vapor chamber with a leaf-shaped liquid-absorbing core that promotes efficient condensate return for cooling. Shitong Chai [21] demonstrated that symmetric leaf-vein structures outperform their asymmetric counterparts in heat transfer characteristics, emphasizing the potential of biomimetic design for optimizing fluid channel configurations in engineering applications. Yi Peng [22] designed a vapor-chamber wick structure inspired by the fractal architecture of leaf veins, featuring polygonal loops formed by Y-shaped bifurcations. This study demonstrated that the novel design outperforms parallel configurations in terms of permeability and resistance, highlighting its effectiveness in enhancing heat transfer and fluid transport efficiency. Xinyu Wan [23] investigated the flow field of a proton-exchange membrane fuel cell (PEMFC) by incorporating a Y-shaped tree-like fractal structure based on Murray's law. The research revealed that an optimal branching angle of 75° resulted in a 26.7% increase in maximum output power density compared to a parallel flow field, showcasing the advantages of tree-like fractal flow fields in improving transport characteristics and overall PEMFC performance. Chuangbei Ma [24] proposed a bio-inspired fractal microchannel heat sink with a modified secondary structure for efficient heat dissipation. The optimized design achieved a heat flux of 577 W/cm^2, demonstrating effectiveness in enhancing heat dissipation and reducing energy consumption. These findings illustrate the successful application of bionic structures in engineering and validate the potential of leaf vascular branching structures in fluid transport, offering new insights for designing recesses in air bearings.

Drawing on the functional similarities between leaf vascular branching structures and air diffusion processes in aerostatic thrust bearings, this study proposes a bionic-compound recess design. Simulations were conducted on three different types of recesses for aerostatic thrust bearings using CFD methods. A comparative analysis was performed between the bionic-compound recess and the other two structures, focusing on static and dynamic performance evaluation criteria. This study aims to explore this innovative design, elucidate the critical relationship between its performance and structure, and provide new insights for enhancing the performance of aerostatic bearings.

2. Compound Recess Based on the Bionic Concept

2.1. Bionic Recess Design

Certain structures in nature, such as leaf veins, animal lungs, and plant roots, are the result of natural optimization through long-term evolution. They possess inherent advantages in terms of drag reduction, mass transfer, and uniform diffusion. Compared to structures like straight channels, flow channels designed with inspiration from these structures exhibit better fluid distribution and mass transfer properties, ensuring a uniform supply of material to all areas while maintaining a low pressure drop, thus improving overall distribution efficiency. Considering that the airflow in a single-orifice thrust bearing structure exhibits the same point-to-area flow characteristics as the liquid transport in a plant leaf, as shown in Figure 1a. The blue arrows in Figure 1, represent the diffusion of the fluid. The biomimetic-compound recess inspired by the branching structure of leaf veins has been designed to enhance the throttling effect of the recess structure and improve its ability to transport air within the bearing air film, thereby enhancing the overall performance of the bearing.

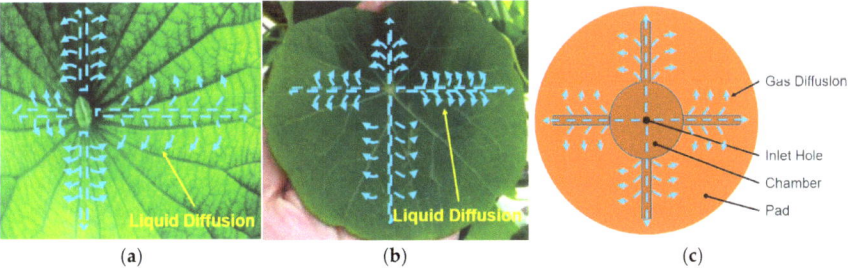

Figure 1. Two similar leaves to single-hole thrust bearings: (**a**) Lotus leaf, (**b**) Nasturtium leaf, (**c**) Schematic diagram of aerostatic bearing.

The branching structure of plant leaf veins, as illustrated in Figure 2, plays a critical role in enabling material exchange and nutrient transport within the leaf. This system ensures the efficient and even distribution of water and nutrients throughout the leaf. The structure is hierarchical, where $k = 1$ and 2 represent the primary and secondary levels, respectively. The hydraulic diameters at each level are $2R_0$ and $2R_1$, with the angle between the levels denoted by θ_1 and the lengths of the main and secondary branches represented by l_1 and l_2. Research indicates that the ratios of these structural parameters at each level significantly influence the hydraulic conductivity of the leaf vein system [25]. Considering the correlation between the structure and function of the leaf blade shown in Figure 1 and the air static-pressure thrust bearing, this paper adopts a bionic approach by introducing the concept of leaf vein branching from Figure 2 into the design of the recess structure, aiming to explore and expand potential pathways for improving the flow performance of air static-pressure bearings. At the same time, considering the requirement of solid balance of bearings at high speeds, recess is designed with a strictly symmetrical branch structure.

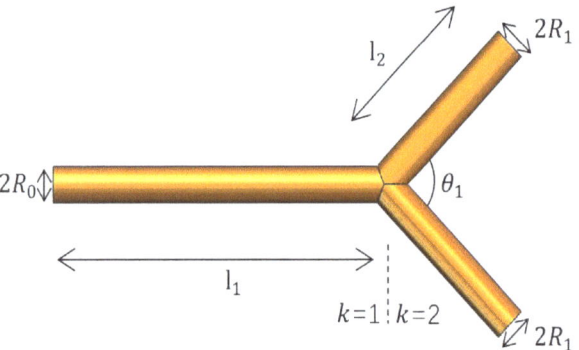

Figure 2. Schematic diagram of bionic branching restrictor structure.

2.2. Physical Models

Aerostatic bearings can be briefly classified into several categories based on the type of recess used. Table 1 [26] illustrates their performance and conducts a longitudinal comparative analysis. Essentially, the bionic-concept recess can be regarded as a variation of simple orifice-type or groove-type recesses. Therefore, to explore the impact of bionic throttle structures on bearing performance, this paper designs three bearings with different recess structures for comparative study. Figure 3 displays the basin diagrams of the bearings with three different recesses, wherein A is the type of bearing with a simple orifice-type recess structure with chambers (SOR); B is the type of bearing with a rectangular-compound recess (RCR); and C is the type of bearing with a bionic-compound recess (BCR). Considering the effects of solid imbalance on bearing operation at high speeds, we have intentionally designed the BCR and RCR structures to be strictly axisymmetric. This design approach ensures uniform mass distribution around the rotational axis, which is essential for maintaining stability and minimizing vibrations at high rotational speeds. The axisymmetric shape facilitates a balanced mass distribution, thereby enhancing the performance and longevity of the bearings in high-speed applications. The three bearings have been designed to maintain the same external diameter D of 40 mm and the same dimensions for the air inlet orifice (diameter D_2 and height H_2). A common range of film thicknesses (from 9 μm to 24 μm) was used. The RCR and BCR have the same pressure-equalizing chamber (diameter D_1 and height H_1) and the same outreach diameter D_0. Meanwhile, we refer to the area within the recess outreach circle as Region I, and the area outside the pure air film section as Region II. In the SOR, we have selected a pressure-equalization chamber with a diameter D_1 ranging from 2 to 30 mm to explore the impact of different chamber sizes on the stability of the bearing. Additionally, within the same range of pressure-equalization chambers, we conduct more comprehensive comparisons of the specific effects of three different structures—A, B, and C—on bearing performance. This systematic research approach will enhance our understanding of how various design parameters affect the overall performance of the bearing. In comparison with the SOR, the RCR with a rectangular structure is of width L on the outside of the pressure-equalizing chamber, and the height is the same as that of the pressure-equalizing chamber. The BCR, on the other hand, introduces a branching structure on the outside of the pressure-equalizing chamber based on the SOR, creating a compound structural design. For the BCR, the branch structure has a primary slot width of L and a secondary slot width of L_1. Additionally, we define the angle formed by the end of the branching structure and the center of the thrust face as the total angle θ in the BCR, where two secondary slots of the branch structure are equal and symmetrical with respect to the axis centra line and the angle between them is called θ_1. The above designs aim to study the hydrodynamic performance of the bearing by adjusting the recess structure, achieving more rational fluid distribution and more stable air-film support.

Table 1. Comparison of various types of restrictors [26].

Restrictor Type		Load Capacity	Stiffness	Stability	Air Consumption
Annular orifice		Low	Low	Fair	Small
Simple orifice		High	High	Poor	Small
Slot		Medium	Medium	Good	Large
Groove		High	High	Good	Medium
Porous		High	High	Excellent	Large

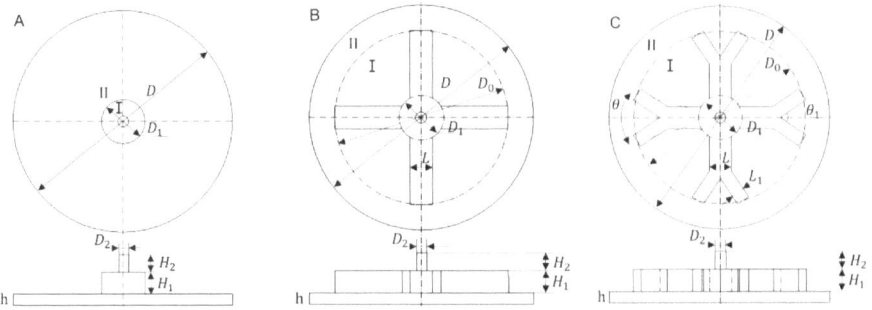

Figure 3. Basin diagram of three different recess structures ((**A**): Bearing with SOR structure; (**B**): Bearing with RCR structure; (**C**): Bearing with BCR structure).

To facilitate the subsequent analysis, the entire thrust face was divided into two regions: Region I, which contains the restrict structure, and Region II, which represents the region without a restrict structure. The ratio of diameter D_0 to the external diameter D is recorded as α_i, where the subscript i denotes the item corresponding to different schemes, each with distinct design parameters. It should be noted that, in the designs of the RCR and BCR, when the width of the groove exceeds the diameter D_1 of the pressure-equalizing chamber, the boundary between the chamber and the groove will disappear, causing the compound structure to lose its complexity and ultimately simplify to a groove recess. The parameters of SOR and RCR bearings are shown in Table 2. In designing the parameters of the branching structure, we have drawn on the work of X. D. Shan [27]. According to their conclusions, when the fluid behaves as a laminar flow, the optimal hydraulic diameter ratio R_0/R_1 is $2^{(1/3)}$, while in a turbulent flow, this ratio is $2^{(7/3)}$. For rectangular cross-section channels, the hydraulic diameter D_h is calculated as follows:

$$D_h = \frac{2WH}{W+H} \quad (1)$$

where W is the width of the rectangular groove structure and H is the height. In this study, according to Equation (1), the width ratios were converted to 0.66 and 0.6, respectively. Since both the simple-orifice and groove recesses are turbulent structures, we chose a width ratio of 0.6 for the BCR to conduct subsequent research.

Table 2. Structure parameters and working conditions.

Calculation Parameters	Value		
	SOR	RCR	BCR
Chamber diameter D_1/(mm)	2~30	2	2
Chamber height H_1/(mm)	1	1	1
Orifice diameter D_2/(mm)	0.2	0.2	0.2
Orifice height H_2/(mm)	0.3	0.3	0.3
Pitch diameter D_0/(mm)	/	20	20
Recess structure ratio α_i	0.05~0.75	0.5	0.5
Primary branch width L/(mm)	/	1	1
Secondary branch width L_1/(mm)	/	/	0.6
Branch angle θ/(°)	/	/	20
Secondary angle θ_1/(°)	/	/	14
Air-film thickness h/(μm)		9~24	
Air temperature T/(K)		293.15	
Density ρ/(kg/m^3)		1.202	
Specific Heat c/(J/(kg/m^3))		1006.43	
Heat Conductivity k/(W/(m·K))		0.0242	
Viscosity μ/(kg/(m·s))		1.7894×10^{-3}	
Molecular weight m_{mo}/(kg/mol)		28.966×10^{-3}	
Supply pressure P_S/(MPa)		0.5/0.6	
Environmental pressure P_O/(MPa)		1.01325	

3. Numerical Modeling

3.1. Governing Equations

The air within the bearing is considered a compressible viscous fluid, assumed to obey ideal air law. Given the complexity of the flow, this paper employs the full Navier–Stokes equations for numerical simulation [28] to more accurately characterize the air state inside the bearing. The following assumptions are made: the lubricating air is treated as a Newtonian fluid with a constant viscosity coefficient μ; inertial forces are neglected; the pressure and air density between the air films remain constant in the direction perpendicular to the bearing; the y-axis is defined as perpendicular to the air-film surface; and the time-dependent elements can be ignored for steady flow conditions. The equations can be expressed as follows:

Continuity equation:

$$\frac{\partial \rho}{\partial t} + \frac{\partial (\rho u)}{\partial x} + \frac{\partial (\rho v)}{\partial y} + \frac{\partial (\rho w)}{\partial z} = 0 \quad (2)$$

where x, y, z are the coordinates; u, v, w are the velocities in the x, y, z directions; and ρ is the density of the air.

Momentum equation:

$$\frac{\partial (\rho u)}{\partial t} + \mathrm{div}(\rho u \mathbf{u}) = -\frac{\partial P}{\partial x} + \mathrm{div}(\mu\ \mathrm{grad}\ u) + S_{Mx} \quad (3)$$

$$\frac{\partial (\rho v)}{\partial t} + \mathrm{div}(\rho v \mathbf{u}) = -\frac{\partial P}{\partial y} + \mathrm{div}(\mu\ \mathrm{grad}\ v) + S_{My} \quad (4)$$

$$\frac{\partial (\rho w)}{\partial t} + \mathrm{div}(\rho w \mathbf{u}) = -\frac{\partial P}{\partial z} + \mathrm{div}(\mu\ \mathrm{grad}\ w) + S_{Mz} \quad (5)$$

where x, y, z are the coordinates, $\mathbf{u}$ is the velocity vector, P is the air pressure, μ is the air kinetic viscosity, and S_{Mx}, S_{My}, S_{Mz} are the momentum sources.

3.2. Computational Domain and Mesh

This study was conducted using CFD simulations with the commercial software ANSYS FLUENT 2022 R1. Given that the thickness of the air film is three orders of magnitude smaller than that of the entire geometry, it is necessary to refine the mesh at the location of the air film to meet the simulation requirements. Meshing the entire geometry, however, would lead to a significant increase in the number of cells. Considering the axisymmetric nature of the flow field in the aerostatic thrust bearing, we employ a quarter model of the entire physical region, as shown in Figure 4a, and apply symmetric boundary conditions for the calculations to minimize computational costs. The model was meshed hexahedrally using ICEM, incorporating 11 mesh layers into the air film, along with refined meshing at critical locations to ensure that the value of y+ remains below 1, as illustrated in Figure 4b. y^+ reflects the resolution of the mesh near the wall, defined by the equation $y^+ = (\rho \tau_W)^{1/2}\, y/\mu$, where y is the distance from the center of the first layer of the mesh adjacent to the wall to the wall, and τ_W is the wall shear stress. Due to the complexity of the fluid motion in the recess during steady-state calculations, vortex effects resulting from turbulent kinetic energy were observed. Therefore, the RNG (k-e) model was employed to more accurately characterize the vortices within the chamber. The calculations utilized the SIMPLE algorithm for pressure–velocity coupling and second-order upwind interpolation for the density, turbulent kinetic energy, and turbulent dissipation rate, as detailed in Table 2.

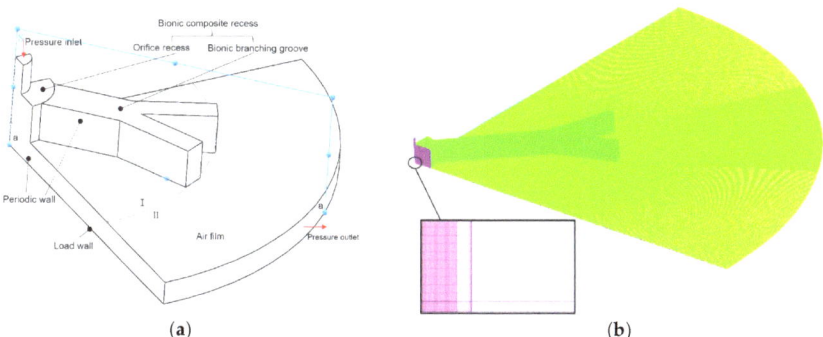

Figure 4. Numerical model of the CFD simulations. (**a**) Computational domain and boundary conditions. (**b**) Illustration of computational meshes.

To validate the CFD simulation methodology employed in this study, the results were compared with the experimental findings of Belfort et al. [29]. The circular-chamber simple-orifice recess model from their work was adopted, the simulation was carried out under the same working conditions using our methodology, and the pressure distribution graph shown in Figure 5 was obtained. The simulation results are largely consistent with the experimental data, with an error margin not exceeding 5%, which is acceptable for engineering applications.

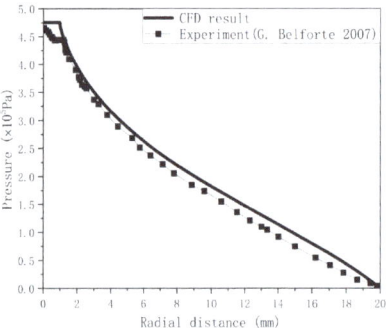

Figure 5. The comparison of radial pressure distribution between CFD result and the existing experimental results [29].

4. Result and Discussion

Qiang Gao et al. [30] demonstrated the pressure distribution in a aerostatic thrust bearing, as illustrated in Figure 6. In the upper diagram of Figure 6, the area outlined by the blue line demonstrates the pressure distribution on the bearing load surface. In the lower diagram in Figure 6, the dark blue area represents the bearing structure, while the light blue area shows the bearing basin. It indicates that for the thrust bearing, when the height of the recess structure significantly exceeds the thickness of the air film, the pressure on the thrust surface within the recess structure (Region I) remains nearly uniform. Conversely, the pressure outside the air outlet (Region II) basically exhibits a linear pressure drop.

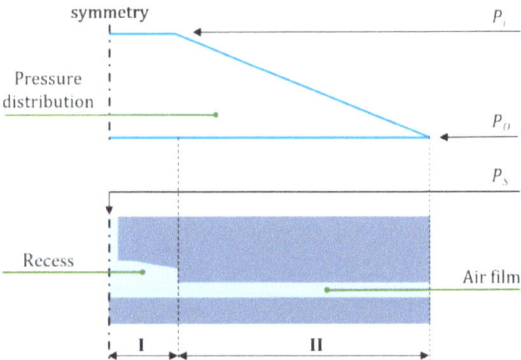

Figure 6. Pressure distribution maps in the aerostatic bearing.

Building on the previous discussion, we can obtain the load-carrying capacity on the thrust surface by individually integrating the pressure over Regions I and II.

$$F_{Ii} = \pi P_i \left(\frac{\alpha_i D}{2}\right)^2 \tag{6}$$

$$F_{IIi} = \int_{\alpha_i L}^{L} 2\pi x(L-x)\frac{P_i}{(1-\alpha_i)L}dx = \pi\left(\frac{D}{2}\right)^2 P_i\left(\frac{1}{3}(1+\alpha_i) - \frac{2}{3}\alpha_i^2\right) \quad (7)$$

$$F_{Ii} + F_{IIi} = F_i = \frac{\pi(D/2)^2 P_i}{3}(1+\alpha_i+\alpha_i^2) \quad (8)$$

where D is the bearing disc diameter; α_i is the size ratio of the recess structure to the bearing; F_{Ii} is the force load on thrust surface by air in Region I; F_{IIi} is the force load on thrust surface by air in Region II; and F_I is the total load on the thrust surface combined from Region I and Region II.

According to the equation, it can be concluded that the pressure P_i inside the recess structure in Region I and the percentage of coverage α_i of the recess are both positively correlated with the bearing's capacity. This means that to increase the actual load capacity of the thrust bearing, it is necessary to simultaneously increase the pressure P_i inside the pressure-equalizing chamber and the percentage of coverage α_i of the recess, which is also the expectation for the bionic recess structure in this study. In fact, P_i is a function of α_i, meaning that the pressure within the recess must change according to the design scheme; however, this does not prevent our ability to qualitatively analyze the load capacity in the equation.

Referring to the literature [29], by analogy with Volt's law of transmission, the resistance of Region I and II of the structure is expressed in terms of R_{Ii} and R_{IIi}, using M'_i to represent the mass flow rate, expressing the relationship between the differential pressure and the resistance and flow as follows:

$$M'_i(R_{Ii} + R_{IIi}) = \Delta P = P_S - P_O \quad (9)$$

where ΔP is the total pressure difference between the air inlet and outlet, and P_S and P_O are pressure defined as the inlet and outlet pressure, as shown in Figure 6. Then, it can be expressed in Region II:

$$M'_i R_{IIi} = \Delta P_{IIi} = P_i - P_O \quad (10)$$

Putting M'_i in (9) on one side of the equal sign and substituting it into (10), we obtain

$$P_i = P_O + \left(\frac{P_S - P_O}{R_{Ii} + R_{IIi}}\right) R_{IIi} \quad (11)$$

To enhance the discussion on the impact of flow resistance in each region on the performance of the thrust bearing. Defining the pressure-drop ratio $E = (P_S - P_i)/(P_S - P_O)$, according to Equations (9) and (10), it can be shown that

$$E = \frac{R_{Ii}}{R_{Ii} + R_{IIi}} \quad (12)$$

Substituting Equation (11) into Equation (10), we obtain

$$P_i = P_O + (P_S - P_O)(1 - E) \quad (13)$$

It is evident that the pressure P_i in Region I increases as the pressure ratio E decreases (smaller R_{Ii}). According to Equation (8), the ideal design can be summarized as having a large proportion α_i of the recess structure and a low flow resistance R_{Ii} of the recess structure in Region I.

The results from [31] indicate that, although the fluid near the inlet of the recess exhibits turbulent flow, it can still be largely considered a laminar flow due to the size of the air film, and the velocity there was lower than 0.1 Ma. Thus, the air flow in Region II can be characterized as the flow of incompressible air within an expansive annular slit. According to Poiseuille's law, we have

$$V(r) = -\frac{h^2}{12\mu}\frac{dP}{dr} \quad (14)$$

where $V(r)$ is the velocity of fluid in radial coordinate, μ is the viscosity of the fluid as shown in Table 2, and dP/dr is the pressure gradient distributed radially along the thrust surface.

Since the fluid is incompressible, the volume flow Q on the surface of the ring is conserved at any radius r, so that

$$Q = 2\pi h \cdot V(r) \quad (15)$$

Substituting Equation (14) into (15) gives

$$Q = -\frac{\pi h^3 r}{6\mu}\frac{dP}{dr} \quad (16)$$

Organizing Equation (16), the radial pressure gradient of the annular slit is obtained as follows:

$$\frac{dP}{dr} = -\frac{6\mu Q}{\pi h^3 r} \quad (17)$$

In order to obtain the corresponding pressure drop in Region II, integration is performed in the radial $\alpha_i L$ to L region

$$P_i - P_O = \frac{6\mu Q}{\pi h^3}\ln(\frac{1}{\alpha_i}) \quad (18)$$

Substituting the above equation into Equation (10) gives the following structural resistance:

$$R_{IIi} = \frac{6\mu}{\pi h^3}\ln(\frac{1}{\alpha_i}) \quad (19)$$

In Equation (19), R_{IIi} will diminish as α_i increases. According to Equations (12) and (13), we need a design that enables E to decline as α_i grows, indicating that R_{II} should decrease more rapidly than R_{IIi}. To verify the effect of α_i on the performance of the aerostatic thrust bearings, the SOR model was first compared with different-sized chamber designs.

4.1. SOR Bearings with Different Recess Ratios

To investigate how the recess structure affects bearing performance, a series of simulations of SOR bearings with different chamber dimensions (size ratio α_i) and air-film thicknesses were conducted and analyzed under the working conditions of P_S = 0.5 MPa.

4.1.1. Load Capacity and Pressure Distribution of SOR Bearings

Figure 7 demonstrates the load capacity of SOR bearings for different values of α_i at different air-film thicknesses. It can be found that, when the selected film thickness is smaller at 9 μm, the overall bearing load capacity increases as the diameter of the chamber increases, though this trend gradually slows down. When the thickness of the air film h > 12 μm, the load capacity of the bearings with increasing chamber diameter shows an initial increase and then decreases, and the peak occurs at the diameter of 20 mm (i.e., α_i = 0.5). Meanwhile, Figure 7 shows that with the increase in air-film thickness, regardless of the type of SOR, the bearing load capacity exhibits a downward trend, and the differentiation also decreases. When the air-film thickness reaches 24 μm, the bearing load capacity tends to be similar across the board, but it can still be observed that larger α_i values are at a disadvantage. This reflects the interaction between pressure drops and the pressure-holding effect for different α_i values under compromised conditions.

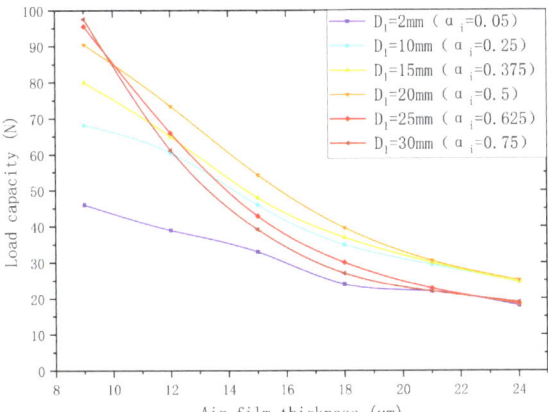

Figure 7. Bearing load capacity for different α_i at different air-film thicknesses.

Pressure distributions in the radial direction of the SOR with varied α_i under the film thickness of 15 μm are shown in Figure 8. It can be observed that the bearing load capacity primarily relies on the pressure-holding effect of the recess structure, which refers to its ability to maintain the diffusion of the central high-pressure region outward, along with the pressure drop occurring at different values of α_i.

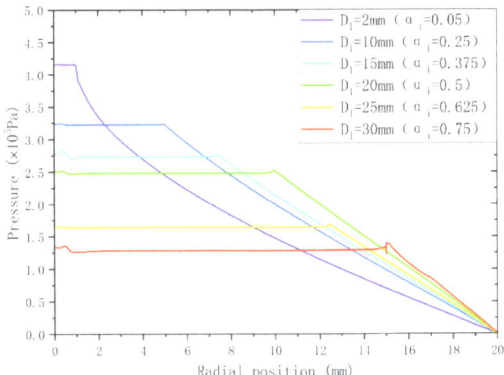

Figure 8. Pressure distribution under different α_i.

Figure 9 shows, from top to bottom, the S-profile velocity streamlines for different α_i values of ROS at P_S = 0.5 MPa, h = 15 μm. It is observed that, except for the smaller chamber, other schemes exhibit obvious vortex structures downstream of the recess structure. As the air flows out of the chamber into the pure air-film region at the boundary between Region I and II, the drastic change in dimensions leads to a change in the velocity and direction of the flow, resulting in the generation of vortices and the emergence of low-pressure regions. However, the vortex in the small α_i scheme is suppressed by the high pressure near the inlet, while the low-pressure region in the large α_i scheme expands in the opposite direction to the flow. Therefore, despite having a lower structural resistance (flow is not impeded), the SOR scheme fails to achieve an increase in load capacity with increasing α_i. It is judged that there is a limitation of α_i for SOR structures, and thus a balance is needed between lowering structural resistance and limiting the expansion of the vortex at the demarcation to meet the expectations of Equation (8).

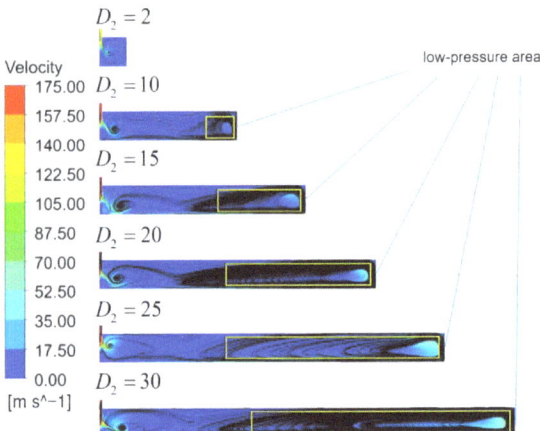

Figure 9. Velocity streamlines at different α_i.

Figure 10 shows the pressure-drop ratios E of the SOR under different-sized ratios of α_i. The data show that E increases with the enlargement of the recess, i.e., the throttling effect gradually decreases with increasing chamber size. As α_i increases, the bearing pressure-drop ratio of the SOR basically shows a linear growth trend.

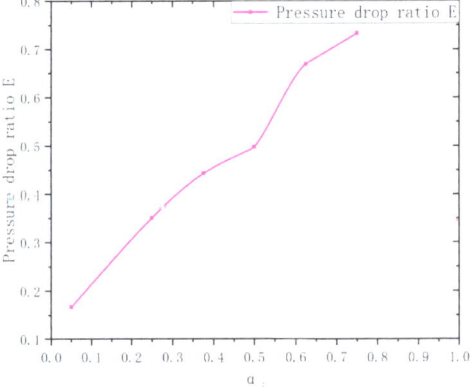

Figure 10. Pressure-drop ratio E with α_i.

4.1.2. Stiffness and Mass Flow Rate of SOR Bearings

Figure 11 shows the stiffness and mass flow rate of different-sized ratios of α_i SOR bearings for different air-film thicknesses. Figure 11a shows that the stiffness of all chambers, except the 25 mm and 30 mm chambers, shows a tendency to increase and then decrease with increasing air-film thickness. The 10 mm, 15 mm, and 20 mm size chambers reach their maximum at an air-film thickness of 12 μm, while the 2 mm chamber reaches its maximum at an air-film thickness of 15 μm. When the air-film thickness $h \leq 12$ μm, the bearing stiffness increases with the increase in the sized ratio α_i. In structures of 25 mm and 30 mm, stiffness decreases rapidly with the air-film thickness, gradually falling below that of the other chamber sizes, and is lower than all structures when the air-film thickness is 21 μm. From the mass flow distribution in Figure 11b, it is evident that the mass flow rate increases with both the air-film thickness and the chamber diameter. Additionally, under different sizes of recess structures, the bearing mass flow rate exhibits varying increases as the air-film thickness rises—the larger the size ratio, the easier it is to reach the critical flow rate as the air-film thickness increases. When the thickness of the air film is 24 μm, the mass flow rates of different sizes of chambers are basically the same, indicating that the throttling effect is essentially negligible under a large thickness of air film.

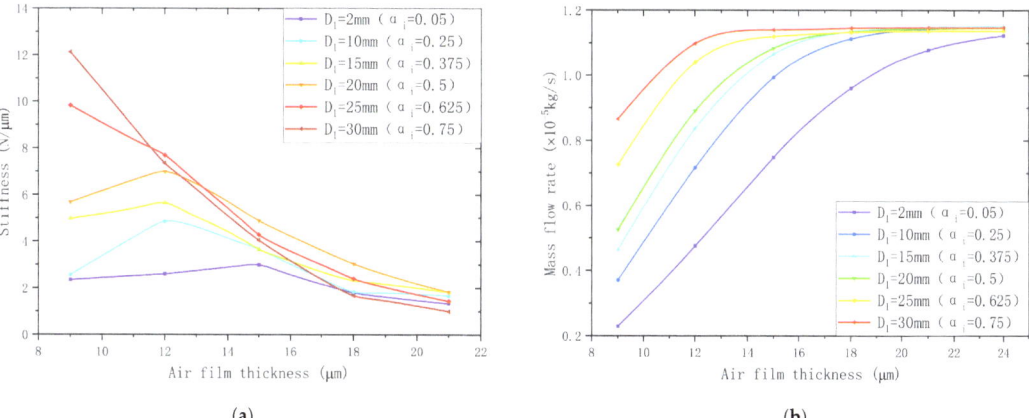

Figure 11. Stiffness and mass flow rate for various α_i: (**a**) Stiffness. (**b**) Mass flow rate.

4.1.3. Dynamic Performance of SOR Bearings

Figure 12 illustrates the velocity contours for different SOR bearings with various sized ratios of α_i, under conditions of $P_S = 0.5$ MPa, $h = 15$ μm. There is only one main vortex present within the chamber. As the chamber diameter increases, the small vortex generated by air separation shifts slightly backward, resulting in the formation of a larger vortex—the larger the chamber, the greater the vortex. According to the research by Xuedong Chen et al. [32], vortices in the chamber are one of the contributing factors to micro-vibrations in bearings. As the intensity of the vortex increases, the associated vibration energy also rises accordingly. Moreover, the pressure drop from the vortex edge to the vortex center directly reflects the intensity of the vortex, which can be used to assess the stability of the bearing. Therefore, in order to more accurately assess the effect of different structures on bearing stability, vector and pressure diagrams of the main vortex positions of different structures were extracted to observe the morphology and pressure distribution of the main vortex. As shown in Figure 12, the main vortex centers of the different structures are at different heights, though all vortices are symmetric with respect to the S-plane (refer to Figure 4). Accordingly, the vector and pressure distributions of each structure at the main vortex position, shown in Figure 13, were extracted along the S plane. It is clear from Figure 13 that there is a significant pressure drop from the edge of the vortex (where the vortex is no longer clearly visible) to the center of the vortex. The pressure drop generated by the vortex in chambers of different sizes is statistically presented in Table 3. In Figure 13, the behavior of the internal vortex varies under different-sized chambers. When the chamber is small, the pressure fluctuations near the vortex are relatively smooth; however, as the chamber size increases, the pressure fluctuations become more intense. Furthermore, as the chamber enlarges, the vortex also increases in size, and the throttling effect diminishes, causing the high-pressure air at the inlet to directly impact the bottom of the bearing, which poses a potential threat to the stability of the bearing. Additionally, the statistical data in Table 3 regarding the pressure drop by the vortex indicate that the pressure drop in the main vortex region of the bearing tends to increase with the diameter of the chamber. In other words, the intensity of the vortex increases as the chamber section increases. These results demonstrate that an increase in the chamber section affects the stability of the bearing. Therefore, in the SOR structure, a larger chamber volume, represented by the size ratio α_i, results in a less stable bearing.

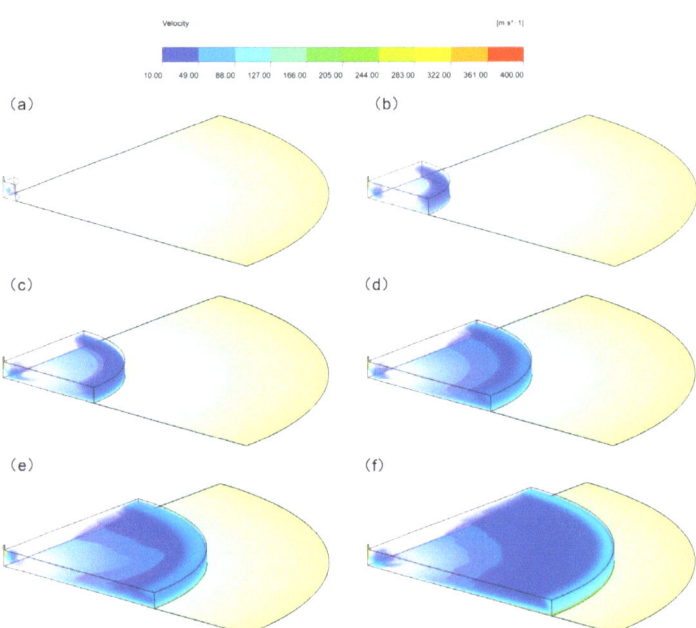

Figure 12. Velocity contours in different-sized chambers. (**a**) D_1 = 2 mm. (**b**) D_1 = 10 mm. (**c**) D_1 = 15 mm. (**d**) D_1 = 20 mm. (**e**) D_1 = 25 mm. (**f**) D_1 = 30 mm.

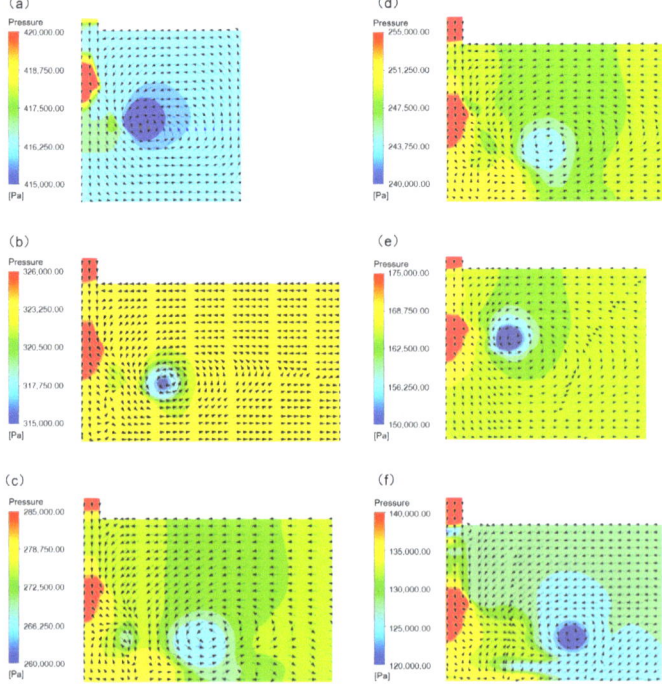

Figure 13. Velocity vector and pressure distribution of different-sized chambers. (**a**) D_1 = 2 mm. (**b**) D_1 = 10 mm. (**c**) D_1 = 15 mm. (**d**) D_1 = 20 mm. (**e**) D_1 = 25 mm. (**f**) D_1 = 30 mm.

Table 3. Pressure drop at different chamber parameters.

Parameters	D_1 = 2 mm (α_i = 0.05)	D_1 = 10 mm (α_i = 0.25)	D_1 = 15 mm (α_i = 0.375)	D_1 = 20 mm (α_i = 0.5)	D_1 = 25 mm (α_i = 0.625)	D_1 = 30 mm (α_i = 0.75)
Pressure drop (Pa)	1250	8023	10,046	15,535	16,028	16,980

In summary, in the SOR structure, the pressure drop and pressure-holding effect caused by the recess are key factors affecting the bearing's load capacity. The ability of the recess to maintain the center pressure is positively correlated with the parameter α_i, while the pressure drop tends to be opposite as α_i increases. Therefore, the key to improving bearing performance is to reduce the ratio of the flow resistance in Region I to that of the entire fluid region as α_i increases. The results indicate that it is crucial to limit the expansion of vortices at the boundary between Region I and II to balance the pressure-holding effect and pressure drop of the recess. Additionally, an increase in α_i enlarges the recess volume, intensifying airflow fluctuations entering the recess structure, which negatively impacts bearing stability. This makes achieving the ideal balance between load capacity and stability more challenging. Therefore, to ensure that the bearing reaches a balance between load capacity and stability, it is necessary to comprehensively consider factors such as the pressure drop caused by the recess structure, the pressure-holding effect and recess volume.

4.2. Effect of Different Recesses on Bearing Performance

The above analyses have been integrated into the design of the RCR and BCR, and the performance of three different types of homogeneous-pressure structure bearings was studied under various working conditions.

4.2.1. Static Performance of Different Types of Recesses

Figure 14 shows the load capacity of various recesses at different air-film thicknesses under a given supply pressure P_S = 0.6 MPa, α_i = 0.5. It can be observed that the compound recess shows a significant increase in the load capacity compared to the SOR for smaller air-film thickness conditions. This phenomenon suggests that the pressure-holding effect of the recess plays a dominant role in the bearing load capacity in this case. In addition, the BCR commonly has a higher load capacity than the RCR. From the pressure distribution and pressure drop under different structures in Figure 15, it can be seen that although the RCR and BCR are also α_i = 0.5, with essentially the same pressure-holding effect, the BCR has a lower pressure-drop ratio, which further results in the load capacity of the BCR exceeding that of the RCR. Figure 15a shows that the main reason for the difference in the pressure-drop ratio is the pressure rise in the branch structure in the sub-stage region, which leads to an increase in the mean pressure and hence a decrease in the pressure-drop ratio. This pressure rise is most likely due to the formation of air accumulation in the sub-stage region as the air flows towards the center on both sides of the branch structure. It is worth noting that the load capacity of all the recesses shows a decreasing trend with increasing air-film thickness. When the air-film thickness is large, there is a significant difference in the variations of pressure drop produced by different recesses, although their pressure-holding effect remains constant. In this case, the pressure-holding effect and the throttle effect balance each other out, so all the results of load capacity tend to be the same.

The pressure-drop ratios of different structures in Figure 15b show that the throttling effect of different types of recesses is roughly divided into three parts, among which the effect of the SOR with a 2 mm chamber is the best, which is much lower than that of other types of bearings. Subsequently, there are SOR bearings and compound recesses with chambers less than 20 mm, and their effect is concentrated in the middle position. Lastly, there are SORs with 25 mm and 30 mm chambers, whose pressure-drop ratio is much higher than that of other types of bearings, and the effect is the worst.

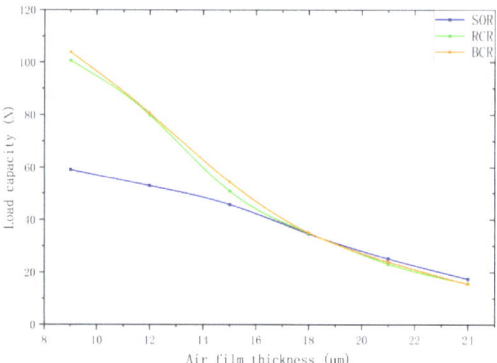

Figure 14. Load capacity of different recesses at different air-film thicknesses.

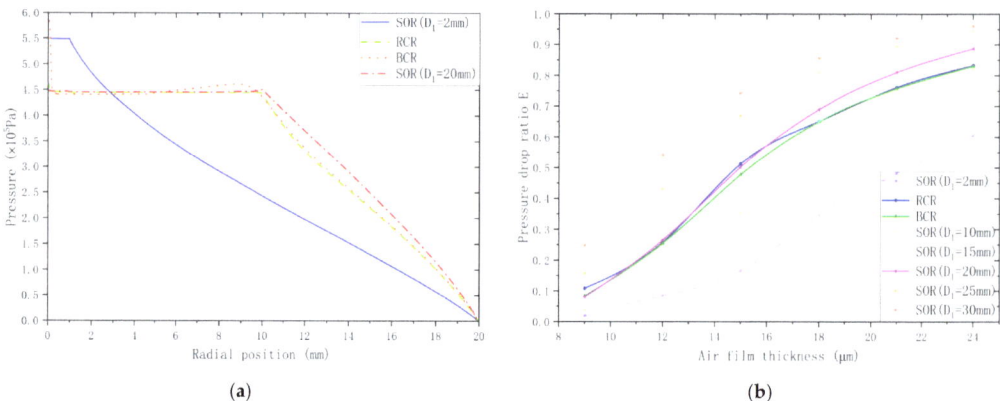

Figure 15. Pressure distribution and pressure-drop ratio of different recesses: (**a**) Pressure distributions at h = 12 μm. (**b**) Pressure-drop ratios for different recesses at different air-film thicknesses.

When RCR, BCR and SOR bearings have the same α_i, the BCR demonstrates the best throttling effect. Regardless of the film thickness, the BCR's pressure-drop ratio is lower than those of the other two structures. Following this, the RCR only exhibits results lower than for the SOR at smaller film thicknesses. As the film thickness increases, the RCR gradually exceeds the SOR and is essentially the same as the BCR at large thicknesses. In addition, the other two recesses were superior to the SOR. Notably, the throttling performance of these structures decreases as the air-film thickness increases.

Bearing stiffness and mass flow plots for different air-film thicknesses, at a supply pressure P_S = 0.6 MPa, of three types of recesses are shown in Figure 16. From Figure 16a, the stiffness of the bearings with different structures exhibits an increasing trend followed by a decrease as the air-film thickness increases. Bearings with compound recesses consistently show greater stiffness than the SOR, indicating that the compound structure at α_i = 0.5 enhances bearing stiffness compared to the small-chamber SOR. Additionally, the SOR bearing achieves its highest stiffness at an air-film thickness of 15 μm. while the bearing with a compound recess peaks at 12 μm. The stiffness trends of the RCR and BCR are essentially similar, with the RCR consistently exhibiting lower stiffness than the BCR, regardless of the air-film thickness. When the air-film thickness exceeds 18 μm, the stiffness of all bearing types converges to a similar level.

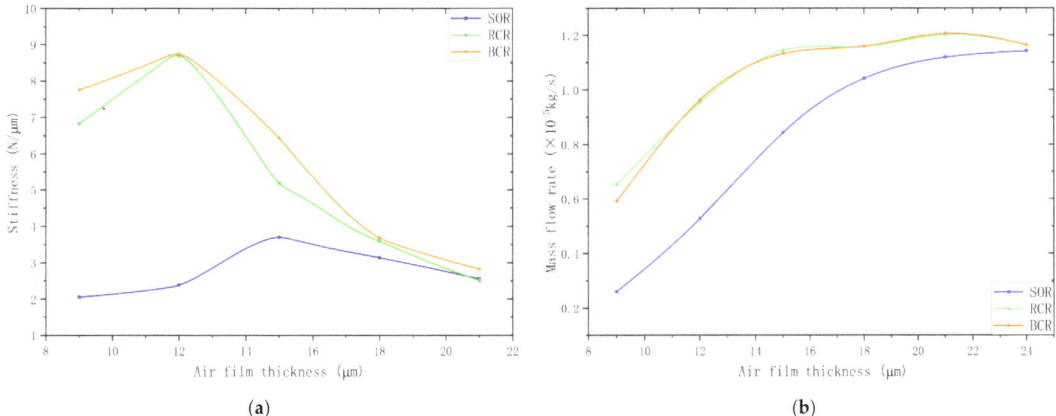

Figure 16. Stiffness and mass flow rate for different types of recesses: (**a**) Stiffness. (**b**) Mass flow rate.

Figure 16b shows the trend of mass flow rate changes with air-film thickness for different structures. Different structures with increasing air-film thickness are basically the same, showing an increasing trend, and the same trend of the SOR with different chamber sizes is shown in Figure 11, which reaches the critical value at the air-film thickness $h = 24$ μm. The air consumption of the RCR and BCR is roughly equivalent, with the RCR's being slightly higher than the BCR's, and both exceed that of the SOR for small chambers.

4.2.2. Dynamic Performance of Different Types of Recesses

Figure 17 presents the velocity contours for various recess bearings at an air-film thickness of 15 μm and a supply pressure of $P_S = 0.5$ MPa. Just like the small-chamber SOR, the compound recess produces only one main cyclone in the chamber. However, compare to compound recess, the SOR with the same chamber diameter of 20 mm ($\alpha_i = 0.5$) produces a larger cyclone near the inlet by sweeping the small cyclone slightly backwards. Figure 18 shows the main vortex vector diagram and pressure distribution at the inlet of different recesses. The figure indicates that, compared to the small-chamber SOR structure, the compound structure has a weaker throttling effect, and as the chamber volume increases, the vortex in the compound structure becomes larger, resulting in a certain impact from the high-pressure air at the inlet on the thrust surface. However, it can be clearly seen that, compared to the large-chamber SOR structure with the same $\alpha_i = 0.5$, the main vortex of the compound structure is more orderly, the vortex is smaller, and the impact area of the high-pressure air flow on the thrust surface is also smaller. Table 4 shows the pressure drop from the edge to the center of the main vortex inside these bearings. It can be observed that for the same $\alpha_i = 0.5$, the BCR exhibits the smallest pressure drop, followed by the RCR, and finally the SOR. The presence of a secondary branch in the BCR bearing increases the flow path of the air within the throttling structure, and the branch angle impedes the expansion of the low-pressure region at the end of the recess structure toward the air inlet. Therefore, the branch structure not only effectively directs the air but also enhances the stability of the bearing.

Table 4. Pressure drop at different types of recesses.

Recess Types (Parameters)	SOC ($D_1 = 2$ mm /$\alpha_i = 0.05$)	SOC ($D_1 = 20$ mm /$\alpha_i = 0.5$)	RCR ($\alpha_i = 0.5$)	BCR ($\alpha_i = 0.5$)
Pressure drop (Pa)	1250	15,535	7961	5470

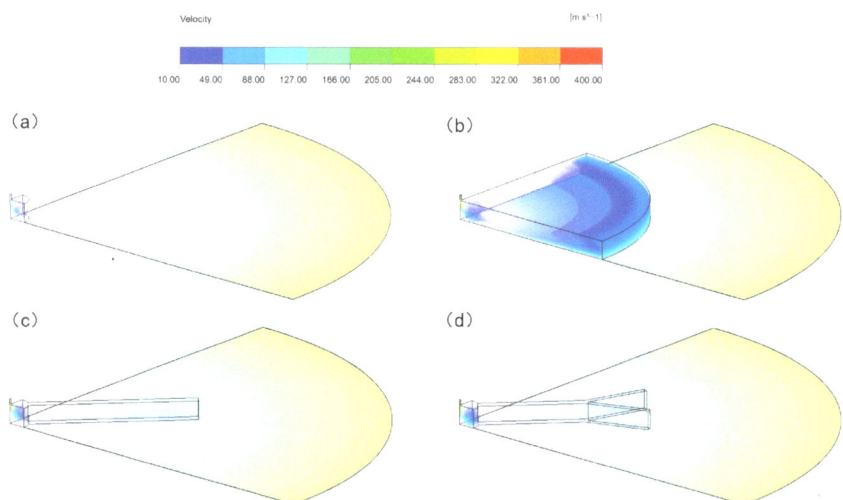

Figure 17. Velocity contours in different types of recesses. (**a**) SOR with $D_1 = 2$ mm. (**b**) SOR with $D_1 = 20$ mm. (**c**) RCR. (**d**) BCR.

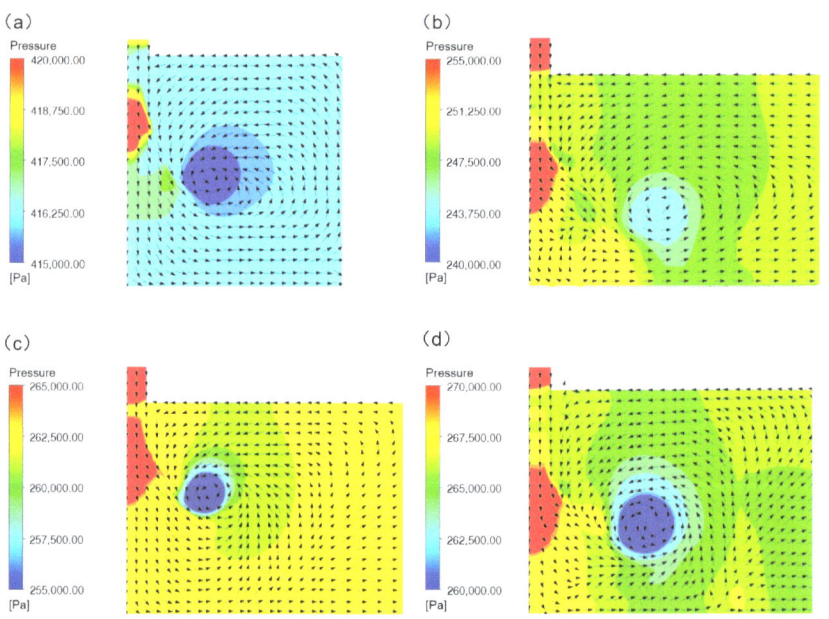

Figure 18. Velocity vector and pressure distribution along the s-plane for different types of recesses. (**a**) SOR with $D_1 = 2$ mm. (**b**) SOR with $D_1 = 20$ mm. (**c**) RCR. (**d**) BCR.

In summary, the compound recess has significant advantages in load capacity and stiffness compared to the small-chamber SOR. Moreover, due to its smaller projected area and air capacity, it also shows significant stability compared to SOR bearings under the same value of α_i. The compound structure not only ensures the pressure-holding distance but also suppresses the expansion of vortices, effectively balancing the bearing's static and dynamic performance. Additionally, thanks to the excellent fluid distribution capability of the bionic structure, the BCR outperforms other structures in both static and dynamic performance.

4.3. Effect of Different Partition Ratios of BCRs on Bearing Performance

The above study found that under the same conditions of α_i, the RCR and BCR achieved significant improvements in stiffness and stability while maintaining similar load capacities, thereby balancing static and dynamic performance. It is noted that the overall performance of the BCR surpasses that of the RCR, which confirms the application potential of bionic structures. To further investigate the effects of the bionic-compound structure on bearing performance, this study varied the proportions of chamber and branch structures in the compound structure while keeping the length ratio, width ratio, and inclusion angle of the branch structure constant. For the sake of convenience in discussion, as shown in Figure 3, we define CR as the proportion of the chamber section of the recess structure in Region I as follows:

$$CR = \frac{D_1}{\alpha_i D} \tag{20}$$

4.3.1. Static Performance BCR Bearings

Figure 19 illustrates the pressure distribution, load capacity, and pressure-drop ratio of BCR bearings with different structural dimensions at a supply pressure of P_S = 0.5 MPa and an air-film thickness of h = 15 μm. From the pressure distribution graphs of different BCR bearings in Figure 19a, it can be observed that except for the structure where CR equals 1 (recess is a chamber without branches), the pressure distribution curves of the other bionic-compound recesses are essentially the same, showing only minor differences at the end. They both exhibit good pressure-holding effects, meaning that the pressure can be maintained after an initial drop until it reaches the 20 mm OD circle. At the same time, the pressure-drop ratio graphs in Figure 19b clearly show the throttling effects of different structures. The E value decreases and then increases with an increase in the CR ratio, indicating that the throttling effect first increases and then decreases. Obviously, bionic-compound recesses with branch structures outperform those with only chambers in terms of effect. In Figure 19b, it is also shown that the load capacity of different BCR bearings at h = 15 μm slightly increases with the increase in the CR ratio. By analyzing the top-down pressure field of the thrust surface in Figure 20, the pressure inside the recess remains consistent, while pressure decay occurs outside the recess (in the air film between the branches).

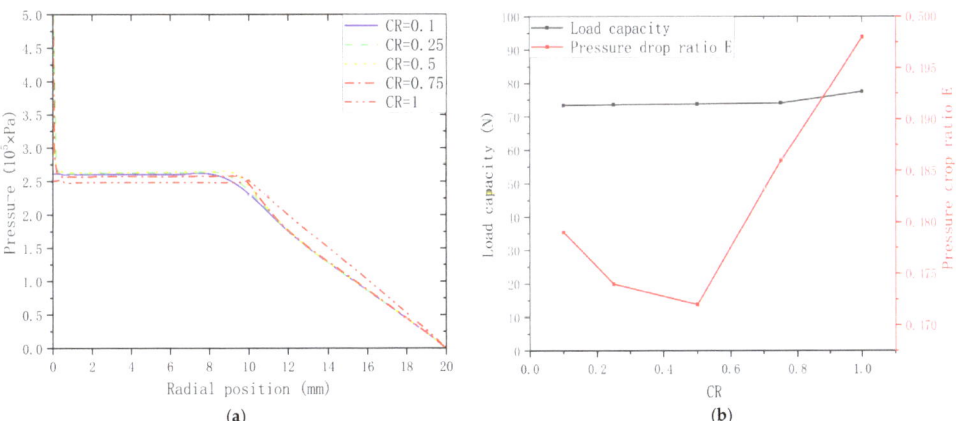

Figure 19. Pressure distribution, pressure-drop ratio and load capacity of bearings with different CR value structures: (**a**) Pressure distribution. (**b**) pressure-drop ratio and load capacity.

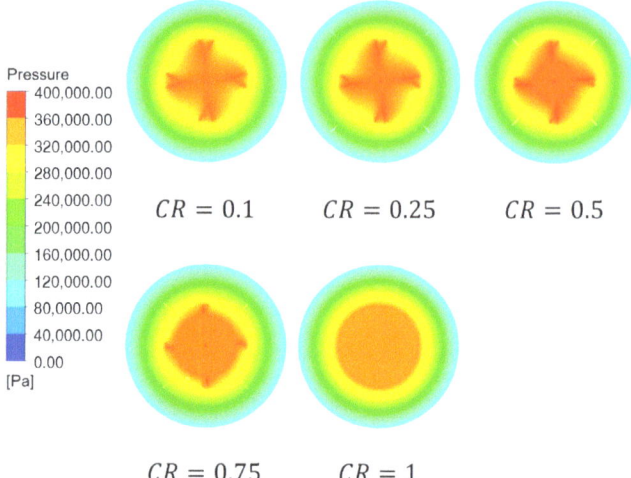

Figure 20. Contours of pressure distribution on bearing surface with different CR values.

4.3.2. Dynamic Performance of BCR Bearings

Figure 21 shows the velocity contours of BCR bearings with different CR ratios under the conditions of $P_S = 0.5$ MPa and air-film thickness $h = 15$ μm. Similar to the results of SOC bearings, when the chamber occupies a small proportion of the homogeneous pressure structure, only one main cyclone is generated in the chamber, and then the air diffuses outward through the branch structure. As the chamber section becomes larger, the increased air capacity inside causes the air flowing from the main cyclone at the inlet of the recess to sweep backward, forming a larger cyclone in the chamber. However, this cyclone primarily remains within the chamber, with no obvious vortices in the slot.

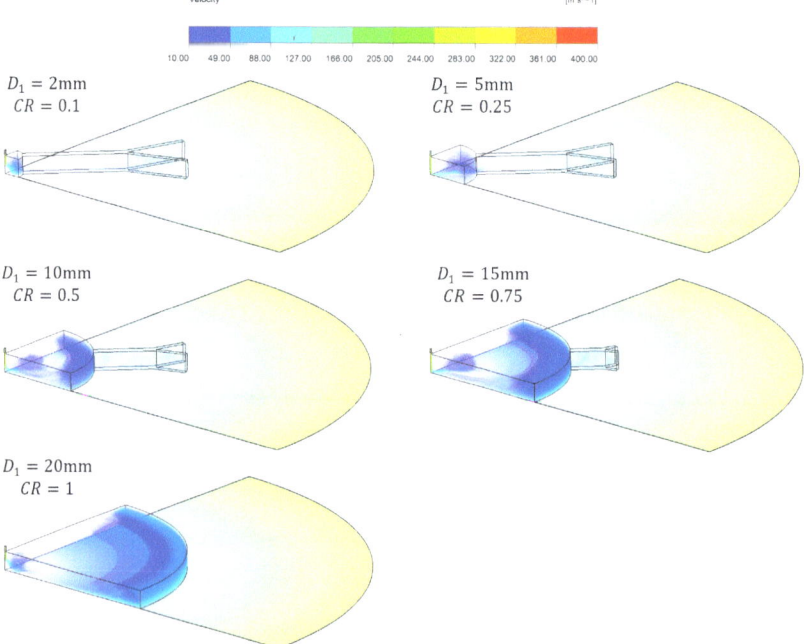

Figure 21. Velocity contours with different CR values.

Figure 22 illustrates the main vortex vector diagram and pressure distribution around the inlet for different CR values. As the CR value increases and the proportion of the chamber section within the recess structure grows, the size of the main vortex at the inlet expands significantly. Additionally, the turbulence intensity within the vortex increases, indicating a more chaotic and irregular airflow pattern. This suggests that as the chamber section grows, the fluid near the inlet becomes more unstable, potentially affecting the pressure distribution and flow behavior, thereby influencing the overall system performance. Meanwhile, Table 5 shows the pressure drop from the edge to the center of the main vortex for different structures. It can be observed that as the void structure increases, the pressure drop tends to rise. This indicates that the proportion of the branch compound structure has a significant impact on the stability of the bearing. As the proportion of the chamber section in the compound recess (i.e., CR) increases, the structural stability gradually declines.

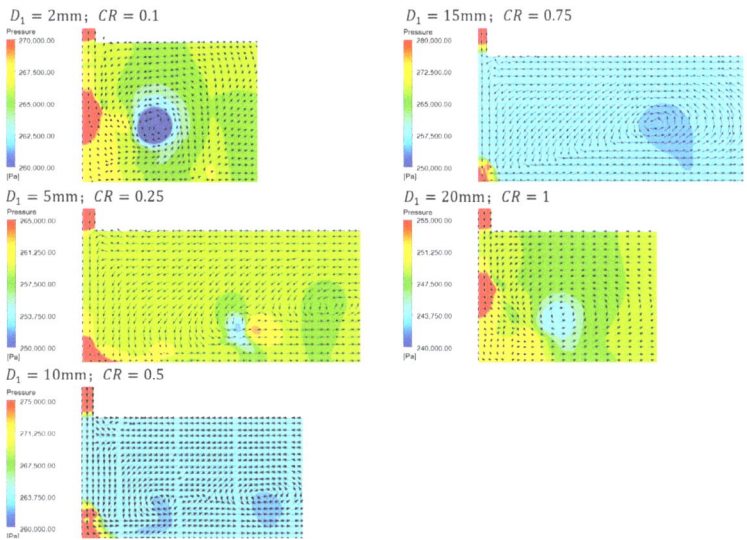

Figure 22. Velocity vector and pressure distribution along the s-plane for different CR values.

Table 5. Pressure drop at different CR values.

Parameters	CR = 0.1	CR = 0.25	CR = 0.5	CR = 0.75	CR = 1
Pressure drop (P_a)	5470	9041	10,623	13,786	15,535

5. Conclusions

This study explores the impact of recess structures on the static and dynamic performance of air static-pressure thrust bearings. Considering the effect of solid imbalance on bearings at high speeds, three structures—SOR, RCR, and BCR—have been designed to strictly adhere to axial symmetry. A series of calculations were conducted through simulation analysis on three representative structures, and the validity of the simulations was verified by comparison with existing experimental data, leading to the following main conclusions:

(1) The throttling effect and pressure-holding effect of the recess are the primary factors determining the static performance of the bearing, and they exhibit opposite trends as the recess's radial ratio α_i changes.

(2) An effective strategy to enhance the bearing's performance is reducing the flow resistance of the recess structure, ensuring that the pressure drop in the throttling region rises slowly, remains stable, or even decreases as α_i increases. This approach strikes an optimal compromise between the pressure-holding effect and the pressure drop.

(3) The compound recess balances both the throttling effect and pressure-holding effect, significantly improving the static performance of the bearing. Moreover, due to its smaller air capacity compared to the traditional simple-chamber recess, it offers a clear advantage in stability, achieving an effective balance between the static and dynamic performance of the bearing.

(4) The BCR equipped with a branch structure has excellent fluid distribution capability. It effectively controls the flow resistance in the recess while ensuring that the recess structures' ratio α_i is sufficiently large. Additionally, the branch structure reduces the impact of sudden changes in the structural dimensions at the boundary between the recess and the air film on the pressure distribution at the air inlet. As a result, it performs exceptionally well in both static and dynamic aspects.

Author Contributions: Conceptualization, F.Y. and H.X.; Data curation, F.Y. and B.C.; Formal analysis, F.Y., H.X., J.Z. and Y.W.; Funding acquisition, H.X. and J.Z.; Investigation, F.Y., H.X., J.Z., B.C. and G.C.; Methodology, F.Y., H.X. and J.Z.; Project administration, F.Y. and H.X.; Resources, H.X. and J.Z.; Software, F.Y., H.X., G.C. and X.Z.; Supervision, H.X., B.C. and G.C.; Validation, F.Y., H.X., J.Z. and B.C.; Visualization, F.Y.; Writing—original draft, F.Y. and B.C.; Writing—review and editing, F.Y., H.X., B.C. and Y.W. All authors have read and agreed to the published version of the manuscript.

Funding: This research was funded by the Chongqing Natural Science Foundation, grant number CSTB2022NSCQ-MSX1326.

Data Availability Statement: Data are contained within the article.

Conflicts of Interest: The authors declare no conflicts of interest. The funders had no role in the design of the study; in the collection, analyses, or interpretation of data; in the writing of the manuscript; or in the decision to publish the results.

References

1. Yan, R.; Zhang, H. Analysis of the Dynamic Characteristics of Downsized Aerostatic Thrust Bearings with Different Pressure Equalizing Grooves at High or Ultra-High Speed. *Adv. Mech. Eng.* **2021**, *13*, 168781402110180. [CrossRef]
2. Wang, Y.; Wu, H.; Rong, Y. Analysis of Hydrostatic Bearings Based on a Unstructured Meshing Scheme and Turbulence Model. *Machines* **2022**, *10*, 1072. [CrossRef]
3. Sahto, M.P.; Wang, W.; Sanjrani, A.N.; Hao, C.; Shah, S.A. Dynamic Performance of Partially Orifice Porous Aerostatic Thrust Bearing. *Micromachines* **2021**, *12*, 989. [CrossRef] [PubMed]
4. Hashimoto, H.; Ochiai, M.; Namba, T. Optimization of Groove Geometry for Thrust Air Bearing According to Various Objective Functions. *ASME J. Tribol.* **2009**, *131*, 041704. [CrossRef]
5. Yoshimura, T.; Hanafusa, T.; Kitagawa, T.; Hirayama, T.; Matsuoka, T.; Yabe, H. Clarifications of the Mechanism of Nano-Fluctuation of Aerostatic Thrust Bearing with Surface Restriction. *Tribol. Int.* **2012**, *48*, 29–34. [CrossRef]
6. Chen, Y.; Huo, D.; Wang, G.; Zhong, L.; Gong, Z. Modelling and Performance Improvement of an Annular Grooved Thrust Air Bearing. *ILT* **2023**, *75*, 1116–1124. [CrossRef]
7. Colombo, F. Dynamic Characterisation of Rectangular Aerostatic Pads with Multiple Inherent Orifices. *Tribol. Lett.* **2018**, *66*, 133. [CrossRef]
8. Yu, P.; Lu, J.; Luo, Q.; Li, G.; Yin, X. Optimization Design of Aerostatic Bearings with Square Micro-Hole Arrayed Restrictor for the Improvement of Stability: Theoretical Predictions and Experimental Measurements. *Lubricants* **2022**, *10*, 295. [CrossRef]
9. Zhuang, H.; Ding, J.; Chen, P.; Chang, Y.; Zeng, X.; Yang, H.; Liu, X.; Wei, W. Numerical Study on Static and Dynamic Performances of a Double-Pad Annular Inherently Compensated Aerostatic Thrust Bearing. *J. Tribol.* **2019**, *141*, 051701. [CrossRef]
10. Gao, S.; Cheng, K.; Chen, S.; Ding, H.; Fu, H. CFD Based Investigation on Influence of Orifice Chamber Shapes for the Design of Aerostatic Thrust Bearings at Ultra-High Speed Spindles. *Tribol. Int.* **2015**, *92*, 211–221. [CrossRef]
11. Yu, P.; Zuo, T.; Lu, J.; Zhong, M.; Zhang, L. Static and Dynamic Performances of Novel Aerostatic Bearings with Primary and Secondary Orifice Restrictors. *Lubricants* **2023**, *11*, 518. [CrossRef]
12. Qiao, Y.J.; Luo, R.; Shi, K.J. Analysis on the Influence for the Sectional Shape of Compound Pressure-Equalizing Groove to the Supporting and Bearing Characteristics of Precision Aerostatic Bearing. *AMM* **2014**, *494–495*, 598–601. [CrossRef]
13. Eleshaky, M.E. CFD Investigation of Pressure Depressions in Aerostatic Circular Thrust Bearings. *Tribol. Int.* **2009**, *42*, 1108–1117. [CrossRef]
14. Li, W.; Wang, G.; Feng, K.; Zhang, Y.; Wang, P. CFD-Based Investigation and Experimental Study on the Performances of Novel Back-Flow Channel Aerostatic Bearings. *Tribol. Int.* **2022**, *165*, 107319. [CrossRef]
15. Gao, S.; Shang, Y.; Gao, Q.; Lu, L.; Zhu, M.; Sun, Y.; Yu, W. CFD-Based Investigation on Effects of Orifice Length–Diameter Ratio for the Design of Hydrostatic Thrust Bearings. *Appl. Sci.* **2021**, *11*, 959. [CrossRef]

16. Yan, R.; Wang, L.; Wang, S. Investigating the Influences of Pressure-Equalizing Grooves on Characteristics of Aerostatic Bearings Based on CFD. *ILT* **2019**, *71*, 853–860. [CrossRef]
17. Feng, K.; Wang, P.; Zhang, Y.; Hou, W.; Li, W.; Wang, J.; Cui, H. Novel 3-D Printed Aerostatic Bearings for the Improvement of Stability: Theoretical Predictions and Experimental Measurements. *Tribol. Int.* **2021**, *163*, 107149. [CrossRef]
18. Jing, D.; He, L.; Wang, X. Optimization Analysis of Fractal Tree-like Microchannel Network for Electroviscous Flow to Realize Minimum Hydraulic Resistance. *Int. J. Heat Mass Transf.* **2018**, *125*, 749–755. [CrossRef]
19. Zhu, R.; Li, K.; Wang, D.; Fei, J.; Yan Tan, J.; Li, S.; Zhang, J.; Li, H.; Fu, S. Biomimetic Optimized Concept with Murray Networks for Accelerated Solar-Driven Water Evaporation. *Chem. Eng. J.* **2023**, *467*, 143383. [CrossRef]
20. Zhou, G.; Zhou, J.; Huai, X. High Performance Vapor Chamber Enabled by Leaf-Vein-Inspired Wick Structure for High-Power Electronics Cooling. *Appl. Therm. Eng.* **2023**, *230*, 120859. [CrossRef]
21. Chai, S.; Mei, X.; Xie, Y.; Lu, L. Inspired Biomimetic Design for Efficient Fluid Channel Configuration Based on Leaf Vein Structures. *Numer. Heat Transf. Part B Fundam.* **2023**, *1*, 1–14. [CrossRef]
22. Peng, Y.; Liu, W.; Wang, N.; Tian, Y.; Chen, X. A Novel Wick Structure of Vapor Chamber Based on the Fractal Architecture of Leaf Vein. *Int. J. Heat Mass Transf.* **2013**, *63*, 120–133. [CrossRef]
23. Wan, X.; Cao, J.; Yang, X.; Wang, L.; Chen, Y.; Cheng, B. Research on The Flow Field of The Pemfc Bipolar Plate Based on The Tree-Like Fractal Theory. *Fractals* **2023**, *31*, 2340185. [CrossRef]
24. Ma, C.; Sun, Y.; Wu, Y.; Zhang, Q.; Wang, Y.; Ding, G. A Bio-Inspired Fractal Microchannel Heat Sink with Secondary Modified Structure and Sub-Total-Sub Fluid Transmission Mode for High Heat Flux and Energy-Saving Heat Dissipation. *Int. J. Heat Mass Transf.* **2023**, *202*, 123717. [CrossRef]
25. Sack, L.; Scoffoni, C. Leaf Venation: Structure, Function, Development, Evolution, Ecology and Applications in the Past, Present and Future. *New Phytol.* **2013**, *198*, 983–1000. [CrossRef]
26. Zhang, G.; Huang, M.; Chen, G.; Li, J.; Liu, Y.; He, J.; Zheng, Y.; Tang, S.; Cui, H. Design and Optimization of Fluid Lubricated Bearings Operated with Extreme Working Performances—A Comprehensive Review. *Int. J. Extrem. Manuf.* **2024**, *6*, 022010. [CrossRef]
27. Shan, X.D.; Wang, M.; Guo, Z.Y. Geometry Optimization of Self-Similar. *Math. Probl. Eng.* **2011**, *11*, 421526. [CrossRef]
28. Ashgriz, N.; Mostaghimi, J. *An Introduction to Computational Fluid Dynamics*, 2nd ed.; Pearson Education Limited: London, UK, 2007; pp. 30–33.
29. Belforte, G.; Raparelli, T.; Viktorov, V.; Trivella, A. Discharge Coefficients of Orifice-Type Restrictor for Aerostatic Bearings. *Tribol. Int.* **2007**, *40*, 512–521. [CrossRef]
30. Gao, Q.; Chen, W.; Lu, L.; Huo, D.; Cheng, K. Aerostatic Bearings Design and Analysis with the Application to Precision Engineering: State-of-the-Art and Future Perspectives. *Tribol. Int.* **2019**, *135*, 1–17. [CrossRef]
31. Zhou, Y.; Chen, X.; Chen, H. A Hybrid Approach to the Numerical Solution of Air Flow Field in Aerostatic Thrust Bearings. *Tribol. Int.* **2016**, *102*, 444–453. [CrossRef]
32. Chen, X.; Chen, H.; Luo, X.; Ye, Y.; Hu, Y.; Xu, J. Air Vortices and Nano-Vibration of Aerostatic Bearings. *Tribol. Lett.* **2011**, *42*, 179–183. [CrossRef]

Disclaimer/Publisher's Note: The statements, opinions and data contained in all publications are solely those of the individual author(s) and contributor(s) and not of MDPI and/or the editor(s). MDPI and/or the editor(s) disclaim responsibility for any injury to people or property resulting from any ideas, methods, instructions or products referred to in the content.

Article

Influence and Optimization of Nozzle Position on Lubricant Distribution in an Angular Contact Ball Bearing Cavity

Baogang Wen [1], Yuanyuan Li [1], Yemin Li [1], Meiling Wang [2,*] and Jingyu Zhai [3,4]

1 School of Mechanical Engineering and Automation, Dalian Polytechnic University, Dalian 116034, China
2 Zhan Tianyou Honors College, Dalian Jiaotong University, Dalian 116028, China
3 School of Mechanical Engineering, Dalian University of Technology, Dalian 116024, China
4 Ningbo Institute of Dalian University of Technology, Ningbo 315032, China
* Correspondence: wml_dljt@163.com; Tel.:+86-411-84106065

Abstract: In this paper, the lubrication flow field model for an angular contact ball bearing considering the characteristics of the nozzle position was constructed with CFD methods, and the simulation results were compared and validated with the test results. The research was carried out on the lubricant distribution characteristics in the bearing cavity under different nozzle angles and heights, and the nozzle position was optimized with the response surface methodology. The results show that the lubrication distribution characteristics in the bearing cavity are closely related to the nozzle angle and height. The weighted average of the oil phase volume fraction on the cage guiding surface decreases first and then increases with the increase of the nozzle height and decreases with the increase of the nozzle angle on the ball surface. The optimal nozzle position was determined by finding the maximum value of the regression function in the specified area.

Keywords: angular contact ball bearing; lubricant distribution; nozzle position; test rig

1. Introduction

The lubrication characteristics of angular contact ball bearings, which are crucial support components in aeroengines and machine tools, are directly related to their life and reliability. These characteristics are influenced by the lubrication performance [1]. The position of the oil jet nozzle used for lubrication is vital for ensuring proper distribution of lubricant within the bearing cavity.

About modeling of lubrication flow field in bearings: Yan [2] established a highly precise numerical model with different nozzle distributions and tracked the oil-air interface using the VOF method. Bei Yan [3] used a coupling of the Level Set function and VOF method to track the oil-air interface and obtain the migration and diffusion process of oil droplets in the bearing cavity. Bao [4] used the Coupled Level Set Volume of Fluid (CLSVOF) method to track the oil-air two-phase flow inside the ball bearing with under-race lubrication. Ge [5] studied the flow behavior of oil on the bearing inner ring surface by adding groove structures to the non-contact area of the bearing inner ring surface. Chen [6] established a transient air–oil two-phase flow model in a ball bearing based on computational fluid dynamics (CFD) to investigate the behaviors of oil transfer and air–oil flow under different capillary conditions with speed, surface tension, and viscosity and found that the oil distribution and air–oil flow behaviors in a ball bearing are strongly related to the speed and the ratio of oil viscosity. Shan [7] established the lubrication analysis model of ball bearing, and the evolution of hydrodynamic behavior, including the oil-air distribution, temperature, and flow characteristics during different lubrication states, is displayed. Oh [8] used flow field simulation techniques of computational fluid dynamics to analyze the three-dimensional airflow behavior in a ball bearing. Liu [9] used a fluid–structure coupled simulation model based on the CFD method of air–oil two-phase flow (AOTPF) in the bearing to discuss the lubricating characteristics of an oil jet-lubricated

ball bearing in the gearbox. Peterson [10] used ANSYS FLUENT computational fluid dynamics (CFD) software to develop a full-scale model of a single-phase oil flow in a deep groove ball bearing (DGBB) and measure the frictional torque of oil-lubricated rolling element bearings and compare fluid drag losses with CFD simulations. Aidarinis [11] experimentally studied the flow field inside the bearing external chamber using a laser Doppler anemometry system and carried out at real operating conditions, both for the airflow and for the lubricant oil flow and for a range of shaft rotating speeds, then computed by fluid dynamics (CFD) modeling.

About testing on lubrication flow field inside bearings: Maccioni [12,13] designed a dedicated test rig to perform Particle Image Velocimetry (PIV) measurements on the lubricant inside a tapered roller bearing using a sapphire outer ring. He [14] directly observed the distribution of the lubricant film in a custom-made model-bearing rig, with an outer ring replaced by a glass ring to allow full optical access. Arya [15] studied the oil flow inside an angular contact ball bearing using Bubble Image Velocimetry (BIV). The oil flow inside the cage and bearing was analyzed using a high-speed camera, and the observed oil flow streamlines demonstrated the influence of operating conditions and cage designs on fluid flow. Wen [16] built one measurement system for lubricant distribution that obtained grayscale images to characterize the lubricant distribution, and then the image pixels were evaluated for the characterization of lubricant volume. Chen [17] proposed a novel experimental method that combines synchronized dual-camera imaging with laser-induced fluorescence (LIF) using extra illumination to observe the oil flow in a ball bearing and then observed the distribution of the lubricant film with an outer ring replaced by a glass ring. Takashi Node [18] used X-ray computed tomography (CT), which is one of the non-destructive inspection techniques to visualize remarkable details of grease distribution in a resin ball bearing. Sakai [19] used a neutron imaging technology to perform non-destructive visualization of the grease fluidity and migration inside a ball bearing. Franken [20] applied fluorescence spectroscopy technology to identify oil migration in the bearing. Nitric oxide laser-induced fluorescence imaging methods [21] and PIV (Micro Particle Image Velocimetry) [22,23] to measure the grease velocity vector field.

About optimizing the bearing nozzle: Hu [24] proposed a new nozzle structure, and the three-dimensional simulation model of the internal moving components of the bearing considering the movement of the lubricant at the nozzle outlet was established. The two conventional nozzle structures and the new nozzle structure were compared and analyzed by numerical simulation for the oil-air two-phase flow distribution law under different working conditions. The fluid flow characteristics in the bearing cavity and contact area and its key structural parameters were investigated and improved by Yan [25]. Li [26] studied the influence of different structural parameters on the force of the valve core and obtained the optimal structural parameters under given operating conditions by using computational fluid dynamics (CFD) and the orthogonal design method. Wu [27,28] investigated the air-oil two-phase flow inside an oil jet-lubricated ball bearing through numerical simulations and corresponding tests and obtained the influence of jet velocity and oil flow rate on the average oil volume fraction inside the bearing cavity and then determined the number of nozzles.

The lubrication flow field in bearings has been extensively studied. However, there is a lack of quantitative verification and limited focus on the flow characteristics of rolling elements. As a result, the lubricant distribution rule of the bearing cavity remains undisclosed.

A lubrication flow field model of an angular contact ball bearing was constructed in this paper and validated by experiments. The study investigated the lubricant distribution of the bearing cavity under different nozzle angles and heights, using observation surfaces on both the ball surface and cage guiding surface. Subsequently, the optimal nozzle position under given operating conditions was determined. These findings can provide valuable insights for lubrication flow field analyses and a method for optimizing nozzle position under multiparameter coupling.

2. Modeling and Verification for Lubrication Flow Field

2.1. Oil-Jet Lubricated Ball Bearing and Its Motion Relationship

The orientation of the nozzle plays a crucial role in lubrication design as it facilitates the injection of lubricating oil into the bearing, as illustrated in Figure 1. The oil injection lubrication area is the gap between the cage and the outer ring of the bearing. Specifically, the nozzle angle ϕ refers to the horizontal angle, while the nozzle height h represents the vertical distance from the center of the nozzle to that of the bearing. When the bearing rotates, the outer ring remains stationary while the inner ring rotates in conjunction with the cage propelled by the balls.

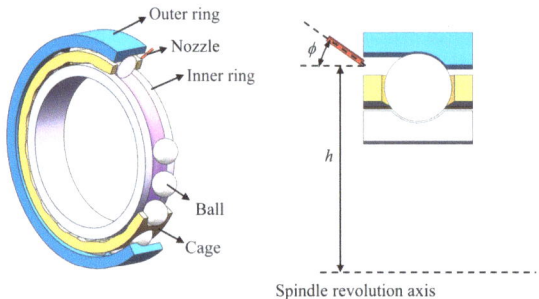

Figure 1. The bearing structure and the spatial orientation of its nozzle.

2.2. CFD Model of the Bearing Cavity Considering the Characteristics of Nozzle Position

A two-phase flow model of the bearing cavity with a nozzle is proposed based on the principles of mass conservation and momentum conservation, incorporating a turbulent model to depict the flow field as depicted in Figure 2.

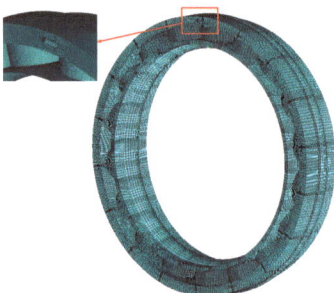

Figure 2. The fluid field of the bearing cavity with the nozzle.

The fluid field within the cavity, along with the nozzle, was discretized into a finite number of non-overlapping units. The differential functions were reformulated as linear expressions comprising of the node values of each variable or its derivatives and the chosen interpolation functions, which were then solved using the weighted residual method. The flow chart depicting the research process is illustrated in Figure 3.

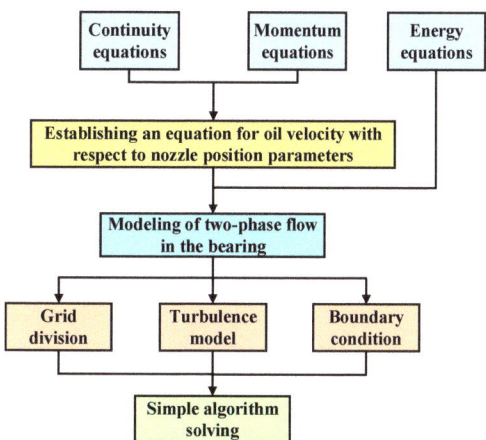

Figure 3. The flowchart of the research.

2.2.1. The Two-Phase Flow Model

The lubrication flow field model for an angular contact ball bearing considering the characteristics of the nozzle position was constructed with the CFD ideal fluid model, which refers to the completely incompressible, non-viscous fluid with no sliding at the interface. The VOF-implicit model in the multi-phase flow model was employed to track the oil-air interface in the bearing cavity with a nozzle [29]. In this context, a cell without oil is denoted as $\varphi_{oil} = 0$, while a cell filled with oil is represented by $\varphi_{oil} = 1$. If $0 < \varphi_{oil} < 1$, is present, it signifies the interface between the oil phase and air phase.

The continuity function for the oil phase volume fraction is given by:

$$\frac{\partial}{\partial t}(\varphi_{oil}\rho_{oil}) + \nabla \cdot \left(\varphi_{oil}\rho_{oil}\vec{v}\right) = S_{\alpha oil} \tag{1}$$

where $\vec{v}$ is the velocity vector and $S_{\alpha oil}$ is the mass source term.

The determination of oil droplet motion as they enter the bearing cavity through the nozzle is essential. Figure 4 illustrates the trajectory of oil droplet motion.

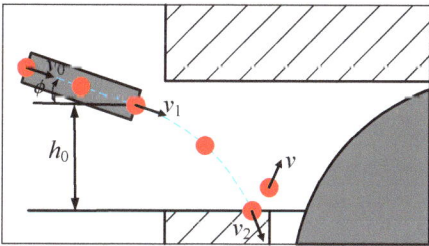

Figure 4. Schematic diagram of an oil droplet motion trajectory.

While oil droplets spray through the nozzle, Bernoulli's function for two interfaces can be obtained as follows:

$$\frac{p_1}{\rho_1} + \frac{v_0^2}{2} = \frac{p_2}{\rho_2} + \frac{v_1^2}{2} \tag{2}$$

where p_1 is the static pressure inside the nozzle, p_2 is the static pressure outside the nozzle, v_1 is the average velocity of the fluid inside the nozzle, v_2 is the average velocity of the fluid outside the nozzle, ρ_1 is the fluid density inside the nozzle, and ρ_2 is the fluid density outside the nozzle.

Assuming oil jet continuity, the continuity function at the interface of the nozzle can be expressed by

$$\rho_1 v_0 A_1 = \rho_2 v_1 A_2 \tag{3}$$

where A_1 is the cross-sectional area of the nozzle inlet and A_2 is the cross-sectional area of the nozzle outlet.

The nozzle angle is ϕ, and its cross-section are circular; meanwhile, the fluid density $\rho = \rho_1 = \rho_2$, $p_1 \gg p_2$, the incident velocity can be obtained using the following:

$$v_1 = \sqrt{\frac{2p}{\rho\left[1 - \left(\frac{1}{\cos\phi}\right)^4\right]}} \tag{4}$$

If the height of oil droplets entering the bearing cavity is h_0 and the velocity is v_2, the energy conservation function can be given by:

$$\frac{1}{2}mv_1^2 + mgh_0 = \frac{1}{2}mv_2^2 \tag{5}$$

The velocity relates to the nozzle angle, and nozzle height can be expressed by:

$$v_2 = \sqrt{\frac{2p}{\rho\left[1 - \left(\frac{1}{\cos\phi}\right)^4\right]} + 2gh_0} \tag{6}$$

The motion of oil droplets after impacting the bearing inner wall must comply with the momentum conservation function as follows:

$$\begin{cases} \rho\left(\frac{\partial u}{\partial t} + u\frac{\partial u}{\partial x} + v\frac{\partial u}{\partial y} + w\frac{\partial u}{\partial z}\right) = \rho F_x + \frac{\partial p_{xx}}{\partial x} + \frac{\partial p_{xy}}{\partial y} + \frac{\partial p_{xz}}{\partial z} \\ \rho\left(\frac{\partial v}{\partial t} + u\frac{\partial v}{\partial x} + v\frac{\partial v}{\partial y} + w\frac{\partial v}{\partial z}\right) = \rho F_y + \frac{\partial p_{yx}}{\partial x} + \frac{\partial p_{yy}}{\partial y} + \frac{\partial p_{yz}}{\partial z} \\ \rho\left(\frac{\partial w}{\partial t} + u\frac{\partial w}{\partial x} + v\frac{\partial w}{\partial y} + w\frac{\partial w}{\partial z}\right) = \rho F_z + \frac{\partial p_{zx}}{\partial x} + \frac{\partial p_{zy}}{\partial y} + \frac{\partial p_{zz}}{\partial z} \end{cases} \tag{7}$$

The continuity function is given by:

$$v_i\frac{\partial \rho}{\partial t} + v_i \text{div}(\rho v) = v_i\left(\frac{\partial \rho}{\partial t} + \text{div}(\rho v)\right) = 0 \tag{8}$$

so it can be obtained using the following:

$$\frac{\partial(\rho v_i)}{\partial t} + \text{div}(\rho v_i v) = \rho\left(\frac{\partial v_i}{\partial t} + v \cdot \text{grad} v_i\right) \tag{9}$$

In the selection of turbulence models, the RNG $k - \varepsilon$ model, which is particularly suitable for the VOF, can better capture the variations in different turbulence scales and more accurately predict various turbulence characteristics compared to the standard $k - \varepsilon$ model. The turbulent energy transport Equation (10) and the energy dissipation transport Equation (11) of the RNG $k - \varepsilon$ model are as follows:

$$\frac{\partial(\rho k)}{\partial t} + \frac{\partial}{\partial x_j}\left(\rho u_j \frac{\partial k}{\partial x_j} - \left(\mu + \frac{\mu_t}{\sigma_k}\right)\frac{\partial k}{\partial x_j}\right) = \tau_{tij}S_{ij} - \rho\varepsilon + \phi_k \tag{10}$$

$$\frac{\partial(\rho\varepsilon)}{\partial t} + \frac{\partial}{\partial x_j}\left(\rho u_j \varepsilon - \left(\mu + \frac{\mu_t}{\sigma_\varepsilon}\right)\frac{\partial \varepsilon}{\partial x_j}\right) = c_{\varepsilon 1}\frac{\varepsilon}{k}\tau_{tij}S_{ij} - c_{\varepsilon 2}f_2\rho\frac{\varepsilon^2}{k} + \phi_\varepsilon \tag{11}$$

where the terms on the right side represent the generation term, dissipation term, and wall term, respectively.

The wall function is used to express the wall term, which is a semi-empirical formula to connect the viscous-affected region between the wall and the fully turbulent region. Among that, the standard wall function uses a dimensionless velocity U^* and a dimensionless distance y^* to describe the law within the boundary layer as follows:

$$U^* = \frac{1}{k}\ln(Ey^*) \tag{12}$$

among the following equations:

$$U^* \equiv \frac{U_p C_\mu^{1/4} k_p^{1/2}}{\tau_w/\rho} \tag{13}$$

$$y^* \equiv \frac{\rho C_\mu^{1/4} k_p^{1/2}}{\mu} \tag{14}$$

where $k = 0.4187$ is the Karman constant, $E = 9.793$ is the empirical constant, and U_p is the velocity at the center of the grid adjacent to the wall, k_p is the turbulent kinetic energy at the center of the grid adjacent to the wall, μ is dynamic viscosity [30].

2.2.2. Solution Method

The finite volume method was employed to discretely solve the governing equations in this study. The central difference scheme was utilized for the diffusion and pressure terms of the momentum equations, while the convection term was treated using a second-order upwind difference scheme [31]. Additionally, a semi-implicit (SIMPLEC) approach was adopted for the velocity-pressure coupling solver, and a convergence criterion of 10^{-4} was set for the residuals of each velocity component and VOF functions.

2.3. Test Rig for Investigating the Lubricant Distribution in the Angular Contact Ball Bearing Cavity

The test rig was constructed to investigate the lubricant distribution characteristics within the cavity of an angular contact ball bearing, as depicted in Figure 5. This setup comprises a drive system, an oil lubrication system, a horizontal two-support rotor system, a tested bearing, and a designated measurement system for assessing lubricant distribution. In addition, the measurement system consists of a high-speed camera, a circular light source, and a shading box. An adjustable nozzle capable of modifying its spatial position by altering both height and angle was adopted in the oil lubrication system. The inner ring and the outer ring were fabricated from a highly transparent resin material, while the balls were made of transparent acrylic, as depicted in Figure 5. The specific structural parameters of the bearing are presented in Table 1.

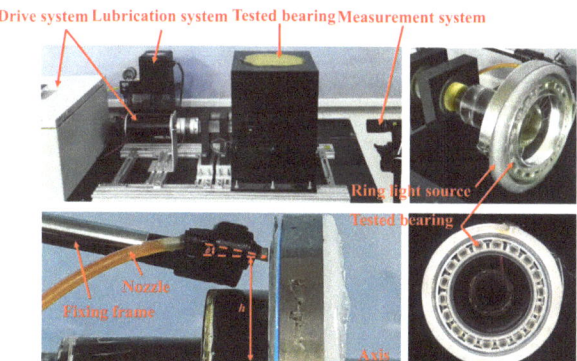

Figure 5. Test rig and tested bearing.

Table 1. Structural dimensions of angular contact ball bearing.

Item	Symbol	Unit	Value
Diameter of inner ring	d	mm	65
Diameter of outer ring	D	mm	100
Diameter of ball	D_w	mm	11
Number of balls	Z	No.	18
Contact angle	α	°	15
Width of ring	B	mm	18

4010 aviation lubricating oil was used, and its physical parameters are shown in Table 2.

Table 2. Oil material parameters.

Parameter	Symbol	Unit	Value
temperature	t	K	313.15
density	ρ	kg/m^3	876
Kinematic viscosity	ν	$[10^{-6}\ \text{m}^2/\text{s}]$	30
Specific heat	c	$[\text{W}/(\text{m}^2 \cdot \text{K})]$	1.96

The tracer material for lubrication was selected as ink. So the shaded part denotes lubricant and clearly represents its distribution in the tested bearing. At a rotational speed of 120 rad/s for the inner ring, a high-speed camera with a resolution of 640 × 370, a frame rate of 1000, and an exposure time of 0.5 ms captured an amplified image of shadows on the ball surface by adjusting the focal length, as depicted in Figure 6a. Subsequently, by adjusting the brightness contrast curve in Photoshop (PS), a clearer image of the lubricant was obtained, as shown in Figure 6b. Finally, pixels were extracted from the grayscale image displayed in Figure 6c using the OTSU method, as shown in red box in Figure 6c.

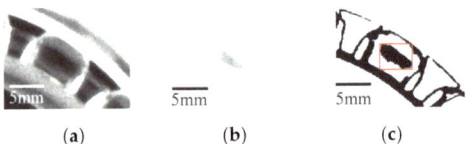

Figure 6. The process of photographing the oil jet lubricated. (a) Original images. (b) PS images. (c) Grayscale images.

2.4. Comparative Verification

Based on the computational fluid dynamics method, a 1/18 two-phase flow model with a nozzle replacing the overall model was established on ANSYS FLUENT in Section 2.2, taking into account the periodic symmetry of the bearing structure as illustrated in Figure 7.

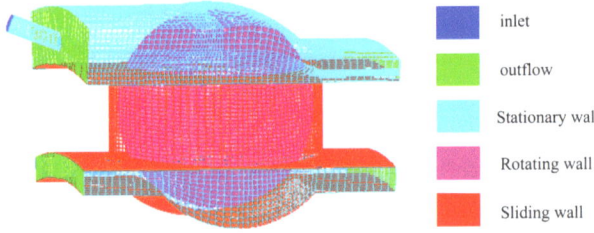

Figure 7. CFD mesh and boundary conditions.

The fluid domain was discretized using a structured mesh, resulting in 158,989 elements and 171,612 nodes. The outer ring of the bearing was designated as a stationary wall, while the front and rear walls were set as pressure outlets. The nozzle served as the velocity inlet with a flow rate of 1 m/s. With an inner ring rotational speed of 120 rad/s, the ball and cage exhibited a revolution speed of 50 rad/s, whereas the ball itself rotated at a speed of 400 rad/s based on previously derived formulas.

The nozzle angle of 20° and the nozzle height of 45 mm, which are consistent with the test conditions, were selected in this section to analyze the internal flow field of the bearing fluid domain. The simulated and experimental images on the ball surface corresponding to four time points at 0.05 s, 0.1 s, 0.15 s, and 0.2 s are depicted in Figure 8.

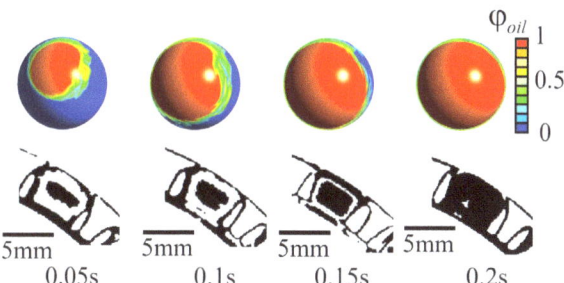

Figure 8. Photographing results of oil distribution on the ball surface.

The oil distribution on the ball surface, obtained from simulation and experiment, was converted into the oil spreading area using CFD-post of Fluent and image recognition algorithms, respectively. The comparison between simulation and experimental results over time revealed the temporal evolution of the oil spreading area on the surface of the ball, as depicted in Figure 9.

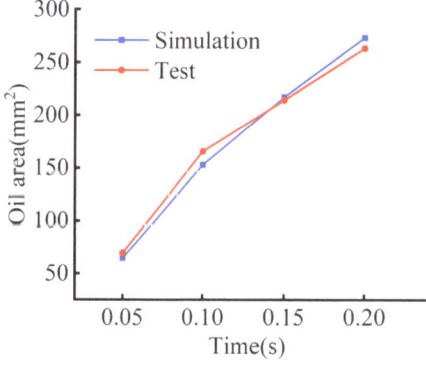

Figure 9. Comparison between simulation and test results.

The oil spreading area on the ball surface demonstrates a consistent trend between simulation and experimental results, indicating the applicability of the proposed model for the lubrication flow field with a nozzle in this study.

3. Influence of Nozzle Positions on Lubricant Distribution and Its Optimization

The ball surface and the cage guiding surface are selected as observation surfaces to comprehensively analyze the lubricant distribution characteristics in the bearing cavity, as shown in Figure 10.

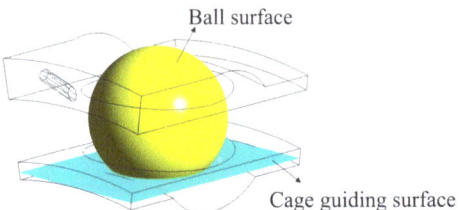

Figure 10. Schematic diagram of observation surfaces in the bearing fluid domain.

The surface weighted average value is calculated by summing up the values of each region on the surface, multiplied by their corresponding weights, and divided by the total weight. If there are n regions on a surface with oil phase volume fractions, respectively denoted as $\varphi_1, \varphi_2, \ldots, \varphi_n$, and their weights are $w_1, w_2, \ldots, w_n$, then the formula is

$$\overline{x} = \frac{\varphi_1 w_1 + \varphi_2 w_2 + \ldots + \varphi_n w_n}{w_1 + w_2 + \ldots + w_n} \tag{15}$$

The value is referred to as the weighted average of the oil phase volume fraction, which provides a more precise depiction of the overall distribution of oil on the surface and then is used to characterize the state of oil distribution.

3.1. Influence of Nozzle Positions on Lubricant Distribution

3.1.1. Influence of Nozzle Height on Lubricant Distribution

The nozzle angle ϕ was set at $35°$ constantly, and the five oil injection heights were equidistantly set from 44.5 mm to 46 mm. The contours of oil distribution about two observation surfaces after stabilization are shown in Figure 11.

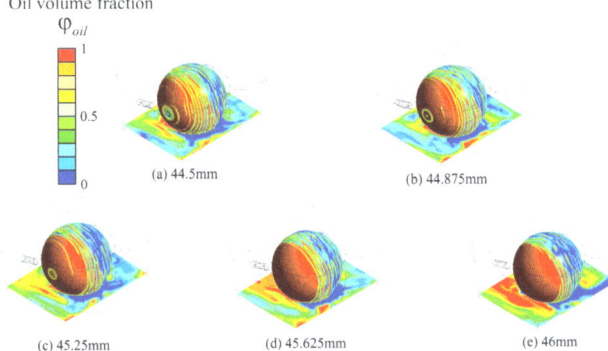

Figure 11. Oil phase distribution maps with different nozzle heights.

The variation trend of the weighted average of the oil phase volume fraction on a ball surface and a cage guiding surface along with the nozzle heights are shown in Figure 12.

The variation trend of the weighted average of the oil phase volume fraction on the ball surface is not statistically significant, as depicted in Figure 12. However, on the cage guiding surface, this trend increases with an increase in nozzle height. As the nozzle height increases, the elevation of oil impacting the ball also rises, resulting in an increased angle at which the oil collides with the ball. Consequently, there is an expansion in the surface area covered by oil on the ball. Simultaneously, as the nozzle height increases, the region where oil impacts transitions from intersecting horizontally with both the ball and cage to a higher position on the ball. This facilitates easier entry of oil into the gap between the ball and cage, ultimately leading to centrifugal force propelling it onto the cage guiding surface.

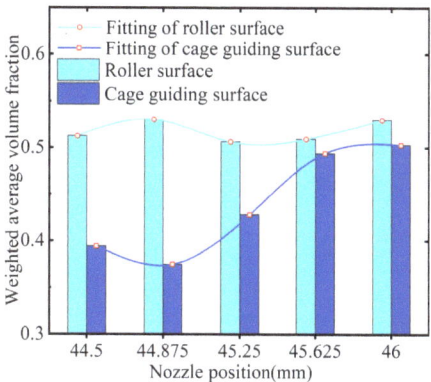

Figure 12. Line graph of the weighted average of the oil phase volume fraction on the ball and the cage guiding surface with different nozzle heights.

3.1.2. Influence of Nozzle Angle on Lubricant Distribution

The nozzle height h was set at 35° constantly, and the five oil injection angles were equidistantly set from 20° to 50°. The contours of oil distribution about two observation surfaces after stabilization are shown in Figure 13.

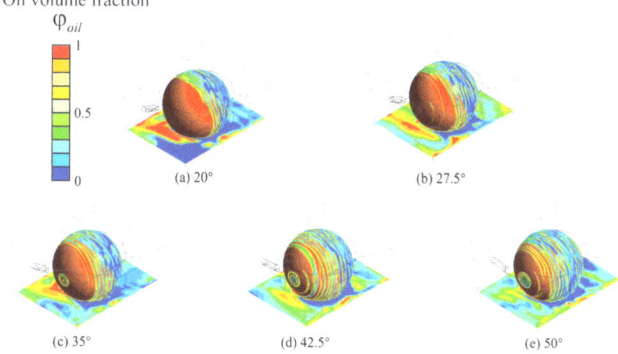

Figure 13. Oil phase distribution maps with different nozzle angles.

The variation trend of the weighted average of the oil phase volume fraction on a ball surface and a cage guiding surface along with the nozzle heights are shown in Figure 14.

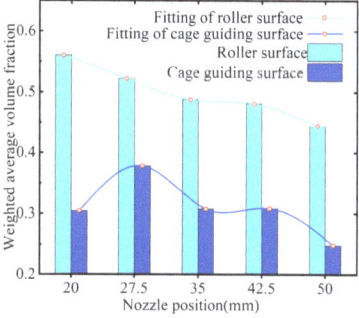

Figure 14. Line graph of the weighted average of the oil phase volume fraction on the ball and the cage guiding surface with different nozzle angles.

The variation trend of the weighted average, in relation to the nozzle heights, is illustrated in Figure 14. The progressive increase in nozzle angle leads to a narrower region between the cage and the ball, resulting in an increased tangential velocity of oil along the ball surface. Furthermore, due to the rotational motion of the ball, oil is expelled along both surfaces—namely, the ball surface and cage guiding surface—leading to a reduction in flow rate on these surfaces. So, both the weighted average of the oil phase volume fraction on the ball surface and the cage guiding surface decrease as the nozzle angle increases.

3.2. Optimization of the Nozzle Position with the Response Surface Methodology

The response surface methodology is a statistical technique that establishes a regression-fitted expression by mathematically fitting limited experimental data using a multivariate quadratic regression function [32]. The equation for the standard quadratic polynomial response surface function [33] is as follows:

$$y = m_0 + \sum_{i=1}^{n} m_i x_i + \sum_{i=1}^{n}\sum_{j=1}^{n} m_{ij} x_i x_j + \sum_{i=1}^{n} m_{ii} x_i^2 \tag{16}$$

where x_i represents the design variable, n is the number of design variables, and m_0, m_i, m_{ii}, and m_{ij} are the coefficients of the quadratic polynomial regression function.

The optimization design indicators include the nozzle angle, nozzle height, and the weighted average of the oil phase volume fraction on the ball surface and the cage guiding surface [34]. A central composite design was established with a 2-factor 3-level table as presented in Table 3.

Table 3. Factor level table.

Factors	Level		
	−1	0	1
Nozzle height h/mm	44.5	45.25	46
Nozzle angle ϕ/°	20	35	50

Nine simulations need to be conducted, and the results are shown in Table 4.

Table 4. Simulation results.

Number	Factors		The Weighted Average of the Oil Phase Volume Fraction on the Ball Surface	The Weighted Average of the Oil Phase Volume Fraction on the Cage Guiding Surface
	h	ϕ		
1	44.5	20	0.52566	0.48531
2	44.5	35	0.51269	0.39439
3	44.5	50	0.4991	0.30932
4	45.25	20	0.50794	0.41052
5	45.25	35	0.50656	0.42806
6	45.25	50	0.43067	0.29146
7	46	20	0.59107	0.37645
8	46	35	0.52987	0.50311
9	46	50	0.44649	0.33966

The dates in Table 5 were analyzed by Design-Expert, and then the regression equations for the weighted average of the oil phase volume fraction on the ball surface and cage guiding surface were obtained.

$$y_1 = 0.48 - 0.041 \times \phi - 0.03 \times h \times \phi + 0.036 \times h^2 \tag{17}$$

$$y_2 = 0.43 + 5.033E - 003 \times \phi - 0.055 \times h + 0.035 \times \phi \times h + 0.025 \times \phi^2 - 0.073 \times h^2 \quad (18)$$

Table 5. Variance analysis table of the regression function for the ball surface.

Source	Sum of Squares	df	Mean Square	F	p-Value
Model	0.016	3	5.441×10^{-3}	23.05	0.0023
ϕ	0.010	1	0.010	43.57	0.0012
$h \times \phi$	3.482×10^{-3}	1	3.482×10^{-3}	14.75	0.0121
h^2	2.557×10^{-3}	1	2.557×10^{-3}	10.83	0.0217
Residual	1.180×10^{-3}	5	2.360×10^{-4}		
Cor Total	0.018	8			

The results of the variance analysis for Formulas (17) and (18) are shown in Tables 5 and 6, respectively.

Table 6. Variance analysis table of the regression function for the cage guiding surface.

Source	Sum of Squares	df	Mean Square	F	p-Value
Model	0.035	5	7.049×10^{-3}	2.87	0.2073
h	1.520×10^{-4}	1	1.520×10^{-4}	0.062	0.8196
ϕ	0.018	1	0.018	7.47	0.0717
$h\phi$	4.844×10^{-3}	1	4.844×10^{-3}	1.97	0.2548
h^2	1.220×10^{-3}	1	1.220×10^{-3}	0.5	0.5318
ϕ^2	0.011	1	0.011	4.35	0.1284
Residual	7.369×10^{-3}	3	2.456×10^{-3}		
Cor Total	0.043	8			

The results in Table 5 indicate that the p-value of the model is $0.0023 \leq 0.05$, suggesting a statistically significant regression relationship between the weighted average of the oil phase volume fraction on the ball surface and the factors considered. Therefore, it can be concluded that Formula (17) is applicable.

The results in Table 6 indicate that the p-value of the model is 0.2073, which is greater than the significance level of 0.05. This suggests a lack of significant regression relationship between the weighted average of the oil phase volume fraction on the cage guiding surface and the factors under consideration. Therefore, it can be concluded that Formula (18) is not applicable.

The two-dimensional influence diagram of the nozzle angle and height on the weighted average of the oil phase volume fraction on the ball surface is obtained from Formula (17), as depicted in Figure 15, and as the color becomes darker, the value increases.

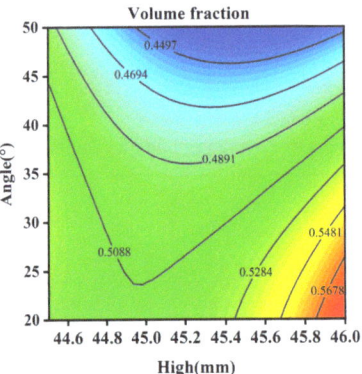

Figure 15. Two-dimensional graph for the weighted average of the oil phase volume fraction on the ball surface.

Unfortunately, the peak value of the function is not within this lubrication area and cannot be used. The maximum weighted average of the oil phase volume fraction on the ball surface is observed when the nozzle height is 46 mm and the nozzle angle is 20, as depicted in Figure 15.

In practical applications, this universal method can also be used to obtain the optimal position for oil-jet lubrication.

4. Conclusions

A two-phase flow model of the angular contact ball bearing was established, taking into account the nozzle, and subsequently validated through experimental results. The lubricant distribution laws of an angular contact ball bearing cavity with a nozzle were analyzed in this paper, followed by the optimization of the nozzle position. The main methods and conclusions can be summarized as follows:

The results indicate that the impact of nozzle height on the ball surface is negligible, whereas it is significant on the cage guiding surface. Additionally, both the ball surface and cage guiding surface are significantly affected by nozzle angle. Regression functions were established using response surface methodology to determine the weighted average of oil phase volume fraction on both surfaces. It was determined that under operating conditions where inner ring speed is 1000 r/min and oil jet velocity is 1 m/s, the optimal nozzle height is 46 mm and the nozzle angle is $20°$.

Author Contributions: Investigation, B.W.; Experiment and simulation, Y.L. (Yuanyuan Li), Y.L. (Yemin Li).; Resources, J.Z.; Writing—review & editing, M.W. All authors have read and agreed to the published version of the manuscript.

Funding: This work was financially supported by the Natural Science Foundation of Liaoning Province (2023-MS-280), the Scientific Research Project of Education Department of Liaoning Province (LJKMZ20220864), and the Key Research and Development Program of Ningbo (2022Z050).

Data Availability Statement: The original contributions presented in the study are included in the article, further inquiries can be directed to the corresponding author.

Conflicts of Interest: The authors declare no conflict of interest.

References

1. Gao, W.; Nelias, D.; Li, K.; Liu, Z.; Lyu, Y. A multiphase computational study of oil distribution inside roller bearings with under-race lubrication. *Tribol. Int.* **2019**, *140*, 105862. [CrossRef]
2. Yan, K.; Wang, Y.; Zhu, Y.; Hong, J.; Zhai, Q. Investigation on heat dissipation characteristic of ball bearing cage and inside cavity at ultra high rotation speed. *Tribol. Int.* **2016**, *93*, 470–481. [CrossRef]

3. Yan, B.; Dong, L.; Yan, K.; Chen, F.; Zhu, Y.; Wang, D. Effects of oil-air lubrication methods on the internal fluid flow and heat dissipation of high-speed ball bearings. *Mech. Syst. Signal Process.* **2021**, *151*, 107409. [CrossRef]
4. Bao, H.; Hou, X.; Lu, F. Analysis of Oil-Air Two-Phase Flow Characteristics inside a Ball Bearing with Under-Race Lubrication. *Processes* **2020**, *8*, 1223. [CrossRef]
5. Ge, L.; Yan, K.; Wang, C.; Zhu, Y.; Hong, J. A novel method for bearing lubrication enhancement via the inner ring groove structure. *J. Phys. Conf. Ser.* **2021**, *1820*, 012092. [CrossRef]
6. Chen, H.; Liang, H.; Wang, W.; Zhang, S. Investigation on the oil transfer behaviors and the air-oil interfacial flow patterns in a ball bearing under different capillary conditions. *Friction* **2022**, *11*, 228–245. [CrossRef]
7. Shan, W.; Chen, Y.; Huang, J.; Wang, X.; Han, Z.; Wu, K. A multiphase flow study for lubrication characteristics on the internal flow pattern of ball bearing. *Results Eng.* **2023**, *20*, 101429. [CrossRef]
8. Oh, I.-S.; Kim, D.; Hong, S.-W.; Kim, K. Three-dimensional air flow patterns within a rotating ball bearing. *Adv. Sci. Lett.* **2013**, *19*, 2180–2183. [CrossRef]
9. Liu, J.; Ni, H.; Xu, Z.; Pan, G. A simulation analysis for lubricating characteristics of an oil-jet lubricated ball bearing. *Simul. Model. Pract. Theory* **2021**, *113*, 102371. [CrossRef]
10. Peterson, W.; Russell, T.; Sadeghi, F.; Berhan, M.T. Experimental and analytical investigation of fluid drag losses in rolling element bearings. *Tribol. Int.* **2021**, *161*, 107106. [CrossRef]
11. Aidarinis, J.; Missirlis, D.; Yakinthos, K.; Goulas, A. CFD modeling and LDA measurements for the air-flow in an aero engine front bearing chamber. *J. Eng. Gas Turbines Power* **2011**, *133*, 082504. [CrossRef]
12. Maccioni, L.; Chernoray, V.G.; Mastrone, M.N.; Bohnert, C.; Concli, F. Study of the impact of aeration on the lubricant behavior in a tapered roller bearing: Innovative numerical modelling and validation via particle image velocimetry. *Tribol. Int.* **2022**, *165*, 107301. [CrossRef]
13. Maccioni, L.; Chernoray, V.G.; Bohnert, C.; Concli, F. Particle Image Velocimetry measurements inside a tapered roller bearing with an outer ring made of sapphire: Design and operation of an innovative test rig. *Tribol. Int.* **2022**, *165*, 107313. [CrossRef]
14. Liang, H.; Zhang, Y.; Wang, W. Influence of the cage on the migration and distribution of lubricating oil inside a ball bearing. *Friction* **2021**, *10*, 1035–1045. [CrossRef]
15. Arya, U.; Peterson, W.; Sadeghi, F.; Meinel, A.; Grillenberger, H. Investigation of oil flow in a ball bearing using Bubble Image Velocimetry and CFD modeling. *Tribol. Int.* **2023**, *177*, 107968. [CrossRef]
16. Wen, B.; Li, Y.; Wang, M.; Yang, Y. Measurement for Lubricant Distribution in an Angular Contact Ball Bearing and Its Influence Investigation. *Lubricants* **2023**, *11*, 63. [CrossRef]
17. Chen, H.; Wang, W.; Liang, H.; Ge, X. Observation of the oil flow in a ball bearing with a novel experiment method and simulation. *Tribol. Int.* **2022**, *174*, 107731. [CrossRef]
18. Noda, T.; Shibasaki, K.; Miyata, S.; Taniguchi, M. X-Ray CT Imaging of Grease Behavior in Ball Bearing and Numerical Validation of Multi-Phase Flows Simulation. *Tribol. Online* **2020**, *15*, 36–44. [CrossRef]
19. Kazumi, S.; Yusuke, A.; Yoshimu, I.; Nobuharu, K.; Yoshihiro, M. Observation of Grease Fluidity in a Ball Bearing Using Neutron Imaging Technology. *Tribol. Online* **2021**, *16*, 146–150.
20. Franken, M.J.Z.; Chennaoui, M.; Wang, J. Mapping of Grease Migration in High-Speed Bearings Using a Technique Based on Fluorescence Spectroscopy. *Tribol. Trans.* **2017**, *60*, 789–793. [CrossRef]
21. Sánchez-González, R.; North, S.W. Nitric oxide laser-induced fluorescence imaging methods and their application to study high-speed flows. In *Frontiers and Advances in Molecular Spectroscopy*; Elsevier: Amsterdam, The Netherlands, 2018; pp. 599–630.
22. Meinhart, C.D.; Wereley, S.T.; Santiago, J.G. PIV measurements of a microchannel flow. *Exp. Fluids* **1999**, *27*, 414–419. [CrossRef]
23. Mastrone, M.N.; Hartono, E.A.; Chernoray, V.; Concli, F. Oil distribution and churning losses of gearboxes: Experimental and numerical analysis. *Tribol. Int.* **2020**, *151*, 106496. [CrossRef]
24. Hu, J.; Xun, B.; Zhang, X.-M.; Zhang, Q.-Y.; Li, G.-W. Design and research of new-type nozzle structure based on oil–air lubrication. *Meccanica* **2023**, *59*, 1–18. [CrossRef]
25. Yan, K.; Zhang, J.; Hong, J.; Wang, Y.; Zhu, Y. Structural optimization of lubrication device for high speed angular contact ball bearing based on internal fluid flow analysis. *Int. J. Heat Mass Transf.* **2016**, *95*, 540–550. [CrossRef]
26. Li, S.; Deng, G.; Hu, Y.; Yu, M.; Ma, T. Optimization of structural parameters of pilot-operated control valve based on CFD and orthogonal method. *Results Eng.* **2024**, *21*, 101914. [CrossRef]
27. Wu, W.; Hu, C.; Hu, J.; Yuan, S. Jet cooling for rolling bearings: Flow visualization and temperature distribution. *Appl. Therm. Eng.* **2016**, *105*, 217–224. [CrossRef]
28. Wu, W.; Hu, J.; Yuan, S.; Hu, C. Numerical and experimental investigation of the stratified air-oil flow inside ball bearings. *Int. J. Heat Mass Transf.* **2016**, *103*, 619–626. [CrossRef]
29. Zeng, Q.; Zhang, J.; Hong, J.; Liu, C. A comparative study on simulation and experiment of oil-air lubrication unit for high speed bearing. *Ind. Lubr. Tribol.* **2016**, *68*, 325–335. [CrossRef]
30. Samokhinv, N. *Mathematical Models in Boundary Layer Theory*; CRC Press: Boca Raton, FL, USA, 2018.
31. Liu, J.; Ni, H.; Zhou, R.; Li, X.; Xing, Q.; Pan, G. A Simulation Analysis of Ball Bearing Lubrication Characteristics Considering the Cage Clearance. *J. Tribol.* **2023**, *145*, 044301. [CrossRef]
32. Hu, J.; Wu, W.; Wu, M.; Yuan, S. Numerical investigation of the air–oil two-phase flow inside an oil-jet lubricated ball bearing. *Int. J. Heat Mass Transf.* **2014**, *68*, 85–93. [CrossRef]

33. Kaur, H.; Rahi, D.K. Response surface methodology-based optimisation of chitin production and its antioxidant activity from Aspergillus niger. *Heliyon* **2024**, *10*, e25646. [CrossRef] [PubMed]
34. Rao, B.R.; Tiwari, R.J. Optimum design of rolling element bearings using genetic algorithms. *Mech. Mach. Theory* **2007**, *42*, 233–250.

Disclaimer/Publisher's Note: The statements, opinions and data contained in all publications are solely those of the individual author(s) and contributor(s) and not of MDPI and/or the editor(s). MDPI and/or the editor(s) disclaim responsibility for any injury to people or property resulting from any ideas, methods, instructions or products referred to in the content.

MDPI AG
Grosspeteranlage 5
4052 Basel
Switzerland
Tel.: +41 61 683 77 34

Lubricants Editorial Office
E-mail: lubricants@mdpi.com
www.mdpi.com/journal/lubricants

Disclaimer/Publisher's Note: The title and front matter of this reprint are at the discretion of the Guest Editor. The publisher is not responsible for their content or any associated concerns. The statements, opinions and data contained in all individual articles are solely those of the individual Editor and contributors and not of MDPI. MDPI disclaims responsibility for any injury to people or property resulting from any ideas, methods, instructions or products referred to in the content.

www.ingramcontent.com/pod-product-compliance
Lightning Source LLC
LaVergne TN
LVHW072320090526
838202LV00019B/2317